ENGINEERING EXPERIMENTATION
Planning, Execution, Reporting

McGraw-Hill Series in Mechanical Engineering

Consulting Editors

Jack P. Holman, *Southern Methodist University*
John R. Lloyd, *Michigan State University*

Anderson: *Computational Fluid Dynamics: The Basics with Applications*
Anderson: *Modern Compressible Flow: With Historical Perspective*
Arora: *Introduction to Optimum Design*
Bray and Stanley: *Nondestructive Evaluation: A Tool for Design, Manufacturing, and Service*
Burton: *Introduction to Dynamic Systems Analysis*
Culp: *Principles of Energy Conversion*
Dally: *Packaging of Electronic Systems: A Mechanical Engineering Approach*
Dieter: *Engineering Design: A Materials and Processing Approach*
Doebelin: *Engineering Experimentation: Planning, Execution, Reporting*
Driels: *Linear Control Systems Engineering*
Eckert and Drake: *Analysis of Heat and Mass Transfer*
Edwards and McKee: *Fundamentals of Mechanical Component Design*
Gebhart: *Heat Conduction and Mass Diffusion*
Gibson: *Principles of Composite Material Mechanics*
Hamrock: *Fundamentals of Fluid Film Lubrication*
Heywood: *Internal Combustion Engine Fundamentals*
Hinze: *Turbulence*
Holman: *Experimental Methods for Engineers*
Howell and Buckius: *Fundamentals of Engineering Thermodynamics*
Hutton: *Applied Mechanical Vibrations*
Juvinall: *Engineering Considerations of Stress, Strain, and Strength*
Kane and Levinson: *Dynamics: Theory and Applications*
Kays and Crawford: *Convective Heat and Mass Transfer*
Kelly: *Fundamentals of Mechanical Vibrations*
Kimbrell: *Kinematics Analysis and Synthesis*
Kreider and Rabl: *Heating and Cooling of Buildings*
Martin: *Kinematics and Dynamics of Machines*
Modest: *Radiative Heat Transfer*
Norton: *Design of Machinery*
Phelan: *Fundamentals of Mechanical Design*
Raven: *Automatic Control Engineering*
Reddy: *An Introduction to the Finite Element Method*
Rosenberg and Karnopp: *Introduction to Physical Systems Dynamics*
Schlichting: *Boundary-Layer Theory*
Shames: *Mechanics of Fluids*
Sherman: *Viscous Flow*
Shigley: *Kinematic Analysis of Mechanisms*
Shigley and Mischke: *Mechanical Engineering Design*
Shigley and Uicker: *Theory of Machines and Mechanisms*
Stiffler: *Design with Microprocessors for Mechanical Engineers*
Stoecker and Jones: *Refrigeration and Air Conditioning*
Ullman: *The Mechanical Design Process*
Vanderplaats: *Numerical Optimization: Techniques for Engineering Design, with Applications*
Wark: *Advanced Thermodynamics for Engineers*
White: *Viscous Fluid Flow*
Zeid: *CAD/CAM Theory and Practice*

Also Available from McGraw-Hill

Schaum's Outline Series in Mechanical Engineering

Most outlines include basic theory, definitions and hundreds of example problems solved in step-by-step detail, and supplementary problems with answers.

Related titles on the current list include:

Acoustics
Continuum Mechanics
Elementary Statics & Strength of Materials
Engineering Economics
Engineering Mechanics
Fluid Dynamics
Fluid Mechanics & Hydraulics
Heat Transfer
Lagrangian Dynamics
Machine Design
Mathematical Handbook of Formulas & Tables
Mechanical Vibrations
Operations Research
Statics & Mechanics of Materials
Strength of Materials
Theoretical Mechanics
Thermodynamics for Engineers
Thermodynamics with Chemical Applications

Schaum's Solved Problems Books

Each title in this series is a complete and expert source of solved problems with solutions worked out in step-by-step detail.

Related titles on the current list include:

3000 Solved Problems in Calculus
2500 Solved Problems in Differential Equations
2500 Solved Problems in Fluid Mechanics & Hydraulics
1000 Solved Problems in Heat Transfer
3000 Solved Problems in Linear Algebra
2000 Solved Problems in Mechanical Engineering Thermodynamics
2000 Solved Problems in Numerical Analysis
700 Solved Problems in Vector Mechanics for Engineers: Dynamics
800 Solved Problems in Vector Mechanics for Engineers: Statics

Available at most college bookstores, or for a complete list of titles and prices, write to: Schaum Division
McGraw-Hill, Inc.
1221 Avenue of the Americas
New York, NY 10020

ENGINEERING EXPERIMENTATION
Planning, Execution, Reporting

Ernest O. Doebelin
The Ohio State University

McGraw-Hill, Inc.
New York   St. Louis   San Francisco   Auckland   Bogotá
Caracas   Lisbon   London   Madrid   Mexico City   Milan   Montreal
New Delhi   San Juan   Singapore   Sydney   Tokyo   Toronto

This book was set in Times Roman.
The editors were John J. Corrigan and James W. Bradley;
the production supervisor was Leroy A. Young.
The cover was designed by Rafael Hernandez.
R. R. Donnelley & Sons Company was printer and binder.

ENGINEERING EXPERIMENTATION
Planning, Execution, Reporting

This book is printed on recycled, acid-free paper
containing 10% postconsumer waste.

1 2 3 4 5 6 7 8 9 0 DOC DOC 9 0 9 8 7 6 5

ISBN 0-07-017339-7

Library of Congress Cataloging-in-Publication Data

Doebelin, Ernest O.
    Engineering experimentation: planning, execution, reporting /
Ernest O. Doebelin.
      p.   cm. — (McGraw-Hill series in mechanical engineering)
    Includes index.
    ISBN 0-07-017339-7
    1. Engineering—Experiments.   2. Experimental design.
    3. Engineering—Statistical methods.   I. Title.   II. Series.
TA 153.D64   1995
620'.00724—dc20   94-21434

ABOUT THE AUTHOR

Ernest O. Doebelin has taught mechanical engineering at The Ohio State University since 1954. Since 1990, he has been professor emeritus and teaches full time for one quarter a year. He received a B.S.M.E. from Case Institute of Technology and M.Sc. and Ph.D. degrees from The Ohio State University. During his career at Ohio State, he originated and developed the curriculum in System Dynamics, Measurement, and Control within mechanical engineering. This includes two undergraduate and two graduate courses in system dynamics, two measurements courses, and two control systems courses. Of these eight courses, seven have laboratories, all of which Prof. Doebelin created, designed, and implemented. He is the author of two control texts, two system dynamics texts, and four editions of a measurements text. He has won many local, regional, and national teaching awards. Recently, he has done volunteer teaching at Otterbein College, a liberal arts school, developing a new course, Understanding Technology, to address the national problem of technology illiteracy.

CONTENTS

3 Measurement System Design and Application 106

4 Experiment Plans 155

PREFACE

Engineering problems are generally solved by some combination of theoretical and experimental work. In preparing students for the experimental portion of this overall effort, we must provide educational materials and school experiences in two major areas. One of these, the study of measurement systems, is the topic of my earlier book, now in its fourth edition. Here, students become familiar with the basic concepts, principles, and analytical tools needed to understand and design the measuring and data processing equipment used in any experimental study. In addition to basic principles, some *familiarity* with existing hardware and methods should be developed. For labs which emphasize set experiments, rather than original projects, this type of information is sufficient.

When, however, students are responsible for carrying out original experimental projects, the above material is still necessary but not really sufficient. This new text is a source for the needed additional material, which may be described as involving planning, execution, and reporting. The two books, taken together, give a rather complete treatment of all major aspects of engineering experimental work. Since the hardware portions of the measurement systems text focus on equipment associated mainly with mechanical, aerospace, electrical, industrial, and civil engineering, that text may require augmentation from other sources when it is used outside these areas. This new text, since it does not deal much with hardware, should be useful for anyone concerned with engineering experimentation, irrespective of the particular field. In fact, many chapters would be useful outside of engineering entirely.

The book can be profitably used with almost any course involving experimentation, but it is particularly appropriate for project laboratories, where students are responsible for the entire process of experimentation, from choice of topic to final reporting. It should prove quite useful to graduate students embarking on an experimental thesis and especially helpful to those who might not have had much lab experience in their undergraduate work. Furthermore, Chaps. 2 and 4 can be the text for a minicourse in applied statistics. Many engineering curriculums outside industrial engineering can benefit from increased emphasis on certain statistical ideas and methods. In our crowded programs, the time for a complete course taught by a mathematics or statistics department may not be available. We have found that a quite respectable student competence can be obtained in about 10 to 20 hours of lecture if the topics are carefully presented with emphasis on application rather than rigorous derivation and proof. If several lab courses are available, this statistics presentation could be apportioned among them. I have given these two chapters a lot of thought and attention in an attempt to develop and present simplified methods that nonspecialists in statistics can understand and use. For example, I show that most problems that are conventionally treated with analysis of variance (ANOVA), an approach most students find confusing, can be treated as well or better with multiple regression. This allows concentration on a small number of basic methods and thus quicker comprehension because of the repetitive use of a few methods rather than sparse use of many different methods. Frequent use of general-purpose statistical software, such as many engineers now have on their personal computers, develops student familiarity and allows efficient treatment of realistic problems. This software also allows one to develop *simulation* as a new teaching tool for statistics, greatly speeding student comprehension of basic concepts and applications.

In terms of student level and maturity, the vast majority of the text material can be comprehended by first-year students; thus one could profitably use the text in an early core course in engineering design concepts that included experimentation as part of the design process. In terms of application, I have made a conscious effort to emphasize the use of experimentation as a valuable adjunct to the design and manufacturing process, rather than concentrating heavily on research studies. Led by earlier Japanese efforts (Taguchi methods, etc.), U.S. industry has recently made many important improvements in quality and productivity through use of systematic experiment design and implementation. I have tried to make many of these techniques accessible to the nonspecialist in statistics by showing simplified approaches to common problems. Considerable emphasis is also given to codes and standards, a much neglected topic in today's engineering curriculums, but an important one for many practicing engineers. Most engineering students leave school almost totally unaware of the significance of the vast body of standard practices used daily in many industries.

While the chapters are all related in the sense that they all discuss some portion of the total process of experimentation, the technical content is often quite self-contained. This allows the instructor to select topics and their sequence to meet the needs of a particular course or lab. Chapter 1 gives a good overview and is easy reading, so it can be used in almost any course. If statistical ideas are to be emphasized and students have no previous exposure to statistics, Chap. 2 must be read before Chap. 4. Measurement system concepts are treated in Chap. 3 for those who have not had a comprehensive course such as might be based on my *Measurement Systems* text. The treatment is necessarily brief but does cover all the basic ideas in a useful way. Chapter 4 is a streamlined treatment of the classical statistical subject of experiment design. Many large books have been written on this subject, and I have packaged and presented the main ideas and methods in a new way, to make these useful tools more accessible to the nonspecialist, with a minimum investment of learning time and effort. Project planning is discussed in Chap. 5 and includes a brief treatment of dimensional analysis as applied to scale modeling and for increasing the efficiency of experimentation.

Apparatus design and construction are covered in Chap. 6. Here, by choice, the hardware coverage relates mainly to mechanical and aerospace types of applications, but many useful organizing concepts are of more general application. Material on the important topic of laboratory automation is included. Some details of experiment execution are covered in Chap. 7, while Chap. 8 discusses statistical, graphical, and numerical data analysis. Since much of the statistical material was presented in Chaps. 2 and 4, the Chap. 8 presentation is mainly a review and recapitulation. Technical communication (reports, papers, articles, and oral presentations) is covered in Chap. 9. This material is largely self-contained and can be presented at an appropriate time in almost any course.

As a firm believer in the importance of laboratory-oriented studies in all engineering curriculums, I hope this text will join my earlier one on measurement systems in encouraging engineering educators to devote significant curricular time to this vital aspect of engineering practice. Taken together, the two books give a comprehensive and organized treatment designed to facilitate student mastery of all aspects of experimental work.

McGraw-Hill and the author would like to thank the following reviewers for their many helpful comments and suggestions: Al Belibahani, University of Cincinnati; Clifford Cremers, University of Kentucky; Jerry Hamelink, Western Michigan University; C. L. Hough, Texas A & M University; T. J. Lawley, University of Texas at Arlington; Barsam Marasli, University of Maryland; Trilochan Singh, Wayne State University; Evan F. C. Somerscales, Rensselaer Polytechnic Institute; Dean Updike, Lehigh University; Raymond P. Vito, Georgia Tech University; and Timothy Wei, Rutgers University.

Ernest O. Doebelin

CHAPTER
1

THEORY AND EXPERIMENTATION IN ENGINEERING

1.1 PROBLEM-SOLVING APPROACHES

In all branches of engineering, from the earliest times to the present and into the future, there exist only two fundamental approaches to solving problems that arise in the discovery of knowledge and its application to society's needs:

1. Theoretical (physical/mathematical) modeling
2. Experimental measurement

In engineering, this is true regardless of the discipline (mechanical engineering, electrical engineering, chemical engineering, etc.) or the engineering function (design, development, research, manufacturing, etc.). While some problems are adequately treated by using only theory, or only experimentation, most require a judiciously chosen mix of these techniques. Even though each specific application will exhibit its own peculiarities, we can identify some important *general* characteristics of the theoretical and experimental methods which will be helpful in deciding on the proper blend when a choice is necessary. Such a discussion also helps us organize our thinking about the whole process.

1

We should first be clear that whenever we choose to describe some device or process with mathematical equations based on physical principles, we always leave the *real* world behind, to a greater or lesser degree. That is, all physical principles, and their mathematical expression, when applied to real-world situations, are approximations of the real behavior. These approximations may, in individual cases, be good, fair, or poor, but *some* discrepancy between modeled and real behavior *always* exists. This fundamental truth will not change tomorrow. We may (and generally *do*) improve the quality of our approximations as time goes by, but perfection is an unreachable (though laudable) goal. We also need to remember that practical engineering (in contrast to "pure" science) labors under (sometimes overriding) constraints of time and money. That is, an engineer may be well aware of a "nearly perfect" theoretical approach to a problem but will consciously choose instead a simpler and less accurate method which is judged "good enough" in terms of overall project objectives. Our first comparison of theory and experiment thus centers on the fact that theories are always approximations involving simplifying assumptions, whereas experiments are run on the actual system and, when properly designed and executed, reveal the *true* behavior. This comparison clearly favors experiment over theory.

Our next comparison redresses the balance by introducing the *generality of results,* a powerful feature in favor of theory. The German scientist Helmholtz, when challenged by the "practical men" of his day for concentrating on what they considered abstract theoretical studies, said, "There is nothing more practical than a good theory." Today, of course, we would agree, based on the evidence of myriad useful products and processes developed with the aid of theory. Perhaps the main reason we prefer theoretical approaches (when their accuracy is adequate) is that theoretical results usually are applicable to a whole *class* of problems whereas experimental results are peculiar to the specific apparatus on which the measurements were made. A simple example, the cantilever beam of Fig. 1.1, illustrates this comparison. Using assumptions characteristic of an introductory course in strength of materials, one can derive the well-known result[1]

$$x = \frac{FL^3}{3EI} \qquad I \triangleq \frac{bt^3}{12} \tag{1.1}$$

Subject to the restrictions implied in the simplifying assumptions, we can use this formula to predict the deflection x of *any* cantilever beam, no

---

[1] The symbol $\triangleq$ means "equal by definition."

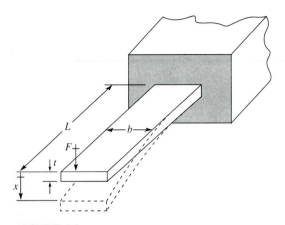

FIGURE 1.1
Cantilever beam experiment.

matter what its dimensions or material might be. If we imagined ourselves for a moment to be the builders of an ancient Roman aqueduct or Egyptian pyramid, engineering wonders constructed long before the theory of the strength of materials was developed, how would we "educate" ourselves about the behavior of cantilever beams? We could certainly construct an actual beam, load it with various weights, and measure the end deflection; however, all this *must* be done with a beam made of a *specific* material and of *specific* dimensions. It is *not* possible to test a "general" beam! If our measurements are carefully made and we don't overstress the beam, a graph of x versus F might lead us to conjecture that x is proportional to F. To discover the effects of material, we could build beams of different materials but identical shape and size. The effects of dimensions L, b, and t could individually be studied in similar fashion; however, the facts that x varies as L^3 and that, for beams of other (circular, triangular, etc.) cross sections, the area moment of inertia I is the significant quantity would be nearly impossible to discover by experiment alone. The ability to solve entire classes of problems with a single analysis gives theoretical methods an efficiency which is their major advantage. (We should at least mention at this point that experimental results may sometimes also be generalized by using techniques such as dimensional analysis.) We have already noted that a significant price paid for the generality of theory is the universal need for simplifying assumptions. Progress in theoretical methods mainly involves, as time goes by, a relaxation of these assumptions, bringing the mathematical model ever closer to the real-world device. This process, by its very nature (inclusion of increasing levels of detail), leads to mathematical formulations of deeper complexity and difficulty. In earlier times, recognition of these mathematical roadblocks stifled the application of complex models, resulting in the use of only simple models, the inaccuracy of which then required extensive experimental fine-tuning. This mathematical intractability of complex (and accurate) models could thus be legitimately considered a negative feature of theoretical approaches.

Today, of course, this problem has been greatly (though not entirely) alleviated by the development of computerized numerical methods (such as the finite element methods) which can "solve" extremely complex models. Note, however, that these methods *sacrifice* generality; they do *not* provide "formula" solutions like our Eq. (1.1) but rather give only numerical answers for individual specific cases, analogous to a physical experiment. Just as the pursuit of accuracy in mathematical modeling leads to undesired complexity, so also does experimental apparatus generally become more intricate when we require higher accuracy in the measured result. The negative effect of complexity, associated with the quest for theoretical accuracy, is counterbalanced by a similar trend in experimentation; thus this comparison shows no obvious or uniform advantage for either side.

Turning now in our comparison to consideration of the "facilities" needed to undertake theoretical or experimental studies, we initially see a considerable advantage on the side of theory. Many theoretical developments have been successfully carried out with little more than a trained engineer (or group) plus pencil and paper. Experimentation, by definition, requires an investment (sometimes considerable) in equipment suitable for the proposed study, plus laboratory space to house the experiment. It thus appears that theoretical studies may be initiated with less commitment of financial resources than would the experimental equivalent. This apparent advantage at least partially disappears when we recall that most *current* theoretical studies will involve *computing facilities* in a significant way. These can easily cost more (initially and/or in maintenance costs) than laboratory equipment, somewhat redressing this imbalance in our comparison. To further show that neither side has a clear superiority here, we need to point out that an investment in computing facilities can be amortized over *many* projects since the computer is a *general-purpose* device, requiring only software modifications to adapt it to an unlimited variety of problems. Much lab equipment and instrumentation, on the other hand, is quite specialized and useful only for a narrow range of tasks.

Our final comparison of theoretical and experimental approaches considers the time required to obtain results. When a "pure scientist" is pursuing knowledge of the basic laws of the universe, the unending nature of such quests allows a certain lack of urgency. In the fast-moving and competitive field of computer microchip development, however, a result which is technically correct but which becomes available a few months after your competitor's engineers have reduced it to practice can have severe economic consequences. We thus see that any technical endeavor has a certain appropriate time scale associated with its success. Deciding what this time scale should be, at the beginning of a new project, can be difficult, but skilled technical managers develop considerable proficiency in this area after years of experience. Once we make our best judgment here, next we can consider whether a theoretical or experimental approach is most likely to realize our timing goals. At first glance, theory seems to offer advantages since one can "start immediately," whereas experimentation requires the building and debugging of an apparatus, purchasing of parts and instruments, and trial runs and calibrations before the first data become available. However, if a chosen theoretical approach turns out to be fruitless and we then have to formulate another, theory begins to lose its advantage. In large manufacturing enterprises, one often finds the same problem being attacked at two levels. In the manufacturing division, where production holdups are catastrophic, quick trial-and-error experimental "fixes" are often the rule. "Anything that works" is accepted without concern about basic understanding. However, to reduce the incidence of such production-line emergencies, a wise management will devote some of its resources to a research and development group, where the less hectic pace allows a more scientific study of the problem.

It should be clear that sometimes a theoretical approach is quickest and at other times experimentation gives the most rapid solution. Judgment, based on extensive experience, is necessary for making such initial choices. In general, though, "simple" problems are more likely to yield quickly to theory, whereas puzzling phenomena may require a strong component of experimentation. A summary of our consideration of the features of theoretical and experimental methods appears in Fig. 1.2.

1.2 FUNCTIONAL TYPES OF ENGINEERING EXPERIMENTS

A careful review of engineering experiments actually performed and reported on reveals that they all can be classified into a small number of different types with regard to their purpose (see Fig. 1.3). A brief examination of these classes will give us a clear appreciation of the scope and importance of experimental methods.

Features of alternative methods of problem solution

Theoretical methods	Experimental methods
Study mathematical models of the real world, which always require simplifying assumptions.	Study the real world, no simplifying assumptions are needed.
Give general results, for a whole class of problems.	Give results specific to the apparatus studied.
Relaxation of assumptions leads to more complex math model.	Higher accuracy measurements require more complex instrumentation.
Facilities needed to commence study can be meager (trained personnel plus "pencil and paper").	Extensive (and expensive) laboratory facilities may be needed.
Study can commence promptly.	Time delays may occur in apparatus construction and debugging.

FIGURE 1.2
Comparison of theoretical and experimental solution methods.

Functional types of engineering experiments

1	Determination of material properties and object dimensions
2	Determination of component parameters, variables, and performance indices
3	Determination of system parameters, variables, and performance indices
4	Evaluation and improvement of theoretical models
5	Product/process improvement by testing
6	Exploratory experimentation
7	Acceptance testing
8	Use of physical models and analogues
9	Teaching/learning through experimentation

FIGURE 1.3
Classification of experiment types.

1.2.1 Determination of Material Properties and Object Dimensions

While the derivation of results like Eq. (1.1) can be purely theoretical, their practical application invariably requires experimentation to obtain numerical values to insert into the formula. That is, the *concept* of Young's modulus E as a relation between stress and strain is a theoretical one; however, numerical values of E for a specific material are obtained by experiment. While physicists pursuing the "fundamental laws of nature" (in terms of ever-more-numerous and smaller basic particles) promise "ultimate explanations" of all physical phenomena, physicists are not, in general, able to predict accurately most important material properties in terms of the basic atomic structure. Thus, when our theory needs numerical values for viscosity, specific heat, elastic modulus, expansion coefficients, etc., we must devise and carry out suitable experiments. This is also clearly the case for physical dimensions such as b, t, and L in Fig. 1.1; they must be experimentally measured. Since all "theoretical" results require numerical data of this sort, this class of experiment is extremely important. Because of the need for accuracy and consistency, in many cases such tests have been codified into standard procedures (which provide explicit instructions for carrying out the experiment) by professional societies such as the American Society for Testing and Materials[1] (ASTM).

1.2.2 Determination of Component Parameters, Variables, and Performance Indices

"One step removed" from tests of material properties and physical dimensions are those experiments aimed at quantifying the behavior of basic components which are to be built into larger systems. Some examples are springs and dampers used in automotive suspension systems; resistors, capacitors, and operational amplifiers used in electronic circuits; and valves, filters, and flow restrictors used in hydraulic systems. For the viscous damper of Fig. 1.4[2] (used in a vibration control system for the *Ranger* spacecraft), the performance attribute of primary interest is the relation between the applied force F and resulting velocity

---

[1] ASTM, 1916 Race St., Philadelphia, PA 19103, phone (215) 299-5585.

[2] E. O. Doebelin, *System Dynamics,* Merrill, Columbus, Ohio, 1972 [most recent printing available from Long's Bookstore, 1836 N. High St., Columbus, OH 43201, phone (614) 294-4674].

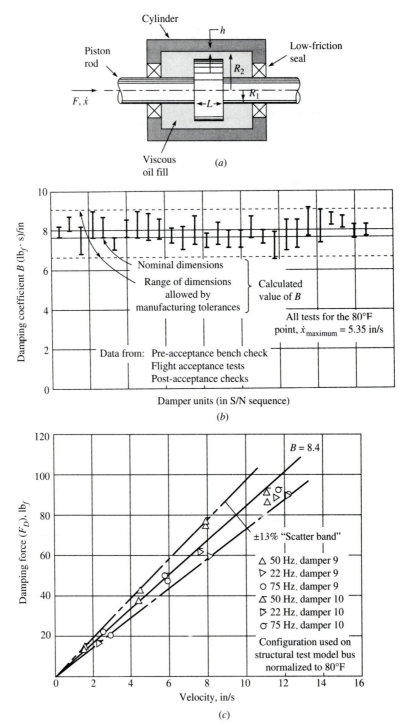

FIGURE 1.4
Experiment to determine the damping coefficient.

v, which ideally is given by $F = Bv$, where B is a constant. Experiments here would be aimed at examining whether a straight-line relation between F and v is closely approximated and, if so, the best numerical value for B. Once confidence in component behavior is established by suitable experiments, design of the larger system can proceed with more assurance that no disastrous surprises will occur.

1.2.3 Determination of System Parameters, Variables, and Performance Indices

Here we immediately confront the ambiguity of the terms *component* and *system*. When we encounter a damper as a graphical symbol in the schematic diagram of a complex vibrating system made up of many masses, springs, and dampers, it is natural to consider the damper as a component. However, when we actually design and build a real damper, which includes seals, bearings, fasteners, cylinder, piston, piston rod, fluid, etc., we might wonder whether this rather complicated assemblage is not a "system." This apparent semantic difficulty needn't present any real problems, however, since the appropriate meaning is usually apparent from the context in which the word is used. For example, when we are discussing the suspension *system* of an automobile, it is clear that a shock absorber (damper) is, in this context, a *component*. Experiments on a suspension system might be directed toward finding the angle of inclination of the roll axis (a parameter), measuring the roll angular velocity (a variable) during a lane-changing maneuver, or determining the ride roughness (a performance index) when traversing a "standard" rough road.

1.2.4 Evaluation and Improvement of Theoretical Models

While the accuracy of computer-aided analysis methods is continually improving, the commitment of a *purely* theoretical design to mass production, without experimental testing of some kind, involves a level of risk most managers are unwilling to take. Much experimentation thus involves the checking of theoretical predictions to establish a desired level of confidence in the calculation methods used. For problems with unfamiliar features, initial theoretical efforts may be quite inaccurate, but can often be significantly improved, by using the guidance supplied by experimental testing programs.

The engineering literature is full of examples of this class; here we quote only one. In support of design studies for electric power generating systems for orbiting space stations, the NASA Lewis Research Center (Cleveland, Ohio) performed theoretical and experimental studies on the dynamic behavior of the condenser in a Rankine-cycle power plant using mercury as the working fluid.[1,2] Since the condenser was only one part of a large system with many interacting components (boiler, turbine, electrical generator, etc.), there was a great incentive to keep the individual models simple, so that the overall power plant model could be kept comprehensible. This philosophy leads to the use of many simplifying assumptions, the validity of which is open to question since much of the technology in this system was unfamiliar. Experimentation was found invaluable in the development of this model and occurred at two different levels. *During* model development, many experiments at the level of basic phenomena were run to establish relations and numerical values for processes such as condensing heat transfer and fluid flow friction which occur in the condenser. When the condenser model was thought to be *complete,* system-level input/output experiments on the entire condenser were used to validate the overall predictions of behavior. The model relates two outputs (condenser pressure and location of the vapor-liquid interface) to three inputs (vapor inflow rate, coolant temperature, and condensate back pressure) through a set of simultaneous ordinary differential equations. Of the six possible input/output responses, two were studied experimentally by using dynamic sinusoidal test methods. Results were quite favorable except for a large inaccuracy in the predicted phase shift between the vapor inflow rate and interface location. This is probably due to a time delay between the formation of condensate droplets and their subsequent migration to join the slug of condensate residing in the downstream end of the condenser tube. No theoretical approach to this phenomenon was known, so it could not be included in the initial model; however, the system-level experiments clearly showed its effect and allowed correction of the final model by addition of a two-second "dead time." The system-level experiments were thus invaluable in validating the majority of the analytical assumptions but also in uncovering an important defect and then suggesting both the form and numerical values of a corrective term.

---

[1] E. O. Doebelin, *System Modeling and Response: Theoretical and Experimental Approaches,* Wiley, New York, 1980, pp. 447–467 (latest printing available from Long's Bookstore, 1836 N. High St., Columbus, OH 43201).

[2] A. A. Schoenberg, *Mathematical Model with Experimental Verification for the Dynamic Behavior of a Single-Tube Condenser,* NASA TN D-3453, 1966.

1.2.5 Product and/or Process Improvement by Testing

The use of experimental testing during the development phase of new products or processes has a long history; however, during the 1980s and 1990s it took on new significance for manufacturing operations in the United States. Japanese manufacturers had earlier developed and implemented efficient methods for *integrating* the design, testing, and manufacturing phases of product development so that the phases occurred more in parallel than in series, leading to economical production of high-quality products with short lead times. Phrases used to describe these methods include *simultaneous engineering, design for manufacture, concurrent engineering,* and *design for assembly.* Systematic experimental test methods, including *Taguchi methods,*[1] are an important part of this process, which we will explain in more detail in later chapters. I consider the introduction of these methods into U.S. manufacturing to be one of the most economically significant engineering developments in many decades, as attested to by numerous in-house short courses developed and presented by companies like AT&T, Ford, and General Electric and by commercial training institutes.[2]

The importance of experimental testing in product development can be appreciated from the following comment by David Packard, a founder of Hewlett-Packard Corp.; "Reliability cannot be achieved by adhering to detailed specifications. Reliability cannot be achieved by formula or analysis. Some of these may help to some extent but there is only one road to reliability. Build it, test it, and fix the things that go wrong. Repeat the process until the desired reliability is achieved."[3]

Some insight into the details of such testing is brought out in the following quote:

> Coincident with the emergence of the prototypes is testing. First the basic concept is examined to see if it addresses all the design constraints. Next, normal operations are explored to find where the design is deficient. (It always is!) Finally the design limits are probed through accelerated tests or abuse testing.

---

[1] M. S. Phadke, *Quality Engineering Using Robust Design,* Prentice-Hall, Englewood Cliffs, NJ, 1989.

[2] *Taguchi Approach to Quality Optimization,* Technicomp, 1111 Chester Ave., 300 Park Plaza, Cleveland, OH 44114, phone 1-800-255-4440.

[3] E. M. Bailey, "Design for Reliability in the HP 256X Family of Line Printers," *Hewlett-Packard Journal,* June 1985, p. 5.

Throughout, the engineer repeatedly cycles through the test-analyze-fix process. Initial testing yields many easily discovered defects, which can be quickly resolved with engineering analysis. Subsequent testing is aimed at improving ruggedness and reliability; these test scenarios usually require many unit-hours of experience, and the conclusions must be reached by careful statistical inference.

Compounding the statistical problem is that of securing representative parts. Much of the initial testing is performed with parts from prototype tools or processes. The design limits are not well understood at this point, and the parts are varying because the process is still unstable and undeveloped. Tolerance analysis of the design is useful, but not sufficient, since the called-for tolerances must be satisfied by a production process. Much effort goes on at this time to allocate the tolerances and allowances between parts and process.

The final phases of testing involve tests under controlled conditions by impartial quality assurance engineers. Here, testing to rigorous HP standards is completed, including temperature and humidity excursions, shock and vibration tests, and transportation and use/abuse tests that seek out the weak links in the design.

Life testing proceeds under accelerated and nonaccelerated conditions, probing the design for deterioration, wearout, and contamination.

In summary, reliable design requires more than theoretical design skills and analysis. Although good first-round designs are an essential foundation, *the bulk of the engineer's efforts go into executing well-thought-out testing programs* whose intent is to stress the design and uncover its limitations so that improvements in the subsystems and the integrated product can be made.[1]

This emphasis on experimental testing is not unique to this product or company. We *often* find that testing efforts comprise more than half of the entire engineering effort.

1.2.6 Exploratory Experimentation

Some experimental work of a tentative and somewhat unfocused nature may be carried out during the early stages of familiarization with a physical phenomenon. Here one wants to keep an open mind, in hopes of observing some unusual or potentially useful behavior which might become the subject of later, more focused investigation. In routine work we usually look for reasons to throw out data points that don't repeat themselves (outliers); in exploratory work the "wild" point may be the

---

[1] J. D. Rhodes, "Managing the Development of the HP Deskjet Printer," *Hewlett-Packard Journal,* October 1988, pp. 51–54.

basis of an important new product or process, so we should not be too quick to dismiss it.

1.2.7 Acceptance Testing

Engineering interfaces with the legal profession in various ways such as product liability, occupational safety and health standards, and fulfillment of contracts. When a purchaser contracts with a supplier for the delivery of a large and expensive machine such as a steam turbine for a central-station power plant, part of the contract consists of detailed performance specifications which the machine must meet before final payment will be authorized. Since large sums of money and important time schedules may be involved, the test procedures to be used (to verify that performance meets that guaranteed) must be precisely spelled out in the contract. Often, standardized test methods promulgated by agencies such as the American Society of Mechanical Engineers (ASME) or ASTM and acceptable to the contracting parties can be used. Experimentation serving this function is called *acceptance testing,* and it is applied both to costly single items (as in our example) and to quantities of components (nuts and bolts, etc.) purchased for assembly into products. When large quantities of a single item are involved, 100 percent testing is rarely economical and sample testing based on statistical principles may be necessary.

1.2.8 Use of Physical Models and Analogues

While computerized *mathematical* models are increasingly the tool of choice for analysis, some use of physical models and analogies persists. A physical model is intended to mimic the behavior of selected aspects of the device under study. It may be constructed of the same, or different, materials and is often of reduced size. Wind-tunnel models of automobiles, used for drag studies, need only duplicate the outside forms of the actual car; other features such as suspension dynamics need not be duplicated. In chemical engineering, however, a pilot plant for a new process often needs to duplicate the full-scale process almost completely, but at reduced scale. Whereas a *model* is considered by most engineers to involve the *same* physical phenomena as the device under study, an *analogue* is defined as a system which obeys the same mathematical equations but involves *different* physical effects. Electronic analog computers (now largely replaced by equivalent digital simulation) were extensively used from 1940 to 1970 to study all kinds of physical systems modeled with ordinary differential equations. The shape of a pressurized soap film stretched across an opening of a certain form is analogous to the torsional stresses and deflections in a rod with that same cross-sectional form. While electronic analog computers and soap bubble measurements are not used much anymore, photoelastic stress analysis remains an important tool. Here, parts made of special transparent plastics are subjected to applied forces and polarized light. The optical response (light and dark fringes at specific spacings) is analogous to the stress response of the actual part and gives numerical values of the stress components.

1.2.9   Teaching and Learning through Experimentation

Most engineering curriculums include experimentation as a portion of the total educational experience. The fraction of the total curriculum devoted to laboratory studies varies from school to school and can take several different forms. Since we remember best those things we actively participate in, hands-on experiments should be part of every lab program. I believe that combined lecture and demonstration also has a place. A carefully designed demonstration apparatus, operated by a skillful presenter, can illustrate a complicated concept in a well-organized sequence, allowing students to comprehend and assimilate it very efficiently. The false starts, misunderstandings, delays, and mistakes characteristic of hands-on experimentation are actually valuable experiences, but they certainly do cause "inefficiency" in terms of concepts learned per unit of class time spent. Thus a mix of demonstrations and hands-on modes allows us to balance these conflicting goals.

Another classification involves set experiments versus original projects. Again, a tradeoff exists; so if there is enough curricular time available, both approaches should be used. Set experiments are carefully designed by the faculty, vary only slightly from year to year, and include rather specific instructions for the students to follow. Such an approach allows little student creativity but is a very efficient way to present a set of important basic concepts, develop familiarity with lab equipment and procedures, and instill good habits in data taking and report writing. Project labs (usually *one* project per term whereas we might have 5 to 15 set experiments) enable students to engage the *entire* process of experimentation, since they must

1. *Select* a problem appropriate to available facilities and time schedule.
2. *Plan* the overall method of attack.
3. *Design* the needed measurement systems.
4. *Build* the apparatus.
5. *Debug* the apparatus and measurement systems.

6. *Execute* the experiment.

7. *Gather and process* the data.

8. *Interpret* the results.

9. *Prepare and present* written and/or oral reports.

When properly managed by the supervising faculty, such a project experience can be one of the most useful parts of the entire curriculum.

A final way of categorizing educational experiments asks whether the lab is *self-contained* or an *adjunct* to a concurrent lecture course. Again, both modes have their place in a well-designed curriculum. In a lab that is an adjunct to a lecture course on, say, fluid mechanics, one would probably emphasize hands-on and demonstration experiences, illustrating and clarifying basic principles with set experiments. Details of instrumentation would play a secondary role, and required reports would be kept short and simple. While alternative approaches are certainly defensible, I like a scheme which uses *two courses devoted specifically to the general area of experimentation as a problem-solving tool.* (These are *not* adjunct to lecture courses in some specific area of engineering science such as heat transfer or electric circuits.) The first course uses set experiments with a combined lecture and lab format to present the general principles underlying the design and use of measurement systems.[1] There is also considerable emphasis on written and oral reporting. This course (together with earlier lab experiences adjunct to several lecture courses) prepares students well for the second experimentation course, a project lab employing a single term-long problem as described above.

1.3 COMPUTER SIMULATION AND PHYSICAL EXPERIMENTATION

We have mentioned several times already that some engineering studies pursued experimentally in the past are today more effectively accomplished by using computerized mathematical models. In addition to *replacing* some types of experimentation, computer simulation is a useful *adjunct* to experimentation. Let's use a simple example to explain this concept. In systems with mechanical moving parts, frictional effects in bearings are sometimes important factors in determining motion. If your experience with friction is mainly through introductory physics and

---

[1] E. O. Doebelin, *Measurement Systems: Application and Design*, 4th ed., McGraw-Hill, New York, 1990.

mechanics courses, you may have a simplistic view of real-world frictional behavior, limited to friction forces which are either constant (Coulomb or dry friction) or proportional to the velocity (viscous friction). Real bearings exhibit more complex behavior which must sometimes be defined by experimental testing. One bearing friction model which is practically useful, but still simple enough for our introductory example, combines dry and viscous friction effects in adjustable proportions, giving a versatile model which can be "fitted" to experimental test results.

Figure 1.5a shows schematically a simple mechanical system including bearing friction models. A simple but useful test gives an initial displacement to the mass, releases the mass with zero velocity, and then measures the curve of displacement x versus time t. Theoretical vibration studies long ago showed that pure dry friction gives a straight-line decay envelope while pure viscous friction results in an exponential (e^{-at}) envelope (Fig. 1.5b). Real bearings may behave according to some *combination* of the two effects, and we can discover the proper combination by matching the measured decay curve with the right member of a family of curves generated with computer simulation.

Although we could write our own simulation programs, ready-made, convenient, and powerful general-purpose packages[1] for physical systems described by ordinary differential equations have been in wide use for many years, most recently in personal computer (PC) versions. (These software products are, of course, useful for many purposes[2] beyond our current example.) We will use CSMP (continuous system modeling program) for our example since I believe it is the easiest to follow for readers who know a little FORTRAN but *don't* know a specific simulation language. (While CSMP is still in use at some facilities, it is no longer supported actively by a software company; thus users new to simulation should choose one of the other packages currently marketed.) *All* these packages are actually quite similar and easy to learn and use. For the system of Fig. 1.5, Newton's law gives

$$M\ddot{x} = -K_s x - B\dot{x} \pm F_{df} \qquad (1.2)$$

Using approximate (but highly accurate) numerical integration, CSMP will solve this equation for the displacement x, so that we can compare it

---

[1] ACSL, Mitchell and Gauthier Assoc., Concord, MA 01742.
CSSL, Simulation Services, Chatsworth, CA 91311.
CSMP, California Scientific Software, Sierra Madre, CA 91024.
SIMULAB, The Math Works Inc., Cochituate Place, 24 Prime Park Way, Natick, MA 01760.

[2] E. O. Doebelin, *System Modeling and Response,* chap. 5.

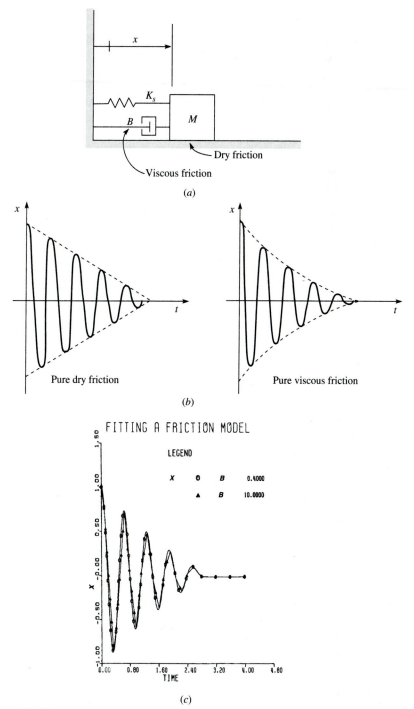

(a)

(b)

FITTING A FRICTION MODEL

(c)

FIGURE 1.5
Simulation used to match model to measured behavior.

with the measured curve. All the details of the solution procedure are automatically taken care of so that the user need only write a few simple but powerful CSMP statements, intermixed with FORTRAN as needed.

1 METHOD RKSFX   Selects a Runge-Kutta fixed-step integrator
2 TITLE FITTING A BEARING FRICTION MODEL
3 PARAM KS = 100., M = 1.0   Sets numerical values
4 PARAM FDF = 4.   Dry friction force magnitude
5 PARAM B = (0.0, .4, .8, 10.)   Multiple values of viscous friction
6 SIGNV = FCNSW(XDOTI, −1, 0.0, 1.)   Algebraic sign of dry
                                                                                     friction
7 NOSORT   Signals upcoming FORTRAN code which must be in
                        sequence
8 10 IF(B − 5.) 40, 40, 200   If $(B − 5.) \leq 0$, go to 40, otherwise 200.
9 40 XDOT2 = (−KS*X − B*XDOT1 − FDF*SIGNV)/M   Newton's
                                                                                          law
10 50 XDOT1 = INTGRL(0.0, XDOT2)   Get velocity by
                                                                  integration.
11 60 X = INTGRL(1.0, XDOT1)   Get displacement by
                                                            integration.
12 70 GO TO 300
13 200 X = AFGEN (CURVE1, TIME)
14 AFGEN CURVE1 = 0.0, 1.0, . . .   Measured values of x
                                                          and t
15 .08, .72, .32, −.86, . . . 4.0, −0.16
16 300 CONTINUE
17 TIMER FINTIM = 4.0, DELT = .002, OUTDEL = 0.004
18 OUTPUT TIME,X   Graph x versus time.
19 PAGE MERGE, XYPLOT, HEIGHT = 5.5, WIDTH = 6.5   Overlay
                                                                        four curves for B values.
20 LABEL FITTING A FRICTION MODEL   Label for graph
21 END
22 PARAM   B = (1.1, 1.5, 1.9, 2.3, 10.)   New B values for a second
                                                                run
23 PARAM FDF = 3.   New FDF value for a second run
24 END

In line 1 we select (from eight available) a numerical integration algorithm (usually this is left to default) suitable for the discontinuous force changes caused by dry friction. Lines beginning with PARAM set numerical values for the spring constant, mass, etc. In any run, *one* parameter may be assigned n multiple values (see line 5). This causes the whole problem to be run n times, once for each parameter value. During graphing, PAGE MERGE (line 19) causes n curves to be overlaid for

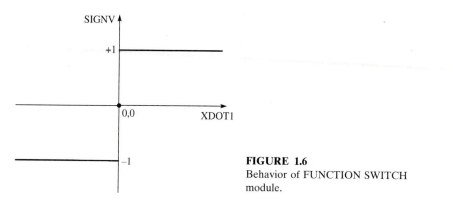

FIGURE 1.6
Behavior of FUNCTION SWITCH
module.

convenient study of the effect of that parameter. Line 6 uses the CSMP module FCNSW (function switch) to give the dry friction force the correct sign as velocity XDOT1 changes sign (see Fig. 1.6).

With few exceptions, the *sequence* of CSMP statements (unlike that of BASIC, FORTRAN, etc.) in a program is not important since CSMP has a built-in *sorting algorithm* which sorts the statements into proper order before processing. If we include in a CSMP program any FORTRAN coding, which *does* require proper sequence, we must warn CSMP of this with a NOSORT statement (line 7). We use this feature in our example to include a graph of the *measured x, t* data in our PAGE MERGE display of several *B* values. The last multiple-*B* value (10.) is used *not* as a value of *B* but rather as a trick to add the graph of measured data to the overlaid graphs for *B* = 0.0, 0.4, and 0.8. In line 8, the IF statement directs program flow through statements 40, 50, 60, 70, and 300 for *B* values less than 5.0. This program path computes acceleration in line 9 and then uses numerical integration in line 10 to get the velocity (initial value 0.0) and again in line 11 to get the displacement (initial value 1.0). When the multiple-value *B* statement reaches 10, the IF statement diverts the program *around* the numerical integration to lines 13 through 15, where we give the experimentally measured *x* versus time curve, using CSMP's arbitrary function generator AFGEN. (In CSMP the independent variable in all differential equations is always called TIME and automatically starts at zero.) Lines 14 and 15 simply list pairs of measured *x, t* values, starting with *t* = 0.0, *x* = 1.0 and continuing until *t* = FINTIM = 4.0 seconds, the end of our problem. We can use as many pairs of points as we wish, and CSMP interpolates linearly between them.

Line 17 sets the physical problem duration FINTIM, the computing increment DELT, and the graphing increment OUTDEL. In line 19, XYPLOT selects a high-resolution electrostatic plotter (rather than the default line printer) while HEIGHT and WIDTH set the graph size. Line 21 signals the end of the first simulation, where three B values and one FDF value are studied. Only two additional lines, 22 and 23, are needed to rerun the entire program for four new values of B and one new value of FDF. This use of END, PARAM, END can be repeated as many times as we wish to explore all desired combinations of dry and viscous friction. Figure 1.5c shows a run where $B = 0.4$ and $FDF = 4$. seem to match the measured response quite well.

Another, more widely known, method of fitting theory to experimental data can be used when we have a *formula* containing the parameters B and FDF, which relates x to t. Then various curve-fitting methods, which we will explore in later chapters, are available for choosing numerical values of B and FDF which fit the theoretical curve to the measured data in some "best" sense, such as least squares. These methods would often be quicker than the simulation scheme we explained above, but are not as versatile. One needs to be familiar with both approaches so as to choose the more appropriate one for each situation.

1.4 LABORATORY FACILITIES IN THE UNITED STATES

Many industrial firms involved in product design and/or manufacture maintain in-house laboratories. In the larger firms there will be many laboratories, each devoted to some special area of research or development. In-house labs not only are convenient but also may be a necessity in new-product development where secrecy must be maintained for competitive advantage. The federal government maintains many large laboratories for commercial and military studies (National Institute for Standards and Technology, the various NASA research centers, Army Tank-Automotive Command, etc.). Some states have also created specialized research centers to encourage economic development of certain industries within the state. Comprehensive research universities will have many specialized labs supported partly by the university and partly by outside projects funded by industry or the federal government. Sometimes federal agencies such as the National Science Foundation will attempt to meet perceived national needs in critical technical and educational fields by supporting development of specialized labs at universities. The Net-Shape Manufacturing Center at The Ohio State University was created in this way.

Smaller companies which cannot afford to maintain a variety of labs (perhaps none at all) need to locate facilities capable of carrying out the studies needed. These will be found mainly at universities; large, comprehensive private research institutes (such as Battelle Memorial

Institute in Columbus, Ohio); or specialized commercial labs. The American Council of Independent Laboratories, Inc.[1] publishes a 385-page guide listing its 329 member labs and describing their capabilities with respect to discipline (acoustic and vibration, biological, chemical, mechanical, etc.) and product or service (abrasives, adhesives, aircraft components, cement, fluid waste disposal, environmental simulation, etc.).

PROBLEMS

1.1. Prepare one-page outlines describing a plan of attack to study the problems listed below, using mainly a theoretical approach. Repeat for a mainly experimental approach.

a. Assembly-line workers using electric screwdrivers are unable to work a complete shift because of excessive hand fatigue and pain.

b. Get a numerical value for the drag coefficient of a newly designed automobile.

c. Predict the failure load of a bolt under a known tension force.

d. Choose the optimum pipe diameter for a pipeline carrying a known flow rate. The pipe purchase price increases in a known way with pipe size, and the operating cost decreases (due to less friction) with pipe size in a known way.

e. Find the steady-state winding temperature of an induction motor under a known steady load.

f. Find the average life (pages printed) for a toner cartridge used in a laser printer.

1.2. Using available libraries, find and briefly describe experiments of the following types, as defined in Fig. 1.3.

a. Type 1	*d.* Type 4	*g.* Type 7
b. Type 2	*e.* Type 5	*h.* Type 8
c. Type 3	*f.* Type 6	*i.* Type 9

BIBLIOGRAPHY

1. J. C. Gibbings, *The Systematic Experiment*, Cambridge University Press, New York, 1986.
2. B. J. Brinkworth, *An Introduction to Experimentation*, American Elsevier, New York, 1968.
3. E. Bright Wilson, Jr., *An Introduction to Scientific Research*, McGraw-Hill, New York, 1952.
4. Hilbert Schenck, Jr., *Theories of Engineering Experimentation*, 2d ed., McGraw-Hill, New York, 1968.
5. Hilbert Schenck, Jr., *An Introduction to the Engineering Research Project*, McGraw-Hill, New York, 1969.

---

[1] Located at 1725 K St., N.W., Washington, DC 20006, phone (202) 887-5872.

CHAPTER
2

ELEMENTARY
APPLIED
STATISTICS

2.1 ENGINEERING APPLICATIONS
OF STATISTICS

Engineers are finding increasing practical usefulness for some of the basic concepts of the branch of mathematics called probability and statistics. This chapter is devoted to the development and application of some elementary ideas from this area. While this is a text devoted to engineering experimentation, the statistical concepts and methods needed there are also applicable to other areas. Our approach in this chapter is thus designed to serve a wide application area, with later chapters showing specific uses in experimentation. This allows use of this chapter as a "minicourse" in applied statistics for students whose curriculum cannot afford the time for a complete course, as taught in a statistics department from one of the standard texts. We have found this approach very satisfactory with not only statistics but also other specialized topics where we feel a practical working knowledge, rather than in-depth specialization, is adequate in our required curriculum. Seeing the practical utility of statistical methods, those students planning careers in statistically sophisticated areas can pursue in-depth treatment in their elective programs. To give you some practically useful statistical tools in a short time, our approach is a pragmatic one which deemphasizes proofs and concentrates on general concepts and specific calculation methods for commonly occurring problems. We begin our discussion by listing and explaining three major application areas where statistical viewpoints and methods provide useful approaches:

1. Failure prevention in machine and process design

2. Experimental engineering analysis

3. Manufacturing quality control

An important phase of the *design* of any machine or process involves proportioning the parts so that failure of one type or another will not occur. Failure occurs when a "load" of some sort exceeds the resistance of the material to that load. Statistics enters the picture when we realize that in real situations (rather than contrived textbook problems) both the applied loads and the strength of the machine part are not precisely known or perfectly predictable, i.e., they are statistical variables. For example, when designers specify SAE 4340 steel with a breaking stress of 100,000 psi, they realize that the particular piece of steel which is used for any given machine part will *not* have a breaking stress of 100,000 psi, but rather will break in a range of, say, 85,000 to 115,000 psi. Furthermore the strength of the given part *cannot* be found without destructively testing that very part, and so the breaking strength will always be somewhat uncertain and statistics may be helpful in deciding the *probability* of the strength falling below a chosen value.

Similarly, the dimensions of parts, and the applied loads are also subject to uncertainty. If one specifies a dimension as 1.25 ± 0.05 inch, the machine shop will accept parts in the range of 1.20 to 1.30 in. This uncertainty in dimension must, of course, produce an associated uncertainty in the strength of the part. Also, although the loads in some machine parts may be quite accurately calculated, many practical loading situations exhibit considerable variability. For example, no one can predict beforehand the history of shock loads sustained by an automotive suspension spring due to road bumps. Statistics can describe such a situation quantitatively in terms of average values and dispersion or "scatter" around this average.

From the above discussion one can see that a design procedure may be formulated in which one compares the probability of the applied stress (a function of loads and dimensions) exceeding a chosen value with the probability of the material strength falling below that value. The probability of the simultaneous occurrence of both these events is the probability of failure. Most design engineers today do not apply these statistical techniques to "everyday" design problems since the basic data needed may be excessively expensive to gather; it is cheaper to simply "overdesign" the parts. However, these techniques *are* used in critical applications, and even in the everyday problem they give a useful way of qualitatively reaching decisions even if the techniques are not applied numerically. Figure 2.1 illustrates some of these statistical concepts as applied to design engineering.

Turning to the area of *experimental engineering analysis*, we find that statistical methods have many important applications such as specifying the uncertainty of measured data, planning experiments to get the maximum amount of significant data with the least effort and expense, and rationally testing hypotheses to decide the probability of their being true or false. Every measurement we make has *some* inaccuracy, and this can be broken down into so-called systematic and random errors.

Systematic errors repeat themselves if the measurement is repeated and may often be removed by calibrating the instrument against a more accurate standard. *Random errors* do not behave in this way. Rather, they exhibit scatter, and the best that we can do is to statistically estimate what their largest values might reasonably be. We can then quote the measurement as, say, 1.76 ± 0.03 to indicate that we are, say, 99 percent sure that the true value is somewhere between 1.73 and 1.79 (see Fig. 2.2). Whole books have been written on the use of statistical methods in planning efficient experiments. Many of these methods were originally developed for use in the biological and social sciences, where the interrelationships between variables are not nearly as clear-cut as they often are in physics and engineering. That is, in biological systems, it is often difficult or impossible to accurately apply principles such as Newton's law or Ohm's law to develop a rational model of some situation, and one is thus forced to run experiments to try empirically to discover relationships which might exist. Of course, physical scientists and engineers also need to run experiments, but their variables are often subject to much more precise control than are those in the biological and social sciences. Statistics has been found to be very useful, indeed necessary, in sorting out the subtle effects of various interacting variables in situations of this sort. As engineering is becoming more involved with biological and social systems, the need of engineers for some statistical background increases. Even in purely technical problems, statistics has been very helpful in giving a rational basis for reaching judgments when the interaction of variables is subtle. The study and improvement of

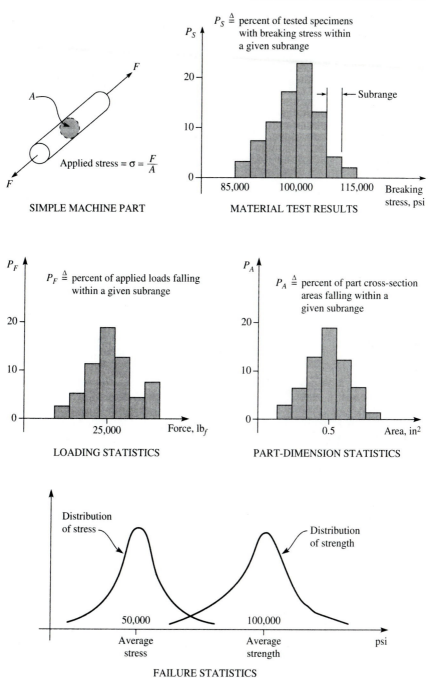

FIGURE 2.1
Statistical interpretation of machine part failure.

Trial number	True value	Measured value
1	1.760	1.53
2	1.760	1.49
3	1.760	1.54
4	1.760	1.52
.	.	.
.	.	.
.	.	.
48	1.760	1.54
49	1.760	1.50
50	1.760	1.51

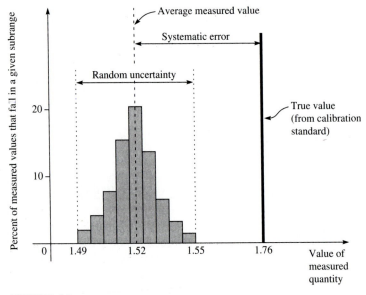

FIGURE 2.2
Statistical view of measurement accuracy.

manufacturing processes (many of which are very complex and poorly understood) is a prime example of such statistical applications.

In recent years, world competition from countries such as Japan, where manufacturing quality has been heavily emphasized and highly developed, has forced a renewed concentration on this area by U.S. manufacturers. Since all manufacturing processes, no matter how carefully designed and operated, exhibit random variability in product quality, statistical methods are the heart of *manufacturing quality control*. A simple, but important, example from this area is the quality control

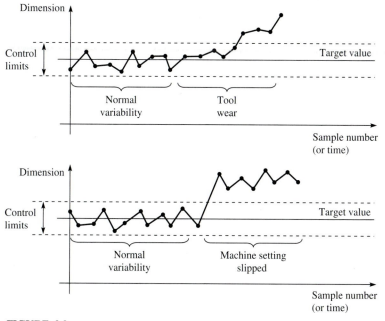

FIGURE 2.3
Quality control charts.

chart (see Fig. 2.3). Here some quality attribute, such as the diameter of a shaft produced by an automatic lathe, has been studied statistically for a sufficient period to reliably establish its "normal" variability, which is then used to establish "control limits" around the target value. By measuring each part produced (or perhaps a *partial* sampling known to be statistically reliable) and plotting the ongoing values, one can detect gradual changes, such as tool wear, or sudden events, such as slippage of settings, and take timely corrective action. Since *some* random variation is normal even in a perfectly functioning process, we need statistical criteria to properly distinguish between normal variability (which we accept without any attempt at correction) and abnormal happenings, which require adjustments to the process.

2.2 BASIC PROBABILITY CONCEPTS AND DEFINITIONS

Probability is a number between 0 and 1 related to a given event. If the event is absolutely certain, its probability is 1, and if the event is impossible, its probability is zero. Three different definitions of probability are in common use, each one being appropriate for certain types of applications.

First is the *classical* definition: If an event can occur in N equally likely and mutually exclusive ways, and if n of these ways have an attribute A, then the probability $P(A)$ of the occurrence of A is defined to be n/N. This definition is appropriate in studying games of chance, for example. If there are 4 aces in a 52-card deck, the probability of drawing an ace in one try is $\frac{4}{52} = 0.077$.

Second is the *frequency* definition: If an experiment is carried out N times and an event A occurs n times, then if we let N approach infinity, the limit of n/N is defined to be the probability $P(A)$ of event A. This is perhaps the most popular definition. It allows treatment of many practical problems in which the classical definition could not be applied. For example, there is no way to compute theoretically the values of n and N if the attribute A is that a 40-year-old U.S. male citizen's weight falls between 150 and 160 lb. To apply the frequency definition to this problem, we would need to take a large (and random) sample of 40-year-olds and tabulate their weights. Since we are usually unable (or it may be technically or economically not feasible) to study an infinite (or even an *entire* finite) population, we must decide on a sample size large enough to be reliable but small enough to be economical. Fortunately, statistical theory provides guidelines for such decisions.

Third is the *subjective* definition: Here the probability $P(A)$ of proposition A is a measure of the "degree of belief" one holds in the proposition. This is perhaps the broadest definition and is necessary, e.g., when one is applying statistics to areas such as national defense strategy. That is, suppose we are trying to choose between two alternative strategies, each of which is likely to produce certain results. Since events such as dropping (or not dropping) of bombs on certain "enemy targets" cannot be experimentally tried out to see how the enemy would respond, we must rely on "expert judgment" to assign a numerical probability in such cases. A similar situation exists in deciding on odds for a horse race or predicting who will win a football title. Here again the classical and frequency interpretations are useless, and a subjective judgment is necessary.

Irrespective of the *interpretation* one puts on probability, once a number has been chosen, any further calculations are independent of the definition.

2.3   PROBABILITY LAWS FOR COMBINATIONS OF EVENTS

We next state, without proof, two basic laws of probability useful in dealing with *combinations* of random events. These laws apply to many practical situations, but we will emphasize their implications for engineering design with respect to

1. Simplicity versus complexity in system design

2. Redundant design to achieve high reliability

We first consider the *multiplication law,* which provides a mathematical basis for the intuitive concept of increasing system reliability through the use of redundancy. Redundant design is no more than the provision of one or more backup devices which provide continuous system operation when failure of a single device occurs. For example, in connecting an electric power source at the front of a military helicopter to power-using electric devices at the rear, instead of running a single pair of wires we might run three parallel-wired pairs, each by a different path, through the fuselage. Thus if enemy fires cuts one pair of wires, the other two pairs remain operative. While such a design philosophy clearly improves reliability, it is desirable to have a more quantitative measure of the degree of improvement.

The multiplication law states that if $A, B, C, \ldots$ are *independent events* (independence means that the occurrence of one event has no effect on the probability of occurrence of any of the others), then the probability that *all* events occur, called their *joint probability* $P(ABC \cdots)$, is the product of their respective probabilities. That is,

$$\text{Joint probability} \triangleq P(ABC \cdots) = P(A \text{ and } B \text{ and } C \text{ and } \ldots)$$

$$= P(A)P(B)P(C) \cdots \qquad (2.1)$$

Thus in the helicopter wiring example, if the probability of a single pair of wires being cut by enemy fire is 0.01 per combat hour, then with a triply redundant design (three pairs of wires) the failure probability drops to

$$P(\text{all three pairs cut}) = (0.01)(0.01)(0.01) = 10^{-6} \qquad (2.2)$$

Redundant design, of course, increases equipment costs and may entail other penalties (such as added weight, important in a helicopter), so its benefits must be shown to outweigh its costs, if we are to justify its use in a particular case.

When the individual events are *not* independent, the multiplication law takes a different form. Whether events are independent or not is not always obvious or clear-cut in practical applications. In the helicopter wiring example, we ran the wires by "different paths," but what does this really mean? If we ran two pairs of wires "right next" to each other and one pair got shot up, then the other pair would almost certainly also, definitely *not* independent events. As we separate the two pairs of wires by greater distances, the two failure events get "more independent." At what separation would it be acceptable to assume independence? What separation is physically possible in the actual helicopter?

As another example where questions of independence might arise, consider the structural design of offshore oil-drilling rigs. These must be designed to withstand loads produced by the action of wind, waves, and earthquakes, all random effects which are difficult to predict and which could occur singly or in combination. When combinations are being considered, we might reason as follows: High winds certainly do not cause earthquakes, nor do earthquakes result in high winds, so these pairs of events seem to be independent. High waves, however, are usually associated with strong winds, so these two events are quite likely to occur together and we should not treat them as independent. While high waves do not cause earthquakes, earthquakes *do* sometimes cause tidal waves (called tsunamis), so there is some dependence here. When the correct interpretation is not obvious after brief consideration, more detailed theoretical and/or experimental studies may be helpful in reaching a decision.

For two events A and B which are *not* independent,

$$P(A \text{ and } B) \triangleq P(AB) = P(A)P(B/A) = P(B)P(A/B) \qquad (2.3)$$

where

$$P(B/A) \triangleq \textit{conditional probability of } B \text{ with respect to } A$$
$$\triangleq \text{probability of } B \text{ occurring, assuming } A \text{ has occurred} \qquad (2.4)$$

$$P(A/B) \triangleq \textit{conditional probability of } A \text{ with respect to } B$$
$$\triangleq \text{probability of } A \text{ occurring, assuming } B \text{ has occurred} \qquad (2.5)$$

In the helicopter example, suppose that, for the wiring separation possible in the actual helicopter, studies have shown for a doubly redundant system that the probability of the second cable being cut by the same burst of fire which cuts the first is 0.03. Then

$$P(\text{failure}) = P(A)P(B/A) = (0.01)(0.03) = 3 \times 10^{-4} \qquad (2.6)$$

While Eq. (2.1) holds for any number of events, extension of Eq. (2.3) to an arbitrary number of events is available in the literature,[1] but is too complex for inclusion in this text.

We turn now to the *addition law,* which states

$$P(A \text{ and/or } B) = P(A) + P(B) - P(AB) \tag{2.7}$$

This is the probability that *one or more* of the events A and B occur. Note that A and B need *not* be independent, so long as we know their joint probability $P(AB)$. If events A and B are *mutually exclusive* (this means that if one of them happens, the other *cannot*), then $P(AB) \equiv 0$ and $P(A \text{ and/or } B) = P(A) + P(B)$. Equation (2.7) can be generalized to any number of events by a process of "continued reapplication," e.g.,

$P(A \text{ and/or } B \text{ and/or } C)$

$$= P(A) + P(B) + P(C) - P(AB) - P(AC) - P(BC) + P(ABC) \tag{2.8}$$

As an example consider the welded structure of Fig. 2.4. It is clear that failure of the structure occurs if any one or more of the three welds fails. If the probability of failure of an individual weld is 0.001 and we assume independence, then

$$P(\text{part failure}) = 0.001 + 0.001 + 0.001 - 10^{-6} - 10^{-6} - 10^{-6} + 10^{-9} \tag{2.9}$$

$$P(\text{part failure}) = 0.003 - 0.000003 + 0.000000001 \approx 0.003 \tag{2.10}$$

Actually, it might be more correct to assume these three events to be mutually exclusive, since any single failure *unloads* the structure, precluding further overstress. In this case $P(\text{part failure}) = 0.003$, not much different from the independent-events assumption. In either case, note that the probability of system failure is very close to three times that for a single weld. A conclusion important for system design *in general* is that,

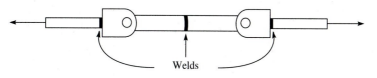

FIGURE 2.4
Use of addition law for weld failure.

---

[1] A. M. Mood, *Introduction to the Theory of Statistics,* McGraw-Hill, New York, 1950, p. 30.

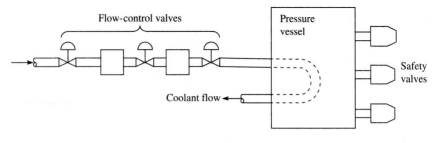

FIGURE 2.5
Addition and multiplication laws applied to pressure vessel.

other things being equal, simple systems (those with few parts) will be more reliable than those with many parts. Even though individual parts may be highly reliable, if a system includes enough of them, the *system* reliability will be poor. Some systems, however, are *unavoidably* complex; we can then improve system reliability (even if component reliability cannot be improved beyond a certain point) by use of redundant design (components in parallel).

Figures 2.5 and 2.6 show some final examples. In Fig. 2.5 a nuclear power plant includes a coolant loop with three valves and a pressure vessel with three safety valves. A loss-of-coolant accident will occur if any one or more of the three valves fails shut, so we can apply the addition law here. The use of three safety valves is a triply redundant design. Since destructive overpressure can occur only if *all* valves fail, the multiplication law shows an increase in reliability. Figure 2.6 shows an abstract system with three stages, all three of which must be successful for system success:

$$P(\text{system success}) = P(1)P(2)P(3) \qquad (2.11)$$

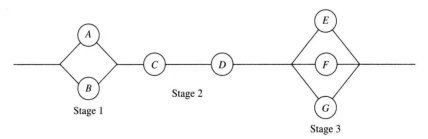

FIGURE 2.6
Multistage system.

Suppose that all events are independent and that the individual probabilities of success are

$$P(A) = 0.7 \qquad P(C) = 0.9 \qquad P(E) = 0.6$$
$$P(B) = 0.7 \qquad P(D) = 0.8 \qquad P(F) = 0.6 \qquad (2.12)$$
$$P(G) = 0.6$$

We then have

$$P(1) = P(A \text{ and/or } B) = P(A) + P(B) - P(AB)$$
$$= 0.7 + 0.7 - 0.49 = 0.91$$
$$P(2) = P(C \text{ and } D) = (0.9)(0.8) = 0.72$$
$$P(3) = P(E \text{ and/or } F \text{ and/or } G) = 0.6 + 0.6 + 0.6$$
$$- 3(0.36) + (0.6)^3 = 0.936$$
$$P(\text{system success}) = P(1)P(2)P(3) = (0.91)(0.72)(0.936) = 0.613$$

$$(2.13)$$

Our overall system thus has a 61.3 percent chance of being successful. Analyses of this type are quite useful in deciding which components are contributing most to the overall unreliability and thus which ones need to have their reliability improved by redesign or development.

2.4 HISTOGRAMS AND PROBABILITY DENSITY FUNCTIONS FOR CONTINUOUS RANDOM VARIABLES

Many practical applications of statistics involve *continuous random variables,* i.e., random variables which are capable of taking on *any* numerical value whatever, between the lowest and highest possible. Dimensions of manufactured parts fit this category as do most material properties, such as the yield strength, modulus of elasticity, density, and viscosity. (Note that if a machined part is *measured* only to the nearest 0.001 in, then the *measured* dimension cannot take on the value 1.79634, but the *actual* dimension can.)

To develop the general concepts of this section, we will use the specific numerical data of Fig. 2.7, which relate to a machine part loaded in compression, that can fail in a buckling mode. The values in the table are the result of tests on 100 "identical" specimens subjected to increasing compressive loads until buckling failure occurred. These data

1	1171.	923.		56	1393.	1209.
2	1042.	924.		57	1152.	1216.
3	1218.	931.		58	1094.	1217.
4	1141.	939.		59	1399.	1218.
5	1298.	1020.		60	1209.	1218.
6	1083.	1021.		61	1185.	1225.
7	1225.	1028.		62	1492.	1231.
8	939.	1040.		63	1233.	1233.
9	1077.	1042.		64	1235.	1233.
10	1275.	1042.		65	1081.	1235.
11	1186.	1051.		66	1302.	1246.
12	1110.	1051.		67	1482.	1249.
13	1181.	1055.		68	1254.	1250.
14	1040.	1055.		69	924.	1254.
15	1185.	1058.		70	1146.	1258.
16	1197.	1062.		71	1150.	1264.
17	1095.	1065.		72	1170.	1270.
18	1124.	1077.		73	1158.	1273.
19	1065.	1081.		74	931.	1273.
20	1464.	1083.		75	1162.	1274.
21	1264.	1090.		76	1249.	1275.
22	1192.	1094.		77	1028.	1285.
23	1273.	1095.		78	1160.	1289.
24	1217.	1106.		79	1361.	1290.
25	1051.	1110.		80	1274.	1298.
26	1146.	1124.		81	1338.	1302.
27	1051.	1133.		82	1258.	1303.
28	1200.	1136.		83	1233.	1312.
29	1141.	1141.		84	1270.	1314.
30	1133.	1141.		85	1333.	1316.
31	1205.	1141.		86	1368.	1327.
32	1196.	1146.		87	1341.	1333.
33	1020.	1146.		88	1141.	1338.
34	1175.	1150.		89	1216.	1341.
35	1218.	1152.		90	1156.	1361.
36	1231.	1152.		91	1290.	1368.
37	1250.	1156.		92	1152.	1393.
38	1058.	1158.		93	1312.	1399.
39	1416.	1160.		94	1246.	1406.
40	1208.	1161.		95	1285.	1416.
41	1316.	1162.		96	1327.	1437.
42	1406.	1163.		97	1106.	1449.
43	1042.	1170.		98	1062.	1464.
44	1273.	1171.		99	1289.	1482.
45	1303.	1175.		100	1090.	1492.
46	923.	1181.				
47	1021.	1185.				
48	1449.	1185.				
49	1055.	1186.				
50	1314.	1192.				
51	1437.	1196.				
52	1161.	1197.				
53	1136.	1200.				
54	1163.	1205.				
55	1055.	1208.				

In testing sequence

In rank order

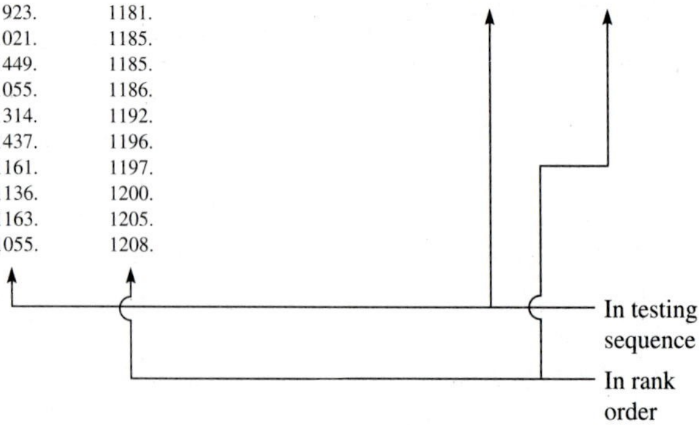

FIGURE 2.7
Buckling-load data set (buckling loads in pound-force).

are typical of the results of many experiments in that they exhibit a random "scatter" around a central value. Before applying any more sophisticated statistical techniques, one always computes two simple parameters; one which locates the "center" of the data and another which measures the spread (dispersion, scatter) on either side. While one could define various different measures useful for these tasks, experience has shown that the ordinary *average (mean) value* is generally the best indicator of central tendency, while the *standard deviation* gives the best measure of dispersion. These two parameters are, in fact, so common that many readers of this text (who generally have not had any formal courses in statistics) will already be familiar with them:

$$\text{Average value} \stackrel{\Delta}{=} \frac{\sum_{i=1}^{n} x_i}{n} \stackrel{\Delta}{=} \bar{x} \tag{2.14}$$

where

$$x_i \stackrel{\Delta}{=} \text{individual value} \tag{2.15}$$

$$n \stackrel{\Delta}{=} \text{number of data values in sample} \tag{2.16}$$

$$s \stackrel{\Delta}{=} \text{standard deviation} \stackrel{\Delta}{=} \left[\frac{\sum_{i=1}^{n} (x_i - \bar{x})^2}{n-1} \right]^{1/2} \tag{2.17}$$

Note that $x_i - \bar{x}$ is the "distance" of an individual point from the average and thus a measure of the scatter. To measure the "total" scatter, we need to add the scatter of all the individual points, but if we sum $x_i - \bar{x}$, this quantity is "as much positive as negative" and in fact sums to zero. To "accumulate" a measure of total scatter, we must prevent this cancellation of negative and positive values. Again, several ways of doing this could be found, but it turns out that squaring $x_i - \bar{x}$ is the most useful statistically. Furthermore, we really want a measure of the *average* scatter, not the total, so we need to divide the sum of $(x_i - \bar{x})^2$ by the number of points. Here the formula seems to be wrong since it uses $n - 1$; however, detailed statistical analysis (beyond our scope here) shows that $n - 1$ is preferred, giving a so-called unbiased estimate of the true standard deviation. To make s a measure of distance, rather than distance squared, we also finally take the square root (this also gives s the same *physical units* as x). We also will sometimes be interested in the *variance* of the sample, which is equal to s^2. It is often useful to put the dispersion on a percentage basis, to make clear whether the dispersion is actually small or large, *relative* to the average value. That is, if two sets of data each have $s = 100$ but one has $\bar{x} = 200$ and the other has $\bar{x} = 2000$, is the scatter of the two sets really equal? Obviously, the data set with $\bar{x} = 2000$ has its points clustered much more closely around the mean than does the set with $\bar{x} = 200$, even though their s values are identical. The dimensionless *coefficient of variation* C_v has been defined to deal with this situation:

$$C_v \triangleq \text{coefficient of variation} \triangleq \frac{s}{\bar{x}} \qquad (2.18)$$

While Eqs. (2.14) and (2.17) allow us to calculate $\bar{x}$ and s for any set of data, we should be clear that they are only *estimates* of the true values. These estimates will get more reliable as we use larger sample sizes (larger n). For small samples, the computed $\bar{x}$ and s will be quite unreliable. While we would intuitively guess that reliability increases with sample size, we need statistical *theory* to give us a quantitative measure of the degree of improvement. Such theory, in the form of *confidence intervals,* is available, but we prefer to put off this discussion until a little later.

We next want to develop the concept of the *histogram,* which will lead us to the important *probability density function.* Let us take the data sample of Fig. 2.7 and put it in ascending order of numerical values (called *rank ordering*). (This rank ordering, computation of $\bar{x}$ and s, and many more statistical operations are of course available on various common statistical computer programs such as SAS and MINITAB.) Once the data are rank-ordered, it is easy to divide the total range from lowest to highest into a number of equal subranges, say, 10. We can then count how many data values fall within each subrange and plot a bar-graph display called a *histogram.* The height h of the bar for each subrange is defined as follows:

$$h \triangleq \frac{\text{probability of finding variable within subrange}}{w} \qquad (2.19)$$

where $w \triangleq$ width of subrange. Note that with such a definition: (1) the area hw of each bar is the probability of finding the variable within that subrange and (2) the total area of all the bars must be 1.00 (100 percent probability that the variable is found between the lowest and highest values). The probabilities for each bar are simply the percentage (decimal) of data values that fall within that subrange.

For example, if 8 values fall within a subrange and there are a total of 100 values, the probability for that subrange is 0.08 (8 percent). To avoid possible ambiguities as to which subrange borderline values actually belong in, it is conventional to define the subranges by using one more significant digit than was used in measuring the variable. In the buckling-load data, the load was measured to the nearest pound, so we define typical subranges as 999.5 to 1099.5, 1099.5 to 1199.5, etc., rather

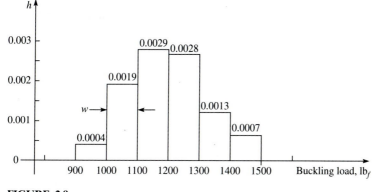

FIGURE 2.8
Definition of histogram.

than 1000 to 1100, 1100 to 1200, etc. Then if we get a data value of exactly 1100, it *definitely* falls in the 1099.5-to-1199.5 range, rather than "on the borderline" between the 1000-to-1100 and 1100-to-1200 sub-ranges. Figure 2.8 shows the histogram for the buckling-load data of Fig. 2.7. There are no rigid rules for deciding on the number and size of subranges when we are constructing a histogram. If we use too few subranges, the histogram does not clearly show the shape of the distribution of probabilities. A large number of subranges means that each will be small and thus some will contain *no* data values, giving undesirable gaps in the histogram, which again are misleading with respect to histogram shape.

To pass from the histogram, which changes shape somewhat depending on our choice of subranges, to the probability density function, which is a unique mathematical curve (function), we postulate a hypothetical experiment for generating the basic data. This experiment is a "mental" one; we could not actually carry it out in the real world since it requires two things:

1. The random variable (buckling load in this case) must be measured with infinitesimal resolution; i.e., the measured value has an infinite number of significant digits.
2. We must test an infinite number of parts.

If these two requirements were satisfied, we could make the subranges smaller and smaller and more and more numerous without any of them being empty and the steplike histogram would, in the limit, become a unique smooth curve, a function f of the random variable x. This $f(x)$ is called the *probability density function*. The subrange width now becomes the infinitesimal dx, but the "bar" area $f(x)\,dx$ still has the same meaning, the probability of finding x in dx; thus

$$P(a < x < b) = \int_a^b f(x)\,dx \qquad (2.20)$$

If we know the probability density function $f(x)$ for some random variable, many useful calculations can be made:

$$P(x < a) = \int_{-\infty}^a f(x)\,dx \qquad (2.21)$$

$$P(x > b) = \int_b^\infty f(x)\,dx \qquad (2.22)$$

Note that

$$P(x = c) = \int_c^c f(x)\,dx = 0 \qquad \text{for } any\ c \text{ and } any \text{ continuous } f(x) \qquad (2.23)$$

and
$$P(-\infty < x < \infty) = \int_{-\infty}^\infty f(x)\,dx = 1.0 \qquad (2.24)$$

Figures 2.9 and 2.10 illustrate these concepts. The result of Eq. (2.23) is sometimes perplexing when first encountered, since in real-world experiments, specific numerical values *do* occur and thus their individual probabilities are not zero. This seeming contradiction is unavoidable and a consequence of the limiting process necessary to define a continuous

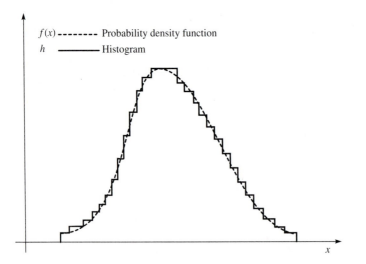

FIGURE 2.9
Definition of probability density function.

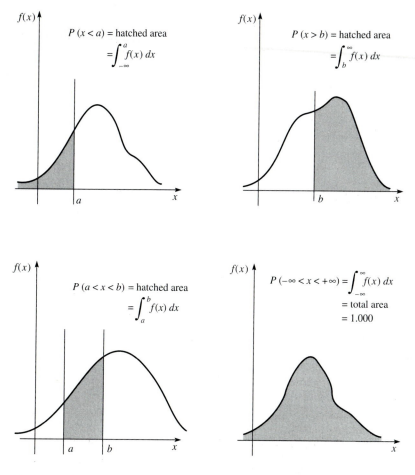

FIGURE 2.10
Graphical interpretations of probabilities.

function $f(x)$. Acceptance of Eq. (2.23) may be strengthened by noting that if $P(x = c)$ were not zero but some definite number, we could "slice" $f(x)$ as finely as we pleased and attach a nonzero probability to *each* x value; then $P(a < x < b)$ could easily be made greater than 1.0, which makes no sense. Another viewpoint would be that in a real-world experiment (like our buckling loads) the random variable is measured only to the nearest pound, so the probability of getting, say, 1213.54 *is* zero. These arguments should make clear that for continuous random variables the probability computations are always associated with some *range* of values.

For each probability density function $f(x)$ there is an associated function $F(x)$, called the *cumulative distribution function*, defined by

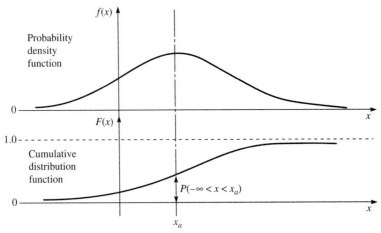

FIGURE 2.11
Definition of cumulative distribution.

$$\text{Cumulative distribution function} \triangleq F(x) \triangleq \int_{-\infty}^{x} f(x)\,dx \qquad (2.25)$$

We see that this is just the probability that x is less than some chosen value x_a:

$$P(-\infty < x \le x_a) = \int_{-\infty}^{x_a} f(x)\,dx \qquad (2.26)$$

Figure 2.11 shows these relationships.

2.5 MATHEMATICAL MODELS FOR CONTINUOUS PROBABILITY DENSITY FUNCTIONS: THE GAUSSIAN (NORMAL) DISTRIBUTION

We defined the probability density function in terms of a histogram based on real-world data. We now wish to introduce some specific functions which have been found to be good (but never perfect) models for many (but not all) classes of real-world data. As with any other mathematical models, the *reasons* for using them include the generalization they provide plus the added speed and convenience of routine calculations. When we are searching for mathematical functions to serve as probability density functions, only a few basic criteria must be satisfied:

1. Our candidate $f(x)$'s must be nonnegative. Why?

2. $\displaystyle\int_{-\infty}^{\infty} f(x)\,dx = 1.0.$

3. The $f(x)$ must be a good curve fit for the real-world data we wish to study. [Actually the cumulative distribution $F(x)$ is often used to check the fit.]

Any functions which meet these requirements are potentially useful mathematical models for probability density functions.

We introduce first the *gaussian probability density function.* This is also called the *normal probability density function,* but we prefer the term *gaussian* since calling one distribution normal seems to imply that others are somehow "abnormal." The term *normal* is not entirely inappropriate, however, because it certainly *is* the most important and commonly used model for many kinds of real-world data. In addition to its goodness of fit for practical data, there are some theoretical reasons (including the famous central-limit theorem) for its importance, but we prefer not to pursue these here. By definition, for the gaussian distribution

$$f(x) \triangleq \frac{1}{\sqrt{2\pi}\,\sigma} e^{-(x-\mu)^2/(2\sigma^2)} \qquad -\infty < x < \infty \qquad (2.27)$$

where

$$\sigma \triangleq \text{standard deviation} \qquad (2.28)$$

$$\mu \triangleq \text{average value (mean value)} \qquad (2.29)$$

We may choose any positive value for σ and any real $(+, -, 0)$ value for μ, giving sufficient "adjustability" to fit many sets of practical data. For any allowed values of σ and μ, the curve is symmetric about μ and has a total area of precisely 1.0. The value of σ determines the *spread* of the curve while μ locates its *center*: see Fig. 2.12. The cumulative distribution $F(x)$ shown must be found by *numerical* integration since no one has ever been able to integrate Eq. (2.27) analytically. Fortunately this integration need be done *only once* (rather than for each σ, μ combination) since we can define the *standard gaussian variable z* by

$$z \triangleq \frac{x - \mu}{\sigma} \qquad (2.30)$$

giving

$$f(z) = \frac{1}{\sqrt{2\pi}} e^{-z^2/2} \qquad (2.31)$$

Note that z has a zero mean value and a standard deviation of exactly 1.0. Figure 2.13a shows these relations, and Fig. 2.13b is the table needed to compute probabilities for any gaussian distribution. This table need only include positive values of z, because of the known symmetry about $z = 0$. The table entries are the probabilities of z being less than the selected value. Since the total area is 1.0, the probabilities of z being *greater* than

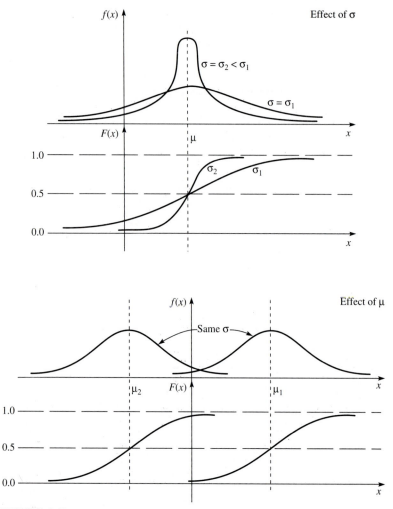

FIGURE 2.12
Gaussian distribution: Effects of σ and μ.

the selected value need not be tabulated—we get them by subtracting the table entry from 1.0.

Use of the gaussian table is best explained with specific examples. Let us point out first that for *any* variable that follows a gaussian distribution,

$$68.3\% \text{ of values fall within } \pm 1\sigma \text{ of } \mu$$

$$95.4\% \text{ of values fall within } \pm 2\sigma \text{ of } \mu \qquad (2.32)$$

$$99.7\% \text{ of values fall within } \pm 3\sigma \text{ of } \mu$$

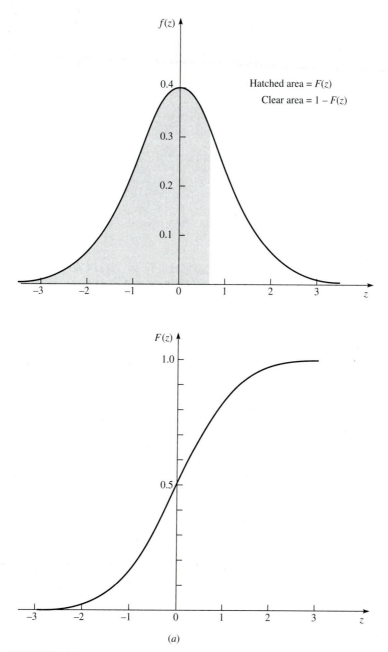

Hatched area = $F(z)$
Clear area = $1 - F(z)$

(a)

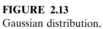

FIGURE 2.13
Gaussian distribution.

Cumulative normal distribution

$$F(z) = \int_{-\infty}^{z} \frac{1}{\sqrt{2\pi}} e^{-z^2/2} dz$$

z	.00	.01	.02	.03	.04	.05	.06	.07	.08	.09
.0	.5000	.5040	.5080	.5120	.5160	.5199	.5239	.5279	.5319	.5359
.1	.5398	.5438	.5517	.5478	.5557	.5596	.5636	.5675	.5714	.5753
.2	.5793	.6838	.5871	.5910	.5948	.5987	.6026	.6064	.6103	.6141
.3	.6179	.6217	.6255	.6293	.6331	.6368	.6406	.6443	.6480	.6517
.4	.6554	.6591	.6628	.6664	.6700	.6736	.6772	.6808	.6844	.6879
.5	.6915	.6950	.6985	.7019	.7034	.7088	.7123	.7137	.7190	.7224
.6	.7257	.7291	.7324	.7357	.7389	.7422	.7454	.7488	.7517	.7349
.7	.7580	.7611	.7642	.7673	.7704	.7734	.7764	.7794	.7823	.7852
.8	.7881	.7910	.7939	.7967	.7995	.8023	.8031	.8078	.8106	.8133
.9	.8139	.8186	.8212	.8238	.8264	.8289	.8315	.8340	.8365	.8389
1.0	.8413	.8438	.8461	.8485	.8508	.8531	.8554	.8577	.8599	.8521
1.1	.8643	.8665	.8686	.8708	.8729	.8749	.8770	.8790	.8810	.8830
1.2	.8649	.8869	.8888	.8907	.8925	.8944	.8962	.8930	.8997	.9015
1.3	.9032	.9049	.9068	.9082	.9099	.9115	.9131	.9147	.9162	.9177
1.4	.9192	.9207	.9222	.9236	.9251	.9265	.9279	.9292	.9306	.9319
1.5	.9332	.9345	.9357	.9370	.9382	.9394	.9406	.9418	.9429	.9441
1.6	.9452	.9463	.9474	.9484	.9495	.9505	.9515	.9525	.9535	.9545
1.7	.9554	.9564	.9573	.9582	.9591	.9599	.9608	.9616	.9625	.9633
1.8	.9641	.9649	.9656	.9664	.9671	.9678	.9686	.9693	.9699	.9706
1.9	.9713	.9719	.9726	.9732	.9738	.9744	.9750	.9756	.9761	.9767
2.0	.9772	.9778	.9783	.9788	.9793	.9798	.9803	.9808	.9812	.9817
2.1	.9821	.9826	.9830	.9834	.9838	.9842	.9846	.9850	.9854	.9857
2.2	.9861	.9864	.9868	.9871	.9875	.9878	.9881	.9884	.9887	.9890
2.3	.9893	.9896	.9898	.9901	.9904	.9906	.9909	.9911	.9913	.9916
2.4	.9918	.9920	.9924	.9925	.9927	.9929	.9931	.9932	.9934	.9936
2.5	.9938	.9940	.9941	.9943	.9945	.9946	.9948	.9949	.9951	.9952
2.6	.9953	.9955	.9956	.9957	.9959	.9960	.9961	.9962	.9963	.9964
2.7	.9965	.9966	.9967	.9968	.9969	.9970	.9971	.9972	.9973	.9974
2.8	.9974	.9975	.9976	.9977	.9977	.9978	.9979	.9979	.9980	.9981
2.9	.9981	.9982	.9982	.9983	.9984	.9984	.9985	.9985	.9986	.9986
3.0	.9987	.9987	.9987	.9988	.9988	.9989	.9989	.9989	.9990	.9990
3.1	.9990	.9991	.9991	.9991	.9992	.9992	.9992	.9992	.9993	.9993
3.2	.9993	.9993	.9994	.9994	.9994	.9994	.9994	.9995	.9995	.9995
3.3	.9995	.9995	.9995	.9996	.9996	.9996	.9996	.9996	.9996	.9997
3.4	.9997	.9997	.9997	.9997	.9997	.9997	.9997	.9997	.9997	.9998

(b)

FIGURE 2.13
(Continued).

$F(z)$	z	$F(z)$	z
.0001	-3.719	.500	.000
.0005	-3.291	.550	.126
.001	-3.090	.600	.253
.005	-2.576	.650	.385
.010	-2.326		
.025	-1.960	.700	.524
.050	-1.645	.750	.674
		.800	.842
.100	-1.282	.850	1.036
.150	-1.036	.900	1.282
.200	-.842		
.250	-.674	.950	1.645
.300	-.524	.975	1.960
		.990	2.326
.350	-.385	.995	2.576
.400	-.253	.999	3.090
.450	-.126	.9995	3.291
.500	.000	.9999	3.719

(c)

FIGURE 2.13
(Continued).

These numbers are obtained from the z table (Fig. 2.13b) by using the entries for $z = 1$, 2, and 3. (For practice, you should verify these numbers.) We want to now use the buckling-load data given earlier, so let's *assume* that past experience has shown the gaussian distribution to be a reasonable model for this physical process. To use the gaussian table for any calculations, first we must have numerical values for μ and σ, the *true* mean and standard deviation of the "total population" (infinite-size sample) of buckling loads. In practice, these are never available, and we must instead use their *estimators* $\bar{x}$ and s, calculated from the available sample. As mentioned earlier, these estimates $\bar{x}$ and s are more reliable for larger samples, and we shall soon define confidence intervals which allow us to quantify this reliability. For the time being, let's *assume* that $\bar{x}$ and s are adequate estimates of μ and σ and proceed on this basis.

Using Eqs. (2.14) and (2.17), we calculate $\bar{x} = 1199.$ lb$_f$ and $s = 124.7$ lb$_f$. One possible practical question might be, What maximum load can we allow if we accept a failure rate of 0.1 percent (1 failure in 1000 trials)? Using the table of percentiles in Fig. 2.13c, we find $z = -3.090$ for 0.001 probability of failure. Using Eq. (2.30) gives

$$x = \sigma z + \mu = (124.7)(-3.090) + 1199. = 814 \text{ lb}_f \qquad (2.33)$$

This result can be used both to illustrate the advantages of mathematical models and to warn of possible dangers. Note that with a sample of 100, it is *impossible* to *directly* calculate 0.1 percent probabilities (one needs at least 1000 items), but the model allows such extrapolation. We should, however, use some caution since the 814-lb$_f$ load predicted is *below* the lowest data point (923.), and thus we have not really experimentally "exercised" this range of buckling loads. In fact, such "complete" experimental testing can rarely be justified economically, so we must often be willing to take some risk in using our mathematical model.

For some additional practice, let's compute

- The probability of buckling occurring for loads of 1000 lb$_f$ or less
- The probability of buckling occurring for loads of 1400 lb$_f$ or less

In performing the calculations, we take the following viewpoint. For the first item, when we say "1000 lb$_f$ or less," we *really* are allowing the 1000-lb$_f$ load to be applied in *every* trial. That is, we can rephrase the question in terms of the actual experimental data of Fig. 2.7 as follows: If we applied a 1000-lb$_f$ load to each of the 100 specimens, what percentage would fail? Based on the measured data, this is 4 percent; however, we wish to use the gaussian model, so we compute as follows:

$$z = \frac{1000 - 1199}{124.7} = -1.60 \tag{2.34}$$

$$P(x < 1000) = P(z < -1.60)$$
$$= 1 - P(z > 1.60) = 1 - 0.9452 = 0.0548 = 5.48\% \tag{2.35}$$

For our original statement "1000 lb$_f$ or *less*," the 5.48 percent serves as a (conservative) *upper* bound. That is, if we apply *various* loads, some of 1000 and others less, the failure rate will be *no worse* than 5.48 percent. By a similar line of reasoning for the second item, the measured data themselves show 93 percent while the gaussian model gives

$$z = \frac{1400 - 1199}{124.7} = 1.61 \tag{2.36}$$

$$P(x < 1400) = P(z < 1.61) = 0.9463 = 94.63\% \tag{2.37}$$

As another example, consider the manufacture of a shaft in an automatic lathe. Suppose the diameter d actually produced follows closely a gaussian distribution with $\bar{d} = 1.250$ cm and $\sigma_d = 0.002$ cm. If only parts in the range $1.245 \leq d \leq 1.255$ will assemble properly with a mating part, what percentage of scrap is being produced?

$$z_\mu = \frac{1.255 - 1.250}{0.002} = 2.5 \qquad z_l = \frac{1.245 - 1.250}{0.002} = -2.5 \tag{2.38}$$

$$P(1.245 \leq d \leq 1.250)$$
$$= P(-2.5 \leq z \leq 2.5) = 2(1.0000 - 0.9938) = 1.24\% \tag{2.39}$$

Suppose now we use a new machine which finishes *both* the diameter d *and* the length l of the shaft, both processes being close to gaussian:

$$\bar{d} = 1.250 \text{ cm} \qquad \sigma_d = 0.002 \text{ cm} \qquad \bar{l} = 10.00 \text{ cm} \qquad \sigma_l = 0.016 \text{ cm}$$

$$(2.40)$$

Now, to be acceptable, suppose parts must satisfy *both*

$$1.245 \leq d \leq 1.255 \quad and \quad 9.96 \leq l \leq 10.04 \qquad (2.41)$$

We can use the addition law of probability:

$$P(l \text{ and/or } d) = P(l) + P(d) - P(ld) \qquad (2.42)$$
$$= P(l) + P(d) - P(l)P(d) \qquad \text{if } l, d \text{ are independent}$$

$$(2.43)$$

$$= P(l) + P(d) - P(l)P(d/l) \qquad \text{if } l, d \text{ are dependent}$$

$$(2.44)$$

To rationally decide whether the two machining processes for l and d are or are not dependent would require an experimental study of actual parts where we graphed length deviations versus diameter deviations, to see whether any relation could be discerned (more sophisticated statistical tests are also available to help answer such questions). (One can certainly think of *theoretical* reasons why *some* dependence could exist. For instance, if an individual workpiece has a higher than normal hardness, *both* l and d dimensions might come out a little larger than normal.) To proceed with our numerical example, let's assume that the two processes are essentially independent. We have then for the l dimension

$$z_\mu = \frac{10.04 - 10.00}{0.016} = 2.5 \qquad z_l = \frac{9.96 - 10.00}{0.016} = -2.5 \qquad (2.45)$$

giving

$$P(l) \stackrel{\Delta}{=} P(9.96 \leq l \leq 10.04) = 1.24\%$$

Since the d numbers are the same as in our earlier example,

$$P(l \text{ and/or } d) = 1.24\% + 1.24\% - (0.0124)(0.0124) = 2.47\% \qquad (2.46)$$

If the l and d processes were "totally" dependent (the deviations in l and d for each part were exactly proportional), we would get

$$P(l \text{ and/or } d) = 1.24 + 1.24 - (1.24)(1.0) = 1.24\% \qquad (2.47)$$

That is, the combination of two "statistically identical" processes (we chose the *same* coefficient of variation in making up the example) produces the *same* percentage of scrap as each individual process because the scrap *l* values occur in the *same* pieces as the scrap *d* values. That is, pieces bad due to *l* deviations *don't* add to the total of scrap because they are "already bad" due to their *d* deviation.

2.6 CHECKING REAL-WORLD DATA FOR CONFORMANCE TO SOME THEORETICAL MATHEMATICAL MODEL: PROBABILITY GRAPH PAPER AND THE CHI-SQUARED GOODNESS-OF-FIT TEST

We have seen that many useful calculations are made possible when we assume that data follow the gaussian distribution. How does one decide that a specific set of data *can* be analyzed by using the gaussian model? Since mathematical models *other* than gaussian ones are also regularly needed to fit certain kinds of data, our question really is the more general one, How do we justify use of *any* particular theoretical distribution? It is essential to state clearly at the outset that we can *never* prove that *any* data are *perfectly* modeled by *any* specific mathematical function. There are *always* a number of reasons why this is so. In the case of the gaussian distribution, one obvious reason is the fact that the model extends from $-\infty$ to $+\infty$. For example, in the case of dimensions of manufactured parts, the gaussian model predicts a small but nonzero probability that the dimension will be *negative*—clearly a physical impossibility. Or, for the buckling loads treated earlier, the mathematical model includes negative (tension) loads—again a violation of reality. Thus when we discuss validation of any statistical mathematical model, we always take the practical viewpoint that it will not (and cannot) be perfect. We require it only to be useful for certain purposes. Often this mainly entails convincing ourselves that it is *not* clearly and grossly in error.

By far the most common test for any distribution is a simple graphical comparison of the *cumulative* distribution of the real-world data and that of the ideal model. If these superimposed plots do not show any obvious and large disagreements, we will often accept the model without further investigation. If this procedure seems somewhat casual and unscientific, recall that

1. There is *no hope* of *ever* proving rigorously that real-world data follow *any* particular mathematical model.
2. We shall show shortly that extremely large samples (with hundreds or even thousands of specimens) are required to obtain very high confidence levels in statistical calculations. These can rarely be justified economically.

Turning now to details, we recall the well-known graphical

device of *rectification*. That is, the human eye and brain are ill equipped to judge whether data points follow a certain curve, but are quite expert at evaluating conformance to straight-line behavior. Since it is possible for *any* curvilinear relation, we therefore develop special graph paper with a plotting scale precisely distorted such that perfectly gaussian data will plot as a straight line. There is sufficient market for such paper that it is commercially available (see Fig. 2.14).

Two vertical scales (left and right) are provided on the paper

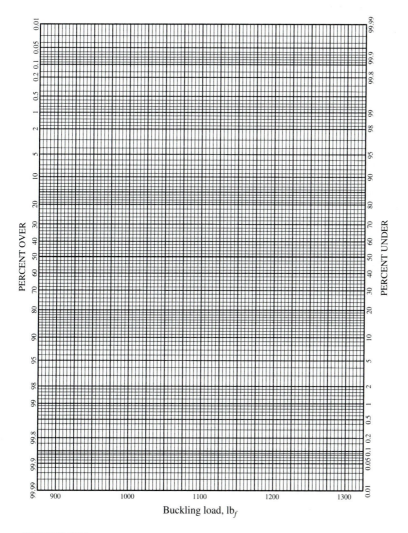

FIGURE 2.14
Gaussian probability graph paper.

shown. We can use either one for plotting purposes, depending on how we define the plotted points. If we (arbitrarily) choose the right-hand scale, then the legend for this axis will read "percentage of data at or below a chosen abscissa value." The abscissa (x axis) is linearly divided and unnumbered since in each case we will choose the numbering to suit the span of the particular data set. Note that it is possible to provide paper covering only *finite*-size samples: for the paper shown, the maximum size is 10,000 since the lowest probability plottable is 0.01 percent, or 1 in 10,000. Let's now explain the use of such graph paper by applying it to the buckling-load data of Fig. 2.7. To get some experience with the effects of sample size, let's pretend we tested only the first 10 specimens, giving this data set:

1171		939	
1042		1042	
1218		1077	
1141		1083	
1298	in-testing sequence	1141	rank-ordered
1083		1171	
1225		1218	
939		1225	
1077		1275	
1275		1298	

Mean = 1147   Standard deviation = 112.6

For our sample, no values are repeated, so our first plotted point is at 939, 10 percent; and each successive point "adds" another 10 percent, with the last plottable point being 1275, 90 percent. The 10th point would be 100 percent which is "off the paper" and thus unplottable. It can be shown that for small samples it is better to plot according to *median ranks.*[1] With this method the percentage at or below is computed from

$$\text{Percentage at or below} = \frac{j - 0.3}{n + 0.4} \qquad (2.48)$$

where $j \triangleq$ rank-order number $(1, 2, 3, \ldots, n)$

$n \triangleq$ sample size

For our example, instead of 9 plottable points at 10, 20, 30, 40, 50, 60, 70, 80, and 90 percent, we now get 10 plottable points at 6.7, 16.3, 26.0, 35.6,

[1] C. Lipson and N. M. Sheth, *Statistical Design and Analysis of Engineering Experiments*, McGraw-Hill, New York, 1973, pp. 16–19.

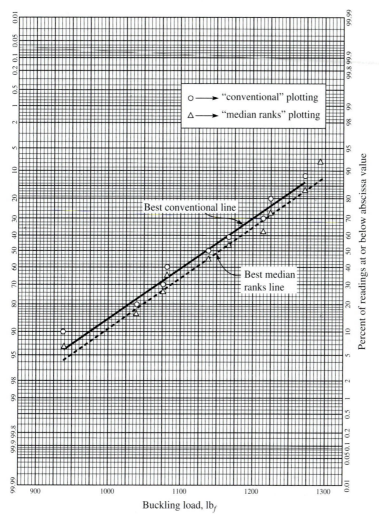

FIGURE 2.15
Comparison of conventional and median-ranks plotting.

45.2, 54.8, 60.6, 74.0, 83.7, and 93.3. Figure 2.15 shows the data plotted both ways. [For samples of about 20 or more, there is little graphical difference between the two schemes, and we could use the simpler scheme with no real penalty. If our data processing is computerized, we might of course use the more complex (and correct) method universally, even where it isn't really needed.]

In Fig. 2.15 the conventional plotting was done first and the "best" (strictly by eye) line was drawn. With such "visual" curve fitting, of course, each individual will come up with a slightly different line; however, it should be clear that these data seem to fit the guassian model pretty well. [In deciding such line fits, one should give less weight to points (like our 10 and 90 percent points) at the "tails" of the distribution, since these areas are always the least reliable and the most likely to be nongaussian.] The median-ranks data were plotted next, and the best line was fitted by eye. The conformance to straight-line (gaussian) behavior again appears quite good, but the visually best line is slightly different, as might be expected, since the plotting scheme is different.

At this point we want to make you aware that the buckling-load data which we have been using is *not* actual experimental data, but was made up. (Many of the example data given in statistics textbooks are similarly "manufactured." Unless the author explicitly states that it is "real" data, you can never be sure.) Not only was our data made up, it also was computer-generated by an algorithm carefully designed to simulate a perfect gaussian distribution. Once you are aware of these facts, you might legitimately wonder why the graphs of Fig. 2.15 did *not* came out as *perfect* straight lines without any scatter, since the data was perfectly gaussian. Actually, the observed behavior is perfectly consistent with that of random gaussian variables; *small samples* taken from *any* distribution will *appear* to deviate from that distribution. Only for very large samples will the observed behavior of the sample approach that of the underlying distribution. Thus we must *expect* such "imperfect" behavior in *any* test method used to check for conformance to a theoretical distribution. This is true even when we "manufacture" *perfect* data, so we should not be surprised to encounter it when we test real-world data. Our purpose here in using the manufactured data was in fact mainly to demonstrate this important point.

Since the ability to generate and manipulate perfect gaussian data is a very useful one, we now show some simple programming using the SAS[1] statistical package. This package performs hundreds of useful statistical operations and is widely used around the world. Here we only touch on some of its simpler capabilities. A program to generate the buckling-load data might go as follows:

DATA STASPL;	Names the upcoming data set STASPL. Use any six-character name you wish. All lines in SAS *must* end with a semicolon.
DO I = 1 TO 100;	Start a DO loop to generate a 100-item sample.

---

[1] SAS Institute, Box 8000, Cary, NC 27512, phone (919) 467-8000.

LOAD = 1200. + 120.*NORMAL (53713);   Generates a normal (gaussian) variable with mean = 1200 and standard deviation = 120. The "seed" of the random number generator is 53713 (explained shortly).

LOAD = ROUND (LOAD, 1);   Rounds off the values just produced (which carry many digits) to the nearest 1 lb, since our buckling loads were measured only to the nearest pound

OUTPUT;   Writes the current data point into the data set STASPL

END;   End of the DO loop

PROC PRINT;   Prints a table showing the data set, in the order generated

PROC SORT;   Sorts the data set into rank order

BY LOAD;   according to values of the variable called LOAD

PROC PRINT;

FORMAT VAR I LOAD 4.0;   Prints a table showing the rank-ordered data set. Variables I and LOAD will be printed in a format four characters wide with zero digits to right of decimal point.

PROC MEANS: Computes and prints average (mean) value, standard deviation, coefficient of variation, and minimum and maximum values of data set.

All random number generators use a *seed value* to start the generation process. In SAS if you let the seed entry be zero [NORMAL(0)], then the generator actually uses the time of day given by the computer clock as the seed. Since this time is never the same twice, using zero as the seed means that a *different* seed is used every time the generator is started. When the seed value is different, the *sequence* of random numbers generated by the DO loop will be different, even though the mean value and standard deviation remain at 1200 and 120 (for our example). Thus if you used the NORMAL statement *twice* in a program, to generate *two* separate sets of random numbers, using a seed of zero in each statement makes the two random variables *independent* (not correlated), since the *sequences* are different. If we *wanted* the two sequences to be the same, we could use the *same* five-digit odd integer as the seed in each, for example, NORMAL (53713). Note that if you ran our example program (which uses 53713 as the seed) *twice,* you would get *exactly* the same sequence of LOAD values each time, which is *not* the behavior of a *truly* random variable; however, this type of defect has no bad effects for most studies, including ours. If you use the seed as zero, then rerunning the identical program will give a different sequence each time, more like a *true* random variable. Using *different* five-digit odd integers as seeds would give different sequences also.

Let's now use the entire (100-item) sample of buckling loads to check graphically for conformance to the gaussian model. Here there will be very little difference between conventional and median-ranks plotting, so we do only the easier conventional method in Fig. 2.16. We now also introduce the concept of the *perfect gaussian line*. For a sample of any size we can always compute $\bar{x}$ and s; they are 1199. and 125. lb_f for our

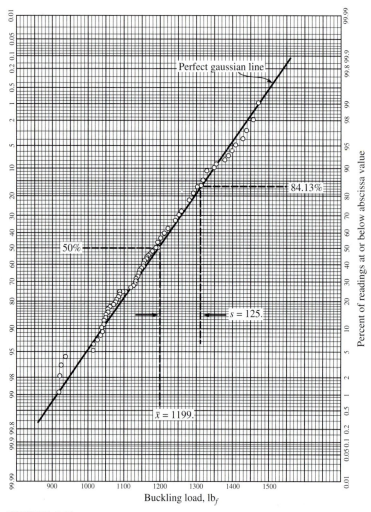

FIGURE 2.16
Definition and use of "perfect gaussian line."

present sample. [Note that, even for a large (100-item) sample, these are close to, but do *not* equal, the 1200. and 120. values requested in the computer model.] By a perfect guassian line we mean that line given by an infinite-size gaussian sample having the $\bar{x}$ and s *measured* on our actual sample. Such a line is always easily plotted on gaussian graph paper since it must go through the points $\bar{x}$, 50 percent and $\bar{x} + s$, 84.13 percent, in our case 1199, 50 percent and 1324, 84.13 percent. The perfect gaussian line may be used in either of two ways. In Fig. 2.16 we have plotted it by using the measured $\bar{x}$ and s, and then we use it to decide whether we will treat the data as gaussian. An alternative approach would be to fit by eye the "best" straight line, treat this as if it were the perfect gaussian line, and use its intercept at 50 percent to *define* the numerical value of $\bar{x}$ and its intercept at 84.13 percent to *define* the numerical value of s, *not* computing $\bar{x}$ and s from the usual formulas at all (or perhaps only as a check calculation). For large samples, the procedure of Fig. 2.16 is probably best. For small samples, the formula-calculated $\bar{x}$ and s values are quite uncertain, and the "visual averaging" performed when we decide on the best line and use *it* to get numbers for $\bar{x}$ and s may actually lead to more reliable values, although we are unable to *prove* this intuitive assertion.

Some analysts try to avoid the visual judgments needed when we use probability graph paper to decide whether data will be treated as gaussian or not, preferring a more numerical criterion. Such a criterion is available through the *chi-squared goodness-of-fit test,* which we now explain. Unfortunately this approach merely substitutes for the visual judgment a numerical judgment which is not really any more effective in my opinion. We thus recommend the graphical approach for most engineering applications. Since the chi-squared test is, however, well known and is widely discussed in textbooks, you probably should at least be familiar with the approach. As with the general graphical method, it can be applied to any distribution, not just a gaussian one. We first divide the rank-ordered sample into N_g groups, where N_g is estimated from

$$N_g = 1.87(n - 1)^{0.4} \qquad \text{for } n \overset{\Delta}{=} \text{sample size} \qquad (2.49)$$

[Specific rules for choosing the number and size of groups are not even mentioned in most statistics texts, since there is no theoretical basis for this. Equation (2.49) was found[1,2] only after "years of searching" and is

---

[1] M. C. Kendal and A. Stuart, *The Advanced Theory of Statistics,* vol. 2, Griffin, London, 1961.

[2] C. A. Williams, Jr., "On the Choice of the Number and Width of Classes for Chi-Square Test of Goodness of Fit," *Journal of the American Statistical Association,* 45, March 1950.

simply offered as the best available guide.] For our 100-item sample $N_g \approx 12$, so we divide the total sample into 12 groups, each containing 8 to 10 items. To provide a little smoothing at the tails, we use the 10-item and 9-item groups at the low and high ends. For the 100-item sample we earlier found $\bar{x} = 1199$ and $s = 125$. The chi-square method compares the number of items *actually* present in each group with the number theoretically predicted to be there if the data followed the given distribution, in our case gaussian with $\sigma = 125$ and $\mu = 1199$. A measure of the "error" between predicted and actual frequencies is formed for each group and then summed over all the groups to get a measure of the "global" (total) error. This global error is compared with tabulated values of a theoretical distribution called the chi-squared distribution. If the global error is too large, we decide that the data are probably not gaussian, with a numerical value for this probability.

The method is best carried out in tabular form, as in Fig. 2.17a, which we now explain. To use the standard gaussian table to calculate the theoretically predicted number of items in each group, we first need to compute z. For the first table entry,

$$z = \frac{x - \mu}{\sigma} = \frac{1042.1 - 1199}{125} = -1.26 \tag{2.50}$$

Group number	Subrange	$z = \dfrac{x-\mu}{\sigma}$	Number of items actually present, n_i	Number of items theoretically present, e_i	$\dfrac{(n_i - e_i)^2}{e_i}$
1	$-\infty$ to 1042.1	−1.26	10	10.38	0.014
2	1042.1 to 1077.1	−0.98	8	5.97	0.690
3	1077.1 to 1124.1	−0.60	8	11.08	0.856
4	1124.1 to 1150.1	−0.39	8	7.40	0.049
5	1150.1 to 1163.1	−0.29	8	3.76	4.781
6	1163.1 to 1192.1	−0.06	8	9.02	0.115
7	1192.1 to 1217.1	0.14	8	7.96	0.000
8	1217.1 to 1246.1	0.38	8	9.23	0.164
9	1246.1 to 1273.1	0.60	8	7.77	0.006
10	1273.1 to 1303.1	0.83	8	7.00	0.143
11	1303.1 to 1368.1	1.35	9	11.48	0.536
12	1368.1 to $+\infty$	∞	9	8.85	0.003

$$\sum \frac{(n_i - e_i)^2}{e_i} \quad \longrightarrow 7.35$$

FIGURE 2.17a
Chi-squared goodness-of-fit test and chi-squared distribution.

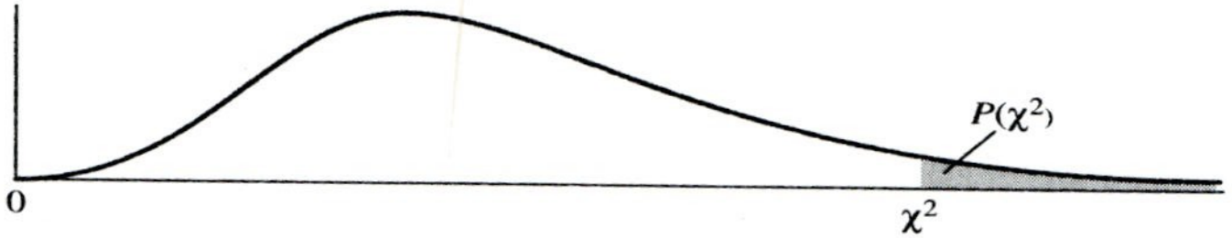

Upper percentage points of the χ^2 distribution

$$P(\chi^2) = \int_{\chi^2}^{\infty} \frac{1}{[(f-2)/2]! \; 2^{f/2}} (\chi^2)^{(f-2)/2} e^{-\chi^2/2} \, d(\chi^2)$$

f \\ $P(\chi^2)$	.995	.990	.975	.950	.900	.750	.500	.250	.100	.050	.025	.10	.005
1	3927×10^{-8}	1571×10^{-7}	9821×10^{-7}	3932×10^{-4}	0.01579	0.1015	0.4549	1.323	2.706	3.841	5.024	6.635	7.879
2	0.01003	0.02010	0.05064	0.1026	.2107	.5754	1.386	2.773	4.605	5.991	7.378	9.210	10.60
3	.07172	.1148	.2158	.3518	.5844	1.213	2.366	4.108	6.251	7.815	9.348	11.34	12.84
4	.2070	.2971	.4844	.7107	1.064	1.923	3.357	5.385	7.779	9.488	11.14	13.28	14.86
5	.4117	.5543	.8312	1.145	1.610	2.675	4.351	6.626	9.236	11.07	12.83	15.09	16.75
6	.6757	.8721	1.237	1.635	2.204	3.455	5.348	7.841	10.64	12.59	14.45	16.81	18.55
7	.9893	1.239	1.690	2.167	2.833	4.255	6.346	9.037	12.02	14.07	16.01	18.48	20.28
8	1.344	1.646	2.180	2.733	3.490	5.071	7.344	10.22	13.36	15.51	17.53	20.09	21.96
9	1.735	2.088	2.700	3.325	4.168	5.899	8.343	11.39	14.68	16.92	19.02	21.67	23.59
10	2.156	2.558	3.247	3.940	4.865	6.737	9.342	12.55	15.99	18.31	20.48	23.21	25.19
11	2.603	3.053	3.816	4.575	5.578	7.584	10.34	13.70	17.28	19.68	21.92	24.72	26.76
12	3.074	3.571	4.404	5.226	6.304	8.438	11.34	14.85	18.55	21.03	23.34	26.22	28.30
13	3.565	4.107	5.009	5.892	7.042	9.299	12.34	15.98	19.81	22.36	24.74	27.69	29.82
14	4.075	4.660	5.629	6.571	7.790	10.17	13.34	17.12	21.06	23.68	26.12	29.14	31.32
15	4.601	5.229	6.262	7.261	8.547	11.04	14.34	18.25	22.31	25.00	27.49	30.58	32.80
16	5.142	5.812	6.908	7.962	9.312	11.91	15.34	19.37	23.54	26.30	28.85	32.00	34.27
17	5.697	6.408	7.564	8.672	10.09	12.79	16.34	20.49	24.77	27.59	30.19	33.41	35.72
18	6.265	7.015	8.231	9.390	10.86	13.68	17.34	21.60	25.99	28.87	31.52	34.81	37.16
19	6.844	7.633	8.907	10.12	11.65	14.58	18.34	22.72	27.20	30.14	32.85	36.19	38.58
20	7.434	8.260	9.591	10.85	12.44	15.45	19.34	23.83	28.41	31.41	34.17	37.57	40.00
21	8.034	8.897	10.28	11.59	13.24	16.34	20.34	24.93	29.62	32.67	35.48	38.93	41.40
22	8.643	9.542	10.98	12.34	14.04	17.24	21.34	26.04	30.81	33.92	36.78	40.29	42.80
23	9.260	10.20	11.69	13.09	14.85	18.14	22.34	27.14	32.01	35.17	38.08	41.64	44.18
24	9.886	10.86	12.40	13.85	15.66	19.04	23.34	28.24	33.20	36.42	39.36	42.98	45.56
25	10.52	11.52	13.12	14.61	16.47	19.94	24.34	29.34	34.38	37.65	40.65	44.31	46.93
26	11.16	12.20	13.84	15.38	17.29	20.84	25.34	30.43	35.56	38.89	41.92	45.64	48.29
27	11.81	12.88	14.57	16.15	18.11	21.75	26.34	31.53	36.74	40.11	43.19	46.96	49.64
28	12.46	13.56	15.31	16.93	18.94	22.66	27.34	32.62	37.92	41.34	44.46	48.28	50.99
29	13.12	14.26	16.05	17.71	19.77	23.57	28.34	33.71	39.09	42.56	45.72	49.59	52.34
30	13.79	14.95	16.79	18.49	20.60	24.48	29.34	34.80	40.26	43.77	46.98	50.89	53.67
40	20.71	22.16	24.43	26.51	29.05	33.66	39.34	45.62	51.80	55.76	59.34	63.69	66.77
50	27.99	29.71	32.36	34.76	37.69	42.94	49.33	56.33	63.17	67.50	71.42	76.15	79.49
60	35.53	37.48	40.48	43.19	46.46	52.29	59.33	66.98	74.40	79.08	83.30	88.38	91.93
70	43.28	45.44	48.76	51.74	55.33	61.70	69.33	77.58	85.53	90.53	95.02	100.42	104.22
80	51.17	53.54	57.15	60.39	64.28	71.14	79.33	88.13	96.58	101.63	106.63	112.33	116.32
90	59.20	61.75	65.65	69.13	73.29	80.62	89.33	98.65	107.56	113.14	118.14	124.12	128.30
100	67.33	70.06	74.22	77.93	82.36	90.13	99.33	109.14	118.50	124.56	129.56	135.81	140.17

FIGURE 2.17b
Chi-squared goodness-of-fit test and chi-squared distribution.

The number e_i of items theoretically present in this first subrange is then

$$e_i = P(z < -1.26)(100) = (1.0000 - 0.8962)100 = 10.38 \qquad (2.51)$$

The other subranges are similarly calculated. Note that for an infinite-size, perfectly gaussian sample, each $n_i - e_i$ would be zero; thus the *smaller* the individual $(n_i - e_i)^2/e_i$, and thus the smaller their total, the closer we are to a gaussian distribution. The question is, *How* small must this "total error" be to make the assumption of gaussian behavior acceptable? This question is answered by a theoretical study which leads to a probability density function called the chi-squared distribution, whose formula and tabulated values appear as Fig. 2.17b. We use the table as follows. The distribution has a single adjustable parameter f, called the *degrees of freedom*. For the problem type under consideration, f is equal to the number of groups minus 3, in our case 9. We now look in the table row for $f = 9$ until we find 7.35, the total of the last column in our calculation table. We find 7.35 bracketed between 5.899 ($P = 75$ percent) and 8.343 ($P = 50$ percent). A rough linear interpolation (suitable for any such problem) gives $P \approx 59$ percent for 7.35. The *meaning* of this is as follows: For a *perfect* gaussian distribution and a sample size of 100, we would get a value as large as 7.35 about 59 percent of the time if we ran such experiments over and over. Thus it is quite likely that our distribution is close to gaussian. If we had gotten a larger number like, say, 20 (rather than 7.35), the table shows that such a large number would occur only about 2.5 percent of the time when the distribution was *really* gaussian. This is quite *unlikely*, so we would hesitate to treat the distribution as gaussian even though there is a small chance that it is. [Recall again that our data *did* come from a "perfect" (computer-generated) gaussian distribution, but we should *not* then expect the number 7.35 to have been zero, just as we don't expect the graph paper test to show a perfect straight line, because the sample size is *not* infinite.]

2.7   DEALING WITH NONGAUSSIAN DATA: TRANSFORMATIONS

When we encounter data that clearly aren't well approximated by a gaussian distribution but we still wish to proceed with some statistical analysis, several options are open to us:

1. Try to "fit" the data to some other statistical mathematical model.

The *Weibull*[1] distribution, e.g., works well for many situations which involve the "life" (time to failure) of various machine components.

2. Try to fit a *gaussian* model, not to the original data, but rather to some suitably chosen *function* of the data. For example, in Fig. 2.15, we might choose to plot on the horizontal axis the *square root* of the buckling load rather than the buckling load itself. If the resulting graph comes out reasonably straight, then this *transformation* was a good choice.

3. Employ "distribution-free" statistical methods, which do not require that the data follow some given distribution function. Since these methods are generally less desirable, we choose this option only when others fail.

In this section we briefly consider only the *transformation* approach. The most common successful example of its use is probably the application to fatigue life testing of standard material specimens. Figure 2.18 is a table of actual test data for 100 specimens of 75S-T aluminum alloy, subjected to ±30,000-psi (completely reversed) bending stress in a rotating-bending fatigue test machine. Even though the material is

---

[1] Lipson and Sheth, op. cit., p. 36.

(1) 33	(19) 92	(37) 188	(55) 279	(73) 514	(91) 1427
(2) 40	(20) 94	(38) 189	(56) 292	(74) 566	(92) 1600
(3) 41	(21) 104	(39) 191	(57) 314	(75) 586	(93) 1601
(4) 42	(22) 115	(40) 196	(58) 317	(76) 592	(94) 1906
(5) 44	(23) 121	(41) 205	(59) 318	(77) 622	(95) 2723
(6) 46	(24) 122	(42) 211	(60) 322	(78) 628	(96) 3013
(7) 50	(25) 124	(43) 221	(61) 330	(79) 688	(97) 3661
(8) 55	(26) 131	(44) 226	(62) 332	(80) 752	(98) 3755
(9) 60	(27) 134	(45) 239	(63) 335	(81) 779	(99) 3781
(10) 91	(28) 137	(46) 244	(64) 343	(82) 807	(100) 4518
(11) 67	(29) 145	(47) 252	(65) 348	(83) 989	
(12) 70	(30) 157	(48) 254	(66) 350	(84) 1082	
(13) 74	(31) 166	(49) 256	(67) 355	(85) 1091	
(14) 81	(32) 168	(50) 261	(68) 413	(86) 1126	
(15) 82	(33) 170	(51) 264	(69) 417	(87) 1146	
(16) 83	(34) 176	(52) 266	(70) 423	(88) 1247	
(17) 84	(35) 183	(53) 268	(71) 453	(89) 1274	
(18) 87	(36) 187	(54) 279	(72) 464	(90) 1399	

FIGURE 2.18
Fatigue specimen lives, 10,000s of cycles.

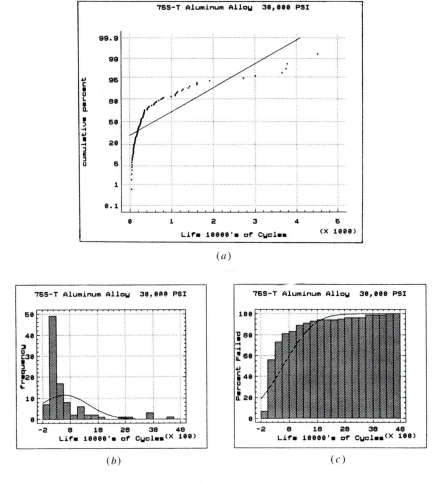

(a)

(b)   (c)

FIGURE 2.19
Proof that fatigue-life data are nongaussian.

carefully selected and the specimens are meticulously machined and "mirror-polished," the specimen cycles to failure exhibit an extreme random scatter, the shortest life being 326,500 cycles and the longest 45,184,200 cycles. When these data are plotted directly (Fig. 2.19a) on gaussian graph paper, they follow a smooth curve quite well, but it is clearly not a straight line and thus not close to gaussian. This is also seen in Fig. 2.19b and c, which shows alternative ways of deciding such questions. In Fig. 2.19b we have plotted a theoretical gaussian curve (with mean and standard deviation the same as those of the actual data) superimposed on a histogram of the actual data. The cumulative

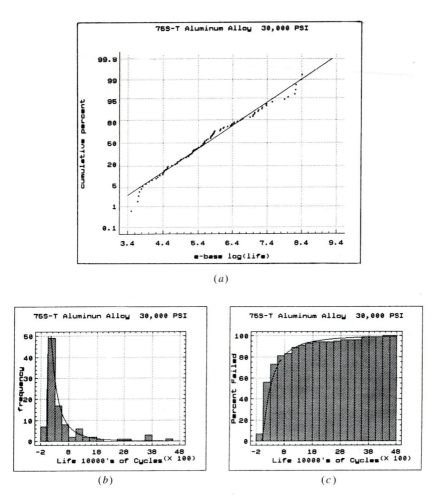

FIGURE 2.20
Log-life transformation makes data gaussian.

distribution graphs are shown in Fig. 2.19c. Clearly, both displays confirm our earlier conclusion from Fig. 2.19a. By simply trying various simple transformations it has been found that if we plot the logarithm (either base e or base 10) of the life (Fig. 2.20a), we now get quite good straight lines. Thus while the life L is *not* a gaussian variable, the quantity $G \triangleq \log L$ can be treated as such. That is, any calculations that are valid for a gaussian variable are valid for G. In the literature, such a distribution is usually called *lognormal*. Figure 2.20b and c compares the theoretical lognormal curves with our actual data, showing good agreement and confirming our decision based on Fig. 2.20a.

2.8 CALCULATION OF MACHINE PART FAILURE PROBABILITY WHEN APPLIED STRESS AND MATERIAL STRENGTH HAVE GAUSSIAN DISTRIBUTIONS

We earlier cautioned that practical applications of statistics to the design of parts with a predictable probability of failure are not common today. However, since the statistical point of view *is* useful to the designer, here we carry through some numerical calculations as an aid to understanding the basic concepts.

In Fig. 2.21 x represents the stress tending to cause failure, and y is the strength of the part. In conventional design, a design factor (safety factor), based on experience, is applied such that the strength exceeds the stress by a certain amount, say 2 to 1. Thus if we estimate the applied stress to be nominally 20,000 psi, we proportion the part such that its

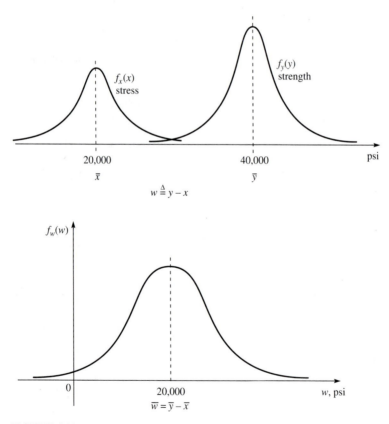

FIGURE 2.21
Failure probability prediction.

strength is 40,000 psi. Looking at this statistically, we recognize that the stress will be a statistical variable, say with probability density function $f_x(x)$, and similarly for strength y, $f_y(y)$.

If we define

$$w \triangleq \text{strength} - \text{stress} = y - x \tag{2.52}$$

then w will also be a statistical variable and has some probability density function $f_w(w)$. Since failure occurs whenever stress is greater than strength $(w < 0)$,

$$\text{Probability of failure} = \text{prob}\,(w < 0) = \int_{-\infty}^{0} f_w(w)\,dw \tag{2.53}$$

Thus if $f_w(w)$ were known, the probability of failure could be computed. It is shown in statistics texts that if $f_x(x)$ is gaussian with mean $\bar{x}$ and variance σ_x^2, while $f_y(y)$ is gaussian with mean $\bar{y}$ and variance σ_y^2, then $f_w(w)$ is gaussian with mean $\bar{w} = \bar{y} - \bar{x}$ and variance $\sigma_w^2 = \sigma_x^2 + \sigma_y^2$. Since $f_w(w)$ is a known gaussian distribution, we can calculate the probability of failure from Eq. (2.53) and the standard gaussian table. As a numerical example, let us take $\bar{x} = 20,000$ psi, $\bar{y} = 40,000$ psi, $\sigma_x = 2000$ psi, and $\sigma_y = 6000$ psi. Then the standardized normal variable z is

$$z = \frac{w - \bar{w}}{\sigma_w} = \frac{w - 20,000}{\sqrt{6000^2 + 2000^2}} = \frac{w - 20,000}{6320} \tag{2.54}$$

Now,

$$\text{Prob}\,(w < 0) = \text{prob}\left(z < \frac{-20,000}{6320}\right) = \text{prob}\,(z < -3.17) \tag{2.55}$$

$$\text{Prob}\,(z < -3.17) = 1 - P(z < 3.17) = 1 - 0.9992 = 0.0008 \tag{2.56}$$

thus the predicted probability of failure is 0.08 percent. Thus if, say, 10,000 such systems were put into service, we would expect eight failures.

Conventional (nonstatistical) design procedures work numerically with the mean values $\bar{x}$ and $\bar{y}$ only, and the variability (taken into account by σ_x and σ_y in the statistical methods) is brought in only in a qualitative sense when one picks a number to use for $\bar{x}$ and $\bar{y}$ and when one decides what safety factor to use. Improvements to be had by reducing variability (such as tightening dimensional tolerances or material specifications) can be *numerically* evaluated with the statistical approach. For example, if a more expensive grade of steel will reduce σ_y above to 2000 psi, we have

$$z = \frac{w - 20,000}{\sqrt{2000^2 + 2000^2}} = \frac{w - 20,000}{2830} \tag{2.57}$$

and

$$\text{Probability of failure} = \text{prob}\,(z < -7.07) = (<1.5 \times 10^{-7})^{\dagger} \quad (2.58)$$

The economic advantage of this greatly reduced failure rate can be compared with the increased cost of the material, in order to judge whether this course of action should be taken. Such judgments can, of course, be made without the use of statistical methods; statistics simply helps to make the judgment more rational.

2.9 CALCULATION OF STATISTICAL VARIABILITY OF QUANTITY WHICH DEPENDS IN A KNOWN WAY ON SEVERAL INDEPENDENT VARIABLES WITH KNOWN STATISTICAL VARIABILITIES

An aircraft inertial guidance and navigation system contains many measuring devices such as gyroscopes and accelerometers and many computing elements. Each of these system elements exhibits statistical variability. When the system calculates the present location of the aircraft, this location is uncertain by some amount. How can we estimate this uncertainty? How is it related to the uncertainty of each system element?

In running acceptance tests on a steam-turbine installation, many individual measurements must be made and then substituted into a complicated formula in order to calculate the thermal efficiency. If our test shows a 33.8 percent efficiency while the manufacturer guarantees 35 percent, what are the chances that the efficiency *really* is 35 percent and we got 33.8 percent because of accumulated errors of measurement? If we can tolerate only 1 percent error in efficiency measurement, how much uncertainty can we tolerate in each individual measuring device?

Questions such as the above regularly arise in engineering practice and can usually be handled by a relatively simple approach, which we now describe. Mathematically, the situation can be cast in the form

$$y = y(x_1, x_2, x_3, \ldots) \quad (2.59)$$

where y is a dependent variable (such as aircraft location or thermal

---

† Available table goes only to $z = -5$. For $z = -5$, the probability equals 1.5×10^{-7}.

efficiency) and the x's are independent variables (such as accelerometer signals or temperature readings). If each independent variable has a nominal value $x_{1,o}$, $x_{2,o}$, etc., and uncertainty (or error, or standard deviation) Δx_1, Δx_2, etc., then if we assume that the Δx's are small percentages of their nominal values (this is usually the case), we can estimate the uncertainty Δy of y by

$$\Delta y \approx \frac{\partial y}{\partial x_1} \Delta x_1 + \frac{\partial y}{\partial x_2} \Delta x_2 + \cdots \tag{2.60}$$

where the partial derivatives are *evaluated* at $x_1 = x_{1,o}$, $x_2 = x_{2,o}$, etc. When the x's are considered random variables with mean values $x_{1,o}$, $x_{2,o}$, etc., and standard deviations σ_{x_1}, σ_{x_2}, etc., the mean value of y may be computed from Eq. (2.59) by substituting $x_{1,o}$, $x_{2,o}$, etc., for x_1, x_2, etc., and the variance σ_y^2 of y may be estimated from

$$\sigma_y^2 \approx \left(\frac{\partial y}{\partial x_1}\right)^2 \sigma_{x_1}^2 + \left(\frac{\partial y}{\partial x_2}\right)^2 \sigma_{x_2}^2 + \cdots \tag{2.61}$$

As an example, suppose we are manufacturing capillary tube flow resistances for use in pneumatic control systems. We purchase capillary tubing from a supplier whose quality control department tells us that the tube inside diameter has a mean value $\bar{d}$ of 0.0100 in and standard deviation σ_d of 0.0005 in. We cut the tube to length in our own cutoff machines, which produce a length with mean value $\bar{l} = 10.00$ in and $\sigma_l = 0.05$ in. To meet specifications, our pneumatic controller can tolerate a flow-resistance standard deviation of no more than 10 percent of the mean value. Will we meet this specification with the values quoted above? The formula for flow resistance is

$$R_{fl} = \frac{p}{q} = \frac{\text{psi}}{\text{in}^3/\text{s}} = \frac{128\,\mu l}{\pi d^4} \tag{2.62}$$

$$\frac{\partial R_{fl}}{\partial l} = \frac{128\mu}{\pi d^4} = \frac{128\mu}{\pi} 10^8 \qquad \frac{\partial R_{fl}}{\partial d} = \frac{(-4)(128\mu l)}{\pi d^5} = \frac{-128\mu}{\pi} \times 40 \times 10^{10}$$

$$\sigma_{R_{fl}}^2 = \left(\frac{128\mu}{\pi}\right)^2 [(10^{16} \times 25 \times 10^{-4}) + (1600 \times 10^{20} \times 25 \times 10^{-8})] \tag{2.63}$$

$$\sigma_{R_{fl}} = \sqrt{\left(\frac{128\mu}{\pi}\right)^2 \underbrace{(25 \times 10^{12}}_{\text{effect of } l} + \underbrace{1600 \times 25 \times 10^{12})}_{\text{effect of } d}} = \frac{128\mu}{\pi}(200 \times 10^6) \tag{2.64}$$

Since we want $\sigma_{R_{fl}}$ as a percentage of R_{fl},

$$\frac{\sigma_{R_{fl}}}{R_{fl}} = \frac{(128\mu/\pi)(200 \times 10^6)}{(128\mu/\pi)(10/10^{-8})} = 0.20 = 20\% \tag{2.65}$$

Note that a σ_d of 5 percent of the mean value of (0.0005 out of 0.01) and a σ_l of 0.5 percent (0.05 out of 10.0) cause a $\sigma_{R_{fl}}$ of 20 percent! The analysis clearly shows that the tube diameter, not length, is the main source of variability because of the d^4 term in R_{fl}. Note from Eq. (2.64) that while σ_d (as a percentage of d) is 10 times σ_l, the *effect* of σ_d on $\sigma_{R_{fl}}$ is 40 times the effect of σ_l, due to the d^4 term.

The above approach is being regularly applied today in the design and manufacture of many products which require the assembly of several or many components into an overall machine or structure. Each part has its own statistical variability in its several dimensions. To assemble properly, there must be a certain geometric compatibility among the various parts; e.g., a shaft must be smaller than the bearing bore into which it fits. A small variation in each of several parts in an assembly can prevent the parts from being assembled easily (or *at all*), thus we must design the parts and operate the manufacturing processes so that this probability is acceptably low. However, we can't just blindly specify super-tight tolerances since this greatly increases the costs. Variations in some dimensions may have a much greater effect than variations in others on the overall ease of assembly, due to the geometry of the assembled mechanism. We need to identify these critical parts and dimensions so that improvement efforts can be concentrated where they are most effective. We also need to study alternative mechanism designs, since some are much more tolerant of individual part inaccuracies than others.

Computer software for efficiently carrying out such studies is available.[1] Figure 2.22, a page from the John Deere sales brochure, gives a few details.

2.10 CONFIDENCE INTERVALS FOR MEAN VALUES AND STANDARD DEVIATIONS OF GAUSSIAN DATA

As mentioned earlier, when we calculate $\bar{x}$ or s for an actual sample, these numerical values are only estimates. When we *use* these estimated numerical values in making practical decisions, it is important to know

---

[1] John Deere Assembly Variation Simulation System (JD/AVSS), Deere & Co., Moline, IL 61265.

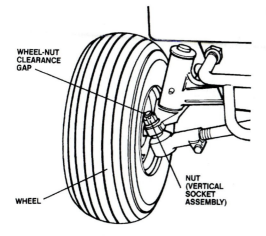

WHEEL-NUT CLEARANCE GAP

NUT (VERTICAL SOCKET ASSEMBLY)

WHEEL

A LET'S TAKE A CLOSER LOOK AT THE JD/AVSS PROCESS

Problem:

Given these parts and features, wheel, wheel-nut clearance gap and nut-socket assembly as shown in illustraton A, what are the answers to the following questions:

▲ Will these parts have proper clearance gaps?
▲ Can the variations, if any, cause interference?

IF SO...

▲ What percentage of assemblies will have it?
▲ Which specifications are major contributors to variation?

B

NOMINAL	MINIMUM	MAXIMUM	MEAN	STD. DEV	RPT NAME(S)
−0.037	−0.038	−0.036	−0.037	0.000	WHEEL.ADC004COSX
0.000	−0.001	0.001	0.000	0.000	WHEEL.ADC004COSY
0.000	−0.001	0.001	0.000	0.000	WHEEL.ADC004COSZ
27.958	22.189	33.661	27.881	1.968	WHEEL.GAP
0.713	0.019	1.468	0.709	0.239	WHEEL.DY
−4.575	−13.294	0.636	−4.668	2.150	WHEEL.DZ
15.065	9.280	20.760	14.988	1.972	WHEEL.RX
−0.100	−0.745	0.626	−0.101	0.234	WHEEL.RY
2.556	−6.224	7.822	2.462	2.171	WHEEL.RZ
BASIC ·					

Use JD/AVSS to get answers to these questions

The AVSS analyser enables the user to statistically analyze the results of simulating your problem. This includes reports of basic statistics for critical dimensions (Figure B). Also included is a histogram analysis (Figure C) that provides percentage of out-of-specification parts, maximum and minimum occurences, and tests for normality. By regression analysis (Figure D), percent contribution and sensitivity of the assembly to variation in individual features can be identified. This provides a means of controlling the most important and sensitive features.

C

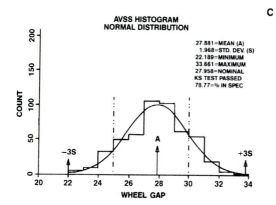

**AVSS HISTOGRAM
NORMAL DISTRIBUTION**

27.881=MEAN (A)
1.968=STD. DEV. (S)
22.189=MINIMUM
33.661=MAXIMUM
27.958=NOMINAL
KS TEST PASSED
78.77=% IN SPEC

COUNT

−3S   A   +3S

20 22 24 26 28 30 32 34

WHEEL GAP

**SEQUENTIAL LINEAR REGRESSION
FOR: WHEEL GAP**

RANK	CONTRIBUTION (%)	SENSITIVITY	VARIANT NAME	
−	−	27.96	CONSTANT TERM	
1	43.79	−0.9942	AM101300.V(	1)
2	37.43	1.010	AM100639.V(	5)
3	7.883	−18.83	AM100639.TP(	1,1)
4	5.055	1.196	AM100639.V(	7)
5	2.808	−10.36	AM100639.TP(	1,2)
REM	3.034	−	NOT ACCOUNTED FOR	

DO YOU WANT ANOTHER ITERATION OF REGRESSION ANALYSIS?

FIGURE 2.22
Software for assembly variation studies.

D JD/AVSS results are determined from computer simulated builds of production assemblies. In this way, proposed design and manufacturing tolerance changes can be quickly evaluated.

JD/AVSS provides designers with a practical problem-solving approach to tolerance assignment during the design cycle. It helps process engineers assess the capability of existing manufacturing processes to meet design specificatons of new products. For current assembly problems, JD/AVSS simulations list major contributors to variation in assembly specifications and help evaluate alternative changes to design specifications required to correct the assembly problem.

how accurate they are. For example, if we are trying to decide whether to buy structural adhesive from vendor A or vendor B, based on the breaking strength of bonds in a standard test specimen, better decisions will be made if we calculate and use a *confidence interval* for $\bar{x}$, not just $\bar{x}$ itself. Suppose we tested, say, 10 specimens each of A and B and computed $\bar{x}_A$ and $\bar{x}_B$. If $\bar{x}_A > \bar{x}_B$, this does *not* necessarily mean that adhesive A is better than B; in fact, they could have exactly the *same* average strength! To convince yourself of this, compute $\bar{x}$ for the first 10 buckling loads of Fig. 2.7 and then for the second 10 loads. We *know* that these two samples were drawn from the *same* gaussian distribution with $\bar{x} = 1200$, yet samples of size 10 will each give *different* estimates of $\bar{x}$. For small samples this difference can be quite great, so it is easy to be misled into a wrong decision if we compare only $\bar{x}$ values. It is *intuitively* obvious that $\bar{x}$ and s estimates improve for larger sample sizes. The confidence interval allows us to *quantify* this improvement.

Confidence intervals for mean (average) values are computed by using a probability density function called the *t distribution*. As usual, we only give a computing method; no proof of its validity is attempted here. We must first decide what confidence level we wish to use. After we compute the confidence interval, we will be able to make the following kind of statement: "My best estimate of $\bar{x}$ is A, and I am 95 percent sure the correct value lies within $\pm R_A$ of A." The number A is merely the value of $\bar{x}$ computed from the given sample in the usual way; the number R_A is obtained from a table of the t distribution, and the number 95 percent is the confidence level. We can choose any value we wish for this; but, for a *given* sample, if we use a higher confidence level (say 99 percent), the number R_A will be *larger*. That is, if we want to be *more sure* that the confidence interval $A - R_A < \bar{x} < A + R_A$ contains the value of $\bar{x}$, then this interval *must* be larger.

It is conventional to define a parameter α such that

$$\alpha \triangleq 1.00 - \text{decimal value of confidence level} \qquad (2.66)$$

For example, $\alpha = 0.05$ for a 95 percent confidence level. The t distribution has an adjustable parameter f, the degrees of freedom. In our current application we always choose $f = n - 1$, where n is the sample size. We look up in our t table (Fig. 2.23) the t value located in the $(n-1)$st row and the $\alpha/2$ column. If $n = 10$ and $\alpha = 0.05$, this value is 2.262. The confidence interval for any such problem is defined by

$$\text{Confidence interval} \triangleq \bar{x} \pm t_{\alpha/2, n-1} \frac{s}{\sqrt{n}} \qquad (2.67)$$

where s is the standard deviation of the sample, computed in the usual

FIGURE 2.23
The t distribution.

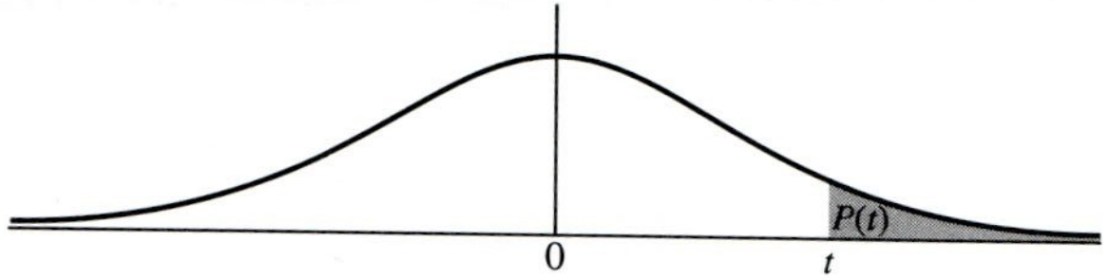

Upper percentage points of the t distribution

f	.40	.30	.25	.20	.15	.10	.05	.025	.01	.005	.0005
1	.325	.727	1.000	1.376	1.963	3.078	6.314	12.706	31.821	63.657	636.619
2	.289	.617	.816	1.061	1.386	1.886	2.920	4.303	6.965	9.925	31.598
3	.277	.584	.765	.978	1.250	1.638	2.353	3.182	4.541	5.841	12.924
4	.271	.569	.741	.941	1.190	1.533	2.132	2.776	3.747	4.604	8.610
5	.267	.559	.727	.920	1.156	1.476	2.015	2.571	3.365	4.032	6.869
6	.265	.553	.718	.906	1.134	1.440	1.943	2.447	3.143	3.707	5.959
7	.263	.549	.711	.896	1.119	1.415	1.895	2.365	2.998	3.499	5.408
8	.262	.546	.706	.889	1.108	1.397	1.860	2.306	2.896	3.355	5.041
9	.261	.543	.703	.883	1.100	1.383	1.833	2.262	2.821	3.250	4.781
10	.260	.542	.700	.879	1.093	1.372	1.812	2.228	2.764	3.169	4.587
11	.260	.540	.697	.876	1.088	1.363	1.796	2.201	2.718	3.106	4.437
12	.259	.539	.695	.873	1.083	1.356	1.782	2.179	2.681	3.055	4.318
13	.259	.538	.694	.870	1.079	1.350	1.771	2.160	2.650	3.012	4.221
14	.258	.537	.692	.868	1.076	1.345	1.761	2.145	2.624	2.977	4.140
15	.258	.536	.691	.866	1.074	1.341	1.753	2.131	2.602	2.947	4.073
16	.258	.535	.690	.865	1.071	1.337	1.746	2.120	2.583	2.921	4.015
17	.257	.534	.689	.863	1.069	1.333	1.740	2.110	2.567	2.898	3.965
18	.257	.534	.688	.862	1.067	1.330	1.734	2.101	2.552	2.878	3.922
19	.257	.533	.688	.861	1.066	1.328	1.729	2.093	2.539	2.861	3.883
20	.257	.533	.687	.860	1.064	1.325	1.725	2.086	2.528	2.845	3.850
21	.257	.532	.686	.859	1.063	1.323	1.721	2.080	2.518	2.831	3.819
22	.256	.532	.686	.858	1.061	1.321	1.717	2.074	2.508	2.819	3.792
23	.256	.532	.685	.858	1.060	1.319	1.714	2.069	2.500	2.807	3.767
24	.256	.531	.685	.857	1.059	1.318	1.711	2.064	2.492	2.797	3.745
25	.256	.531	.684	.856	1.058	1.316	1.708	2.060	2.485	2.787	3.725
26	.256	.531	.684	.856	1.058	1.315	1.706	2.056	2.479	2.779	3.707
27	.256	.531	.684	.855	1.057	1.314	1.703	2.052	2.473	2.771	3.690
28	.256	.530	.683	.855	1.056	1.313	1.701	2.048	2.467	2.763	3.674
29	.256	.530	.683	.854	1.055	1.311	1.699	2.045	2.462	2.756	3.659
30	.256	.530	.683	.854	1.055	1.310	1.697	2.042	2.457	2.750	3.646
40	.255	.529	.681	.851	1.050	1.303	1.684	2.021	2.423	2.704	3.551
60	.254	.527	.679	.848	1.046	1.296	1.671	2.000	2.390	2.660	3.460
120	.254	.526	.677	.845	1.041	1.289	1.658	1.980	2.358	2.617	3.373
∞	.253	.524	.674	.842	1.036	1.282	1.645	1.960	2.326	2.576	3.291

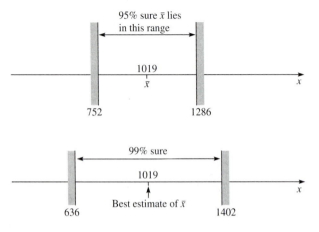

FIGURE 2.24
Confidence intervals for mean value, 95 and 99 percent.

way. Suppose we have a 10-point data set with $\bar{x} = 1019$ and $s = 373$, and we choose a 95 percent confidence level (see Fig. 2.24),

$$\text{Confidence interval} = 1019 \pm 2.262 \,\frac{373}{\sqrt{10}} = 1019 \pm 267 \qquad (2.68)$$

Our best estimate of $\bar{x}$ is thus 1019, and we are 95 percent sure the true value is somewhere between 752 and 1286. By "95 percent sure" we mean that if we use this method routinely in our work, we will be correct in "bracketing" the true $\bar{x}$ values 19 times out of 20. If we want to be 99 percent sure, the interval *widens* (± 383).

In Eq. (2.67) the main effect of increasing the sample size n is found in $\sqrt{n}$; the value of $t_{\alpha/2,n-1}$ varies little with n for n greater than 10. Note that s will also change as n is increased, but unpredictably, so no trend with n can be inferred, except that s should "converge" to a constant value as n gets larger. Thus, as n gets larger, its *main* effect will be to reduce the width of the confidence interval in proportion to $\sqrt{n}$. If we wish to cut the confidence interval in half, we need to quadruple the sample size, for example.

When *comparing* mean values, as in the adhesive strength values for vendors A and B mentioned above, we should always include in such comparisons the confidence intervals of each computed mean value when we make our decision, perhaps using a graphical display as in Fig. 2.25 as an aid. When the two confidence intervals show no overlap, then the decision is fairly clear-cut, except that we *could* confuse it by changing from, say, a 90 percent confidence level to 98 percent. That is, when 90 percent confidence intervals *don't* overlap, it is quite possible that 98

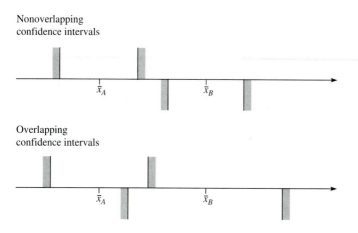

Nonoverlapping
confidence intervals

Overlapping
confidence intervals

FIGURE 2.25
Comparing two mean values by using confidence intervals.

percent intervals *will.* When overlap *does* occur, then the *amount* of overlap becomes critical in making a decision. The larger the overlap, the less likely it is that there is a *real* difference between the two mean values being compared. Note that our statistical calculations do *not* relieve us of the need to use judgment in reaching a conclusion, but they *do* put our decision on a more rational basis.

When the sample size n exceeds about 30, the estimated standard deviation is close enough to the true value σ that the confidence-interval problem can be treated as one where σ is *known,* rather than being estimated from the sample. Statistics texts show that the t distribution for $n \to \infty$ approaches a gaussian one and that the confidence interval for $\bar{x}$ can then be well approximated by

$$\text{Confidence interval} = \bar{x} \pm z_{\alpha/2} \frac{s}{\sqrt{n}} \tag{2.69}$$

We can see the quality of this approximation by making a table:

α	$t_{\alpha/2,29}$	$z_{\alpha/2}$
0.10	1.699	1.645
0.05	2.045	1.960
0.02	2.462	2.326
0.01	2.756	2.576

Instead of using confidence intervals, an alternative approach to practical problems of the type discussed above can be found in the method of *hypothesis testing,* discussed in many statistics textbooks. I have found the confidence-interval viewpoint personally preferable and also have noted that most students seem to understand it more easily, so we do not develop the hypothesis-testing concept here. Details on the relative advantages of the two methods are available in the literature.[1]

Sometimes it is important to obtain more information about the *standard deviation* than just its estimated value, so confidence intervals for s are also of interest. For example, many manufacturing processes can be adjusted to obtain the mean value desired for some critical feature (dimension, density, viscosity, etc.) of the product, but the variability around the mean is more or less fixed. If we are trying to choose between two such alternative processes, then we want to know which has the smaller standard deviation, since it will produce less "scrap" (off-specification) product. Just as with comparing average values, *direct* comparison of the raw estimates of the two standard deviations can easily lead to wrong decisions, whereas inclusion of confidence intervals reduces this likelihood. Confidence intervals on s are computed as follows:

1. Compute s from the sample in the usual way.
2. Choose the desired confidence level. This is often 95 percent, but you *can* use any value. For 95 percent,

$$95 = 100(1 - \alpha)$$
$$\alpha = 0.05 \tag{2.70}$$

3. Look up in a chi-squared table

$$\chi^2_{\alpha/2, n-1} \quad \text{and} \quad \chi^2_{1-\alpha/2, n-1}$$

where n is the sample size.
4. The confidence limits are

$$\left[\frac{(n-1)s^2}{\chi^2_{\alpha/2, n-1}} \right]^{1/2} < s < \left[\frac{(n-1)s^2}{\chi^2_{1-\alpha/2, n-1}} \right]^{1/2} \tag{2.71}$$

For example, if $n = 20$, $s = 11$, and $\alpha = 0.05$, we find

$$\chi^2_{0.025, 19} = 32.85 \qquad \chi^2_{0.975, 19} = 8.907 \tag{2.72}$$

---

[1] Mary G. Natrella, *The Relation Between Confidence Intervals and Tests of Significance,* Special Publication 300, vol. 1, National Bureau of Standards, Washington, 1969, pp. 388–391. G. E. P. Box, W. G. Hunter, J. S. Hunter, *Statistics for Experimenters,* Wiley, New York, 1978, p. 109.

which gives

$$s = 11.0 \begin{array}{c} +5.1 \\ \\ -2.6 \end{array} \qquad 8.4 < s < 16.1 \qquad (2.73)$$

Note that the plus and minus limits are *not* equal as they were for the $\bar{x}$ limits; i.e., the best estimate 11.0 is *not* midway between 8.4 and 16.1. The "probabilities," however, *are* equal. That is, in the above example, the true value will be found 47.5 percent of the time between 8.4 and 11.0 and 47.5 percent of the time between 11.0 and 16.1. Figure 2.26 simplifies the calculation of confidence limits for s.

Actually, in comparing either the mean values or the standard deviations of two data sets representing some product or process characteristics to decide which is better, more direct and convenient methods than the graphical overlap approach suggested above are available. For comparing mean values $\bar{x}_A$ and $\bar{x}_B$, a confidence interval for their difference $\bar{x}_A - \bar{x}_B \stackrel{\Delta}{=} \bar{x}_{A-B}$ is available.[1] We first compute $\bar{x}_A$, s_A, $\bar{x}_B$, and s_B from the two separate data sets in the usual way, assigning the symbol $\bar{x}_A$ to the larger of the two mean values. We will need a value from the table of the t distribution (Fig. 2.23), choosing the table column according to our choice of confidence level; e.g., the column headed 0.025 would be used for 95 percent confidence, 0.05 for 90 percent, etc. The table row (degrees of freedom f) is calculated from

$$f = \frac{\left(\dfrac{s_A^2}{n_A} + \dfrac{s_B^2}{n_B}\right)^2}{(s_A^2/n_A)^2/(n_A - 1) + (s_B^2/n_B)^2/(n_B - 1)} \qquad (2.74)$$

The confidence interval is then given by

$$\bar{x}_{A-B} \pm t_{\alpha/2} \sqrt{\frac{s_A^2}{n_A} + \frac{s_B^2}{n_B}} \qquad (2.75)$$

where $t_{\alpha/2}$ is the value taken from the t table. Interpretation of such results is best explained with a few numerical examples. If in Eq. (2.75) we obtained, say, 5.1 ± 2.5, then it is highly likely that $\bar{x}_A$ really is larger than $\bar{x}_B$; and if we are using 95 percent confidence, then we are 95

[1] R. L. Mason, R. F. Gunsdt, and J. L. Hess, *Statistical Design and Analysis of Experiments*, Wiley, New York, 1989, p. 274.

Confidence intervals for standard deviation								
Sample size	50% confidence			90% confidence			95% confidence		
	Lo	**Hi**	**± %**	**Lo**	**Hi**	**± %**	**Lo**	**Hi**	**± %**
6	0.867s	1.36s	25.	0.672s	2.09s	71.	0.624s	2.45s	91.
11	0.893s	1.22s	16.	0.739s	1.59s	43.	0.699s	1.76s	53.
16	0.907s	1.16s	13.	0.775s	1.44s	33.	0.739s	1.55s	40.
21	0.916s	1.14s	11.	0.798s	1.36s	28.	0.765s	1.44s	34.
41	0.936s	1.09s	7.7	0.847s	1.23s	19.	0.821s	1.28s	23.
61	0.946s	1.07s	6.2	0.871s	1.18s	15.	0.849s	1.22s	18.
101	0.957s	1.05s	4.8	0.897s	1.13s	12.	0.879s	1.16s	14.

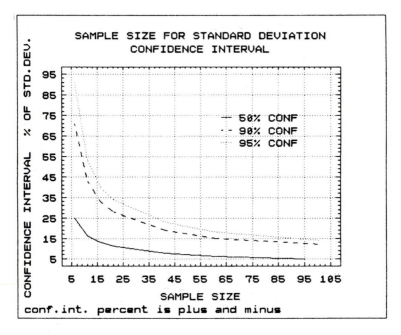

FIGURE 2.26
Choosing the sample size for confidence intervals on standard deviations.

percent sure that $\bar{x}_A - \bar{x}_B$ is between 2.6 and 7.6. If instead we got 5.1 ± 10.3, then $\bar{x}_A - \bar{x}_B$ could be anywhere between -5.2 and $+15.4$, and we have low confidence that $\bar{x}_A$ really is larger than $\bar{x}_B$.

When we wish to compare the standard deviations (variability) of two products or processes to choose the better of the two (smaller variability is generally preferable), we need to use the F distribution (Fig. 2.27). The definition of the confidence interval in this case is given in terms of the ratio σ_A / σ_B of the true standard deviations[1]

$$\frac{1}{F_{\alpha/2, \nu_A, \nu_B}} \frac{s_A^2}{s_B^2} \le \frac{\sigma_A^2}{\sigma_B^2} \le \frac{s_A^2}{s_B^2} F_{\alpha/2, \nu_B, \nu_A} \tag{2.76}$$

Here s_A and s_B are the sample standard deviations computed in the usual way, as is the confidence level (95 percent confidence has $\alpha = 0.05$, etc.). The F distribution requires specification of two different degrees of freedom; that is, $F_{\alpha/2, \nu_1, \nu_2}$ has degrees of freedom

$$\nu_1 \triangleq \text{numerator degree of freedom} = n_1 - 1$$
$$\nu_2 \triangleq \text{denominator degree of freedom} = n_2 - 1 \tag{2.77}$$

where the n's are the sample sizes. Figure 2.27 is for 90 percent confidence; Lipson and Sheth have tables for 90 to 99.9 percent. Note first that for infinite sample size, F becomes 1.0 for every α, and thus in Eq. (2.76)

$$\frac{\sigma_A}{\sigma_B} = \frac{s_A}{s_B} \tag{2.78}$$

and the confidence interval has shrunk to zero width. If both the A and B samples were of size 10, then for 90 percent confidence the two F values in Eq. (2.76) are both 3.18 and

$$0.56 \frac{s_A}{s_B} \le \frac{\sigma_A}{\sigma_B} \le 1.78 \frac{s_A}{s_B} \tag{2.79}$$

If, say, s_A / s_B were 1.5, then $0.84 \le \sigma_A / \sigma_B \le 2.67$, which says that either σ_A or σ_B *could* be the larger of the two. We see again that a simple calculation and direct comparison of only s_A and s_B can be misleading,

[1] Lipson and Sheth, *Statistical Design and Analysis of Experiments*, p. 109.

v_2	Degrees of freedom for the numerator (v_1)																	
	1	2	3	4	5	6	7	8	9	10	11	12	13	14	15	16	17	18
1	161	200	216	225	230	234	237	239	241	242	243	244	245	245	246	246	247	247
2	18.5	19	19.2	19.2	19.3	19.3	19.4	19.4	19.4	19.4	19.4	19.4	19.4	19.4	19.4	19.4	19.4	19.4
3	10.1	9.55	9.28	9.12	9.01	8.94	8.89	8.85	8.81	8.79	8.76	8.74	8.73	8.71	8.70	8.69	8.68	8.67
4	7.71	6.94	6.59	6.39	6.26	6.16	6.09	6.04	6.00	5.96	5.94	5.91	5.82	5.87	5.86	5.84	5.83	5.82
5	6.61	5.79	5.41	5.19	5.05	4.95	4.88	4.82	4.77	4.74	4.70	4.08	4.66	4.64	4.62	4.60	4.59	4.58
6	5.99	5.14	4.76	4.53	4.39	4.28	4.21	4.15	4.10	4.06	4.03	4.00	3.98	3.96	3.94	3.92	3.91	3.90
7	5.59	4.74	4.35	4.12	3.97	3.87	3.79	3.73	3.68	3.64	3.60	3.57	3.55	3.53	3.51	3.49	3.48	3.47
8	5.32	4.46	4.07	3.84	3.69	3.58	3.50	3.44	3.39	3.35	3.31	3.28	3.26	3.24	3.22	3.20	3.19	3.17
9	5.12	4.26	3.86	3.63	3.48	3.37	3.29	3.23	3.18	3.14	3.10	3.07	3.05	3.03	3.01	2.99	2.97	2.96
10	4.96	4.10	3.71	3.48	3.33	3.22	3.14	3.07	3.02	2.98	2.94	2.91	2.89	2.86	2.85	2.83	2.81	2.80
11	4.84	2.98	3.50	3.36	3.20	3.01	2.95	2.90	2.85	2.82	2.82	2.79	2.76	2.74	2.72	2.70	2.69	2.67
12	4.75	3.98	3.49	3.26	3.11	3.00	2.91	2.85	2.80	2.75	2.72	2.69	2.66	2.64	2.62	2.60	2.58	2.57
13	4.67	3.81	3.41	3.18	3.03	2.92	2.83	2.77	2.71	2.67	2.63	2.60	2.58	2.55	2.53	2.51	2.50	2.48
14	4.60	3.74	3.34	3.11	2.96	2.85	2.76	2.70	2.65	2.60	2.57	2.53	2.51	2.48	2.46	2.44	2.43	2.41
15	4.54	3.68	3.29	3.06	2.90	2.79	2.71	2.64	2.59	2.54	2.51	2.48	2.45	2.42	2.40	2.38	2.37	2.35
16	4.49	3.63	3.24	3.01	2.85	2.74	2.66	2.59	2.54	2.49	2.46	2.42	2.40	2.37	2.35	2.33	2.32	2.30
17	4.45	3.59	3.20	2.96	2.81	2.70	2.61	2.55	2.49	2.45	2.41	2.38	2.34	2.33	2.31	2.29	2.27	2.26
18	4.41	3.55	3.16	2.93	2.77	2.66	2.58	2.51	2.46	2.41	2.37	2.34	2.31	2.29	2.27	2.25	2.23	2.22
19	4.38	3.52	3.13	2.90	2.74	2.63	2.54	2.48	2.42	2.38	2.34	2.31	2.28	2.26	2.23	2.21	2.20	2.18
20	4.35	3.49	3.10	2.87	2.71	2.60	2.51	2.45	2.39	2.35	2.31	2.28	2.25	2.22	2.20	2.18	2.17	2.15
21	4.32	3.47	3.07	2.82	2.68	2.57	2.49	2.42	2.37	2.32	2.28	2.25	2.22	2.20	2.18	2.16	2.14	2.12
22	4.30	3.44	3.05	2.84	2.66	2.55	2.46	2.40	2.34	2.30	2.26	2.23	2.20	2.17	2.15	2.16	2.11	2.10
23	4.28	3.42	3.03	2.80	2.64	2.53	2.44	2.37	2.32	2.27	2.23	2.20	2.18	2.15	2.13	2.11	2.09	2.07
24	4.26	3.40	3.01	2.78	2.62	2.51	2.42	2.36	2.30	2.25	2.21	2.18	2.15	2.13	2.11	1.09	2.07	2.05
25	4.24	3.39	2.99	2.76	2.60	2.49	2.40	2.34	2.28	2.24	2.20	2.16	2.14	2.11	2.09	2.07	2.05	2.04
26	4.23	3.37	2.98	2.74	2.59	2.47	2.39	2.32	2.27	2.22	2.18	2.15	2.12	2.09	2.07	2.05	2.03	2.02
27	4.21	3.35	2.96	2.73	2.57	2.46	2.37	2.31	2.25	2.20	2.17	2.13	2.10	2.08	2.06	2.04	2.02	2.00
28	4.20	3.34	2.95	2.71	2.56	2.45	2.36	2.29	2.24	2.19	2.15	2.12	2.09	2.06	2.04	2.02	2.00	1.99
29	4.18	3.33	2.93	2.70	2.55	2.43	2.35	2.28	2.22	2.18	2.14	2.10	2.08	2.05	2.03	2.01	1.99	1.97
30	4.17	3.32	2.92	2.69	2.53	2.42	2.33	2.27	2.21	2.16	2.13	2.09	2.06	2.04	2.01	1.99	1.98	1.96
32	4.15	3.29	2.90	2.67	2.51	2.40	2.31	2.24	2.19	2.14	2.10	2.07	2.04	2.01	1.99	1.97	1.95	1.94
34	4.13	3.28	2.88	2.65	2.49	2.38	2.29	2.23	2.17	2.12	2.08	2.05	2.02	1.99	1.97	1.95	1.93	1.92
36	4.11	3.26	2.87	2.63	2.48	2.36	2.28	2.21	2.15	2.11	2.07	2.03	2.00	1.98	1.95	1.93	1.92	1.90
38	4.10	3.24	2.85	2.62	2.47	2.35	2.26	2.19	2.14	2.09	2.05	2.02	1.99	1.96	1.94	1.92	1.90	1.88
40	4.08	3.23	2.84	2.61	2.45	2.34	2.25	2.18	2.12	2.08	2.04	2.00	1.97	1.95	1.92	1.90	1.89	1.87
42	4.07	3.22	2.83	2.59	2.44	2.32	2.24	2.16	2.11	2.06	2.03	1.99	1.96	1.93	1.91	1.89	1.87	1.86
44	4.06	3.21	2.82	2.58	2.43	2.31	2.23	2.16	2.10	2.05	2.01	1.98	1.95	1.92	1.90	1.88	1.86	1.84
46	4.05	3.20	2.81	2.57	2.42	2.30	2.22	2.15	2.09	2.04	2.00	1.97	1.94	1.91	1.89	1.87	1.85	1.83
48	4.04	3.19	2.80	2.57	2.41	2.29	2.21	2.14	2.08	2.03	1.99	1.96	1.93	1.90	1.88	1.86	1.84	1.82
50	4.03	3.18	2.79	2.56	2.40	2.29	2.20	2.13	2.07	2.03	1.99	1.95	1.92	1.89	1.87	1.85	1.83	1.81
55	4.02	3.16	2.77	2.54	2.38	2.27	2.18	2.11	2.06	2.01	1.97	1.93	1.90	1.88	1.85	1.83	1.81	1.79
60	4.00	3.15	2.76	2.53	2.37	2.25	2.17	2.10	2.04	1.99	1.95	1.92	1.89	1.86	1.84	1.82	1.80	1.78
65	3.99	3.14	2.75	2.51	2.36	2.24	2.15	2.08	2.03	1.98	1.94	1.90	1.87	1.85	1.82	1.80	1.78	1.76
70	3.98	3.13	2.74	2.50	2.35	2.23	2.14	2.07	2.02	1.97	1.93	1.89	1.86	1.84	1.81	1.79	1.77	1.75
80	3.96	3.11	2.73	2.49	2.33	2.21	2.13	2.06	2.00	1.95	1.91	1.88	1.84	1.82	1.79	1.77	1.75	1.73
90	3.95	3.10	2.71	2.47	2.32	2.20	2.11	2.04	1.99	1.94	1.90	1.86	1.83	1.80	1.78	1.76	1.74	1.72
100	3.94	3.09	2.70	2.46	2.31	2.19	2.10	2.03	1.97	1.93	1.89	1.85	1.82	1.79	1.77	1.75	1.73	1.71
125	3.92	3.07	2.68	2.44	2.29	2.17	2.08	2.01	1.96	1.91	1.87	1.83	1.80	1.77	1.76	1.72	1.70	1.69
150	3.90	3.08	2.66	2.43	2.27	2.16	2.07	2.00	1.94	1.89	1.85	1.82	1.79	1.76	1.73	1.71	1.69	1.67
200	3.89	3.04	2.65	2.42	2.26	2.14	2.06	1.98	1.93	1.88	1.84	1.80	1.77	1.74	1.72	1.69	1.67	1.65
300	3.87	3.03	2.63	2.40	2.24	2.13	2.04	1.97	1.91	1.86	1.82	1.78	1.75	1.72	1.70	1.68	1.66	1.64
500	3.86	3.01	2.62	2.39	2.23	2.12	2.03	1.96	1.90	1.85	1.81	1.77	1.74	1.71	1.69	1.66	1.64	1.62
1000	3.85	3.00	2.61	2.38	2.22	2.11	2.02	1.95	1.89	1.84	1.80	1.76	1.73	1.70	1.68	1.65	1.63	1.61
∞	3.84	3.00	2.60	2.37	2.21	2.10	2.01	1.94	1.88	1.83	1.79	1.75	1.72	1.69	1.67	1.64	1.62	1.60

Degrees of freedom for the denominator (v_2)

FIGURE 2.27
The F distribution.

Degrees of freedom for the numerator (v_1)

19	20	22	24	26	28	30	35	40	45	50	60	80	100	200	500	∞	
248	248	249	249	249	250	250	251	251	251	252	252	252	253	254	254	254	1
19.4	19.4	19.5	19.5	19.5	19.5	19.5	19.5	19.5	19.5	19.5	19.5	19.5	19.5	19.5	19.5	19.5	2
8.67	8.66	8.65	8.64	8.63	8.62	8.62	8.60	8.59	8.59	8.58	8.57	8.56	8.55	8.54	8.53	8.53	3
5.81	5.80	5.79	5.77	5.76	5.75	5.75	5.73	5.72	5.71	5.70	5.69	5.67	5.66	5.65	5.64	5.63	4
4.57	4.56	4.54	4.53	4.52	4.50	4.50	4.48	4.46	4.45	4.44	4.43	4.41	4.41	4.30	4.37	4.37	5
3.88	3.87	3.88	3.84	3.83	3.82	3.81	3.79	3.77	3.76	3.75	3.74	3.72	3.71	3.09	3.63	3.67	6
3.46	3.44	3.43	3.41	3.40	3.39	3.38	3.36	3.34	3.33	3.32	3.30	3.20	3.29	2.25	3.24	3.23	7
3.16	3.15	3.13	3.12	3.10	3.09	3.08	3.06	3.04	3.03	3.02	3.01	2.99	2.97	2.95	2.94	2.93	8
2.95	2.94	2.92	2.90	2.89	2.87	2.86	2.84	2.83	2.81	2.80	2.79	2.77	2.76	2.23	2.72	2.71	9
2.78	2.77	2.75	2.74	2.72	2.71	2.70	2.68	2.66	2.65	2.64	2.62	2.60	2.59	2.55	2.55	2.54	10
2.66	2.65	2.63	2.61	2.59	2.58	2.57	2.55	2.53	2.52	2.51	2.49	2.47	2.46	2.43	2.42	2.40	11
2.56	2.54	2.52	2.51	2.49	2.48	2.47	2.44	2.43	2.41	2.40	2.38	2.36	2.35	2.32	2.31	2.30	12
2.47	2.46	2.44	2.42	2.41	2.39	2.38	2.36	2.34	2.33	2.31	2.30	2.27	2.26	2.23	2.22	2.21	13
2.40	2.39	2.37	2.35	2.33	2.32	2.31	2.28	2.27	2.26	2.24	2.22	2.20	2.19	2.16	2.14	2.13	14
2.34	2.33	2.31	2.29	2.27	2.26	2.25	2.22	2.20	2.19	2.18	2.16	2.14	2.12	2.10	2.08	2.07	15
2.29	2.28	2.25	2.24	2.22	2.21	2.19	2.17	2.15	2.14	2.12	2.11	2.08	2.07	2.04	2.02	2.01	16
2.24	2.23	2.21	2.19	2.17	2.16	2.15	2.12	2.10	2.09	2.08	2.06	2.03	2.02	1.09	1.97	1.96	17
2.20	2.19	2.17	2.15	2.13	2.12	2.11	2.08	2.06	2.05	2.04	2.02	1.99	1.98	1.95	1.93	1.92	18
2.17	2.16	2.13	2.11	2.10	2.08	2.07	2.05	2.03	2.01	2.00	1.98	1.96	1.94	1.91	1.89	1.88	19
2.14	2.12	2.10	2.08	2.07	2.05	2.04	2.01	1.99	1.98	1.97	1.95	1.92	1.91	1.88	1.86	1.84	20
2.11	2.10	2.07	2.05	2.04	2.02	2.01	1.98	1.96	1.95	1.94	1.92	1.89	1.88	1.84	1.82	1.81	21
2.08	2.07	2.03	2.03	2.01	2.00	1.98	1.96	1.94	1.92	1.91	1.89	1.86	1.85	1.82	1.80	1.78	22
2.06	2.05	2.02	2.00	1.99	1.97	1.96	1.93	1.91	1.90	1.88	1.85	1.84	1.82	1.79	1.77	1.76	23
2.04	2.03	2.00	1.98	1.97	1.95	1.94	1.91	1.89	1.88	1.86	1.84	1.82	1.80	1.77	1.75	1.73	24
2.02	2.01	1.98	1.96	1.95	1.93	1.92	1.89	1.87	1.85	1.84	1.82	1.80	1.78	1.75	1.73	1.71	25
2.00	1.99	1.97	1.95	1.93	1.91	1.90	1.87	1.85	1.84	1.82	1.80	1.78	1.76	1.73	1.71	1.69	26
1.99	1.97	1.95	1.93	1.91	1.90	1.88	1.86	1.84	1.82	1.81	1.79	1.76	1.74	1.71	1.69	1.67	27
1.97	1.96	1.93	1.91	1.90	1.88	1.87	1.84	1.82	1.80	1.79	1.77	1.74	1.73	1.60	1.67	1.65	28
1.96	1.94	1.92	1.90	1.88	1.87	1.85	1.83	1.81	1.79	1.77	1.75	1.73	1.71	1.67	1.65	1.64	29
1.95	1.93	1.91	1.89	1.87	1.85	1.84	1.81	1.79	1.77	1.76	1.74	1.71	1.70	1.68	1.64	1.62	30
1.92	1.91	1.88	1.86	1.85	1.83	1.82	1.79	1.77	1.75	1.74	1.71	1.69	1.67	1.63	1.61	1.59	32
1.90	1.89	1.86	1.84	1.82	1.80	1.80	1.77	1.75	1.73	1.71	1.69	1.66	1.65	1.61	1.59	1.57	34
1.88	1.87	1.85	1.82	1.81	1.79	1.78	1.75	1.73	1.71	1.69	1.67	1.64	1.62	1.59	1.56	1.55	36
1.87	1.85	1.83	1.81	1.79	1.77	1.76	1.73	1.71	1.69	1.68	1.65	1.62	1.61	1.57	1.54	1.53	38
1.85	1.84	1.81	1.79	1.77	1.76	1.74	1.72	1.69	1.67	1.66	1.64	1.61	1.50	1.55	1.53	1.51	40
1.84	1.83	1.80	1.78	1.76	1.74	1.73	1.70	1.68	1.66	1.65	1.62	1.59	1.57	1.53	1.51	1.49	42
1.83	1.81	1.79	1.77	1.75	1.73	1.72	1.69	1.67	1.65	1.63	1.61	1.58	1.56	1.52	1.49	1.48	44
1.82	1.80	1.78	1.76	1.74	1.72	1.71	1.68	1.65	1.64	1.62	1.60	1.57	1.55	1.51	1.48	1.46	46
1.81	1.79	1.77	1.75	1.73	1.71	1.70	1.67	1.64	1.62	1.61	1.59	1.56	1.54	1.49	1.47	1.45	48
1.80	1.78	1.76	1.74	1.72	1.70	1.69	1.66	1.63	1.61	1.60	1.58	1.54	1.52	1.48	1.46	1.44	50
1.78	1.76	1.74	1.72	1.70	1.68	1.67	1.64	1.61	1.59	1.58	1.55	1.52	1.50	1.46	1.43	1.41	55
1.76	1.75	1.72	1.70	1.68	1.66	1.65	1.62	1.59	1.57	1.56	1.53	1.50	1.48	1.44	1.41	1.39	60
1.75	1.73	1.71	1.69	1.67	1.65	1.63	1.60	1.58	1.56	1.54	1.52	1.49	1.48	1.42	1.39	1.37	65
1.74	1.72	1.70	1.67	1.65	1.64	1.62	1.59	1.57	1.55	1.53	1.50	1.47	1.45	1.40	1.37	1.35	70
1.72	1.70	1.68	1.65	1.63	1.62	1.60	1.57	1.54	1.52	1.51	1.48	1.45	1.43	1.38	1.35	1.32	80
1.70	1.69	1.66	1.64	1.62	1.60	1.59	1.55	1.53	1.51	1.49	1.46	1.43	1.41	1.36	1.32	1.30	90
1.69	1.68	1.65	1.63	1.61	1.59	1.57	1.54	1.52	1.49	1.48	1.45	1.41	1.39	1.34	1.31	1.28	100
1.67	1.65	1.63	1.60	1.58	1.57	1.55	1.52	1.49	1.47	1.45	1.42	1.39	1.36	1.31	1.27	1.25	125
1.66	1.64	1.61	1.59	1.57	1.55	1.53	1.50	1.48	1.45	1.44	1.41	1.37	1.34	1.29	1.23	1.22	150
1.64	1.62	1.60	1.57	1.55	1.53	1.52	1.48	1.46	1.43	1.41	1.39	1.35	1.32	1.26	1.22	1.19	200
1.62	1.61	1.58	1.55	1.53	1.51	1.50	1.46	1.43	1.41	1.39	1.38	1.32	1.30	1.23	1.19	1.15	300
1.61	1.50	1.56	1.54	1.52	1.50	1.48	1.45	1.42	1.40	1.38	1.34	1.30	1.28	1.21	1.16	1.11	500
1.60	1.58	1.55	1.53	1.51	1.49	1.47	1.44	1.41	1.38	1.36	1.33	1.29	1.26	1.19	1.13	1.08	1000
1.59	1.57	1.54	1.52	1.50	1.48	1.46	1.42	1.39	1.37	1.35	1.32	1.27	1.24	1.17	1.11	1.00	∞

Degrees of freedom for the denominator (v_2)

since that tells us s_A is 1.5 times s_B, whereas in reality it is quite possible that s_A is smaller than s_B.

2.11 DISCRETE DISTRIBUTIONS: BINOMIAL AND POISSON

For continuous distributions (gaussian, chi-squared, t, etc.) the probability of the random variable's taking on any *specific* value is zero, as we showed earlier. For those applications where such probabilities are *not* zero, we define and use some kind of *discrete* distribution, binomial and Poisson being two of the most common.

The binomial distribution applies to situations where events are judged on a yes-or-no basis. If you flip a coin, it comes up either heads or tails; there is no in between. Each die of a pair of cast dice will have a specific one of its six sides face upward. Inspection of parts for manufactured objects is used to discover and reject those which are defective before they get built into a complex assembly to cause expensive trouble later. In many cases the inspection uses only two categories: defective and acceptable. Let's use this last application as an example to define and explain the use of the binomial distribution. We assume that our parts come from a population which has a fixed percentage of good and bad items and that this percentage remains the same as we withdraw samples to be tested. (This requires that the population be large enough or else that we *replace* those items withdrawn with others of like characteristics as we draw the sample.) Now let

$$p \triangleq \text{probability of getting a good part} \qquad (2.80)$$

$$q \triangleq \text{probability of getting a bad part} = 1 - p \qquad (2.81)$$

$$n \triangleq \text{number of parts in batch being tested} \qquad (2.82)$$

With the above definitions and assumptions, it can be shown that

$$P_n(x) = p^x q^{n-x} \left[\frac{n!}{x!\,(n-x)!} \right] \qquad (2.83)$$

where

$$P_n(x) \triangleq \text{probability of getting } x \text{ good parts in batch of } n \qquad (2.84)$$

As a numerical example, take $n = 10$ and $p = 0.25$ to get the table and graphs of Fig. 2.28. The *cumulative distribution* $C_n(x)$ is the

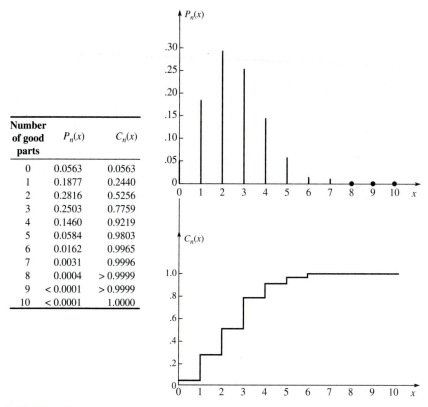

Number of good parts	$P_n(x)$	$C_n(x)$
0	0.0563	0.0563
1	0.1877	0.2440
2	0.2816	0.5256
3	0.2503	0.7759
4	0.1460	0.9219
5	0.0584	0.9803
6	0.0162	0.9965
7	0.0031	0.9996
8	0.0004	> 0.9999
9	< 0.0001	> 0.9999
10	< 0.0001	1.0000

FIGURE 2.28
Example of a binomial distribution.

probability of getting x or less good parts in a batch of n:

$$C_n(x) \triangleq \sum_{x=0}^{x} P_n(x) \qquad (2.85)$$

In practical applications we are often in the position of not knowing the true value of p and trying to estimate it from data on a finite-size sample. Confidence limits for p have been worked out, and the results presented in the form of tables (for $n < 30$) and graphs (for $n \geq 30$).[1] Our Fig. 2.29 shows these graphs. To illustrate the risk of using "nonstatisti-

---

[1] E. Crow, F. Davis, and M. Maxfield, *Statistics Manual,* Dover, New York, 1960, p. 51.

cal" reasoning on small-sample data, consider a case where 30 items are tested and *all* are found to be "perfect." Our natural tendency is to conclude that $p = 1.0$; however, the 95 percent chart shows that p could be anywhere between 1.0 and 0.88. If we want to be 99 percent sure, the range is even wider.

Turning now to the *Poisson distribution,* we introduce it by using one of its common applications, the emission of radioactive particles from a material, as detected by a Geiger counter. These particles are emitted randomly with respect to time; however, if we count them for a "long" time, an *average* rate of emission m in particles per second can be calculated. If we then wish to predict the probability $P(x)$ of counting exactly x particles in a 1-second interval, this phenomenon has been

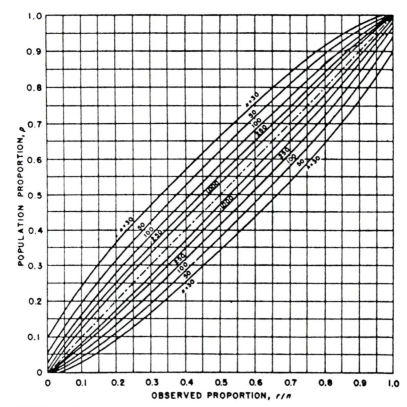

90% CONFIDENCE BELTS FOR PROPORTIONS

FIGURE 2.29
Estimating p for the binomial distribution.

95% CONFIDENCE BELTS FOR PROPORTIONS

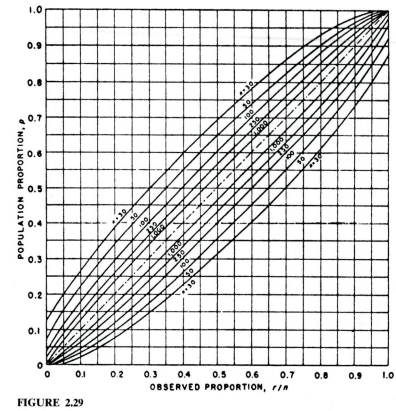

FIGURE 2.29
(Continued).

shown to closely follow the Poisson mathematical model

$$P(x) = \frac{m^x e^{-m}}{x!}$$ (2.86)

where e is the base of natural logarithms. Figure 2.30 is a chart based on Eq. (2.86), useful for many practical problems. Thus if $m = 3$ and we want the probability of getting 1 or more counts in a 1-second period, we use the $x = 0$ curve and find $P(1 \text{ or more}) = 95$ percent and $P(0) = 5$ percent.

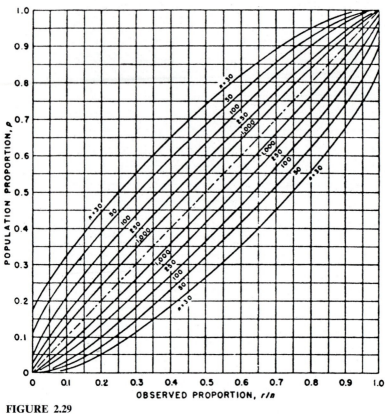

FIGURE 2.29
(Continued).

Some other phenomena for which the Poisson model might be reasonable include

1. The number of phone calls per minute arriving at the emergency road service of the Columbus Ohio Automobile Club. (Would m be the same for a nice summer day compared with a winter snowstorm?)
2. The number of tractor/trailers passing a given point on a highway per quarter-hour.
3. The number of defects per 10,000 ft of tape in a magnetic tape manufacturing line.

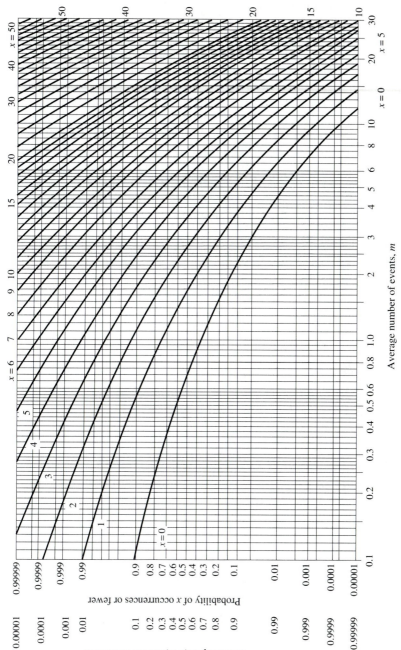

FIGURE 2.30
Poisson distribution graph.

As an example using the Poisson distribution, let's do the following problem. A steel-galvanizing process which has been running for a long time (such that valid data are available for it) includes a machine vision system which keeps track of surface defects. Records taken when the process is running *properly* show that defects occur randomly and that, on average, 0.6 defects are found per 1000 ft of product. It is desired to install a control system which will shut down the process when defects are occurring at an abnormally high rate, so as to reduce the amount of scrap produced. It is suggested that the gage be set to examine successive 1000-ft lengths and to report the number of defects found in each 1000-ft length as it passes through the measuring station. Use the Poisson graph (Fig. 2.30) to answer the following questions.

1. Would it seem reasonable to shut down if 1 or more defects is found in 1000 ft?
2. Repeat question 1, using shutdown criteria of 2 or more, 3 or more, and 4 or more defects.
3. Discuss the advantages and disadvantages of using a 5000-ft sample rather than a 1000-ft one.
4. Do you think there is some "best" sample length which should be used?

Since the average defect rate is 0.6 per 1000 ft, a statistically naive person might think that shutting down when we observe at least 1 defect in any 1000-ft sample is not unreasonable, since 1.0 is almost twice 0.6. However, consideration of the Poisson statistics of such a process, using Fig. 2.30, can give us a better insight into the situation. For the case of 1 or more defects, the curve labeled $x = 0$ is appropriate and we see that, *for a normally running process,* we would observe 1 or more defects in a 1000-ft sample about 46 percent of the time. Thus, if we use a shutdown criterion of 1 or more defects, we will be shut down 46 percent of the time when *nothing* is wrong—clearly not a reasonable procedure. For 2, 3, and 4 defects, this number becomes, respectively, 13, 3, and 0.4 percent, none of which seem unreasonable. Before we impulsively decide to use the 3-defect or 4-defect criterion, we must recall that we can make *two* kinds of mistakes in making such decisions:

1. Shutting down when nothing is wrong
2. Not shutting down when something *is* wrong

Going from 1 toward 4 defects reduces the risk of making the first kind of mistake but *increases* the risk of making the second kind, so there must be a compromise or tradeoff. While the Poisson graph allows us to quantify the first type of risk rather easily, the information to quantify the second type and strike the best compromise generally requires a detailed technical and economic study of the process, which can be very time-consuming and expensive.

Further insight into such matters is obtained when we consider question 3 above, concerning the 5000-ft sample. For a 5000-ft sample the 1 defect per 1000 ft criterion translates into 5 defects per 5000 ft, and $m = 3.0$ rather than $m = 0.6$. Consulting Fig. 2.30, we now get ($m = 3.0$, $x = 4$) a shutdown probability (when nothing is wrong) of 18 percent— much better than 46 percent. For 6 defects per 5000 ft this drops to about 8 percent. Clearly, using the longer sample length improves this aspect of the situation, but again there is a tradeoff. When we use 5000-ft samples and something obvious does go wrong, we produce more scrap than when we use 1000-ft samples. At this point, the answer to question 4 seems clearer. Very short sample lengths are undesirable since the data is statistically unreliable. Very long sample lengths ensure accurate decisions but produce excessive scrap when the process does fail. Thus some intermediate length should be "best" at balancing these conflicting goals; however, finding this optimum length may be difficult. Fortunately, the system overall performance may be nearly the same for a fair *range* of sample lengths (a "broad", rather than "sharp" optimum). Thus we need not locate a theoretical optimum perfectly in many practical situations, making design of the system less complex.

2.12 COMPUTER SIMULATION (MONTE CARLO) METHODS

The facilities of any comprehensive statistical package, such as SAS, allow the practical treatment of statistical problems which have not yielded to theoretical analysis. One can, e.g., use SAS to simulate a real-world experiment involving random variables, take samples of data from this simulated experiment, and process these results to obtain useful conclusions. Such simulations are sometimes called *Monte Carlo methods.*

We will now show a simple example to illustrate the approach. Earlier we discussed several practical situations which involved a random variable defined as the *difference* of two gaussian random variables, such as stress and strength or hole size and shaft size. Theory has been successful in showing that the "difference variable" is also gaussian, with known mean and standard deviation. However, when the two variables are *not* both gaussian, theory is not available to predict the statistical behavior of the difference variable, and if we need such information, a simulation approach may give it to us.

We earlier used SAS's capability for generating samples of gaussian random variables with any desired average value and standard deviation. SAS also has a similar capability for the following distributions:

Uniform   Exponential

Poisson   Gamma

Cauchy   Triangular

Binomial

Our upcoming example will use the *uniform distribution,* and since we have not previously discussed it, Fig. 2.31 gives some basic information. As the word *uniform* implies, variables described by this distribution have a defined range $x_a \leq x \leq x_b$ in which any value has an *equal* probability of occurring. This might be a good approximation for data generated by *selecting* items from a *restricted range* of a gaussian distribution, as in Fig. 2.31*b*. Other physical situations, of course, can also lead to distributions close to the uniform.

For our example we will consider the clearance or interference between a bearing bore and its mating shaft. We assume the shaft diameter DS is gaussian with average 0.999 cm and standard deviation 0.0005 cm, while the bore diameter DH has a uniform distribution over the range 1.000 to 1.002 cm. If during assembly we randomly choose a shaft and bearing, we want this assembly to have a fit which is neither too tight nor too loose. If both variables were gaussian, we would define the clearance as CLEAR = DH − DS and use the standard gaussian tables to calculate the probability of CLEAR being, say, less than 0.001 and greater than 0.003. When DS is gaussian and DH is uniform, no such theoretical approach is available. We can, however, get useful results with a simulation (Monte Carlo) approach. We next show and explain a SAS program for doing this.

```
DATA SAS01;     Names the upcoming data set SAS01
DO I = 1 to 100;     Creates a 100-item data sample
DS = .999 + .0005*NORMAL(53615);     Creates gaussian shaft diameter
DH = 1.000 + .002*UNIFORM (53615);     Creates uniform hole diameter
CLEAR = DH − DS;     Computes clearance
```

CLEARR = ROUND (CLEAR, .000001);   Rounds CLEAR to nearest
.000001 cm

OUTPUT;   Accumulates current items into data set SAS01

END;   Terminates the DO loop

PROC SORT;⎫   Puts data set in rank order
BY CLEARR; ⎭   according to value of CLEARR

PROC PRINT;   Prints the rank-ordered data set SAS01 (see Fig. 2.32)

PROC CHART;   This, plus next line, plots a histogram of DH (see
Fig. 2.33)

VBAR DH/TYPE = PCT MIDPOINTS = .9998 1.0000 1.0002 1.0004
1.0006 1.0008 1.0010 1.0012 1.0014 1.0016 1.0018 1.0020 1.0022;

PROC CHART;   Plots histogram of CLEARR (see Fig. 2.34)

VBAR CLEARR/TYPE = PCT MIDPOINTS = 0.0 .0003 .0006 .0009
.0012 .0015 .0018 .0021, .0024, .0027 .0030 .0033 .0036 .0039 .0042;

PROC CHART;   Plots cumulative distribution of CLEARR (see Fig.
2.35).

VBAR CLEARR/TYPE = CPCT MIDPOINTS = 0.0 .0003 .0006 .0009
.0012 .0015 .0018 .0021 .0024 .0027 .0030 .0033 .0036 .0039 .0042;

PROC UNIVARIATE; ⎫   Computes and prints mean, standard devia-
                 ⎬   tion, and various other statistical parameters
VAR CLEARR;   ⎭   for variable CLEARR (see Fig. 2.36)

DATA SAS011;   Creates a new data set called SAS011

DO I = 1 TO 100;   Creates a 100-item sample

CLRNRM1 = .00203305 + .000740992*NORMAL (53715);   Creates a
gaussian variable with *same* mean and standard deviation as found
for variable CLEARR in PROC UNIVARIATE step above. This
required a *prior* run of the program, to get the numbers .00203305
and .000740992.

CLRNRM = ROUND (CLRNRM1, .000001);   Rounds CLRNRM1

OUTPUT;   Accumulates current item into data set SAS011

END;   Terminates the DO loop

PROC CHART;   Plots distribution of CLRNRM (see Fig. 2.37)

VBAR CLRNRM/TYPE = PCT MIDPOINTS = 0.0 .0003 .0006 .0009
.0012 .0015 .0018 .0021 .0024 .0027 .0030 .0033 .0036 .0039 .0042;

PROC CHART;   Plots cumulative distribution of CLRNRM (see Fig.
2.38)

VBAR CLRNRM/TYPE = CPCT MIDPOINTS = 0.0 .0003 .0006 .0009
.0012 .0015 .0018 .0021 .0024 .0027 .0030 .0033 .0036 .0039 .0042;

PROC UNIVARIATE; ⎫
                 ⎬   Computes and prints mean, standard devia-
                 ⎬   tion, and various other statistical parameters
VAR CLRNRM;   ⎭   for variable CLRNRM (see Fig. 2.39)

/*

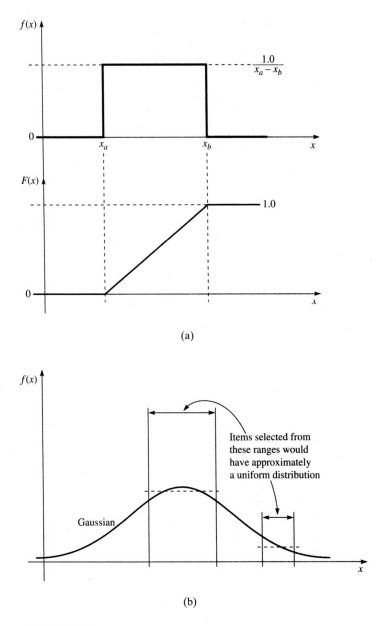

(a)

(b)

FIGURE 2.31
The uniform distribution.

OBS	I	DS	DH	CLEAR	CLEARR
1	11	0.99956	1.00002	0.00046289	0.000463
2	55	0.99952	1.00002	0.00049845	0.000498
3	94	0.99993	1.00050	0.00056986	0.000570
4	34	0.99961	1.00029	0.00067728	0.000677
5	21	0.99940	1.00013	0.00073017	0.000730
6	23	0.99934	1.00012	0.00077939	0.000779
7	76	0.99925	1.00023	0.00098160	0.000982
8	84	0.99947	1.00049	0.00101600	0.001016
9	65	0.99962	1.00067	0.00105030	0.001050
10	44	1.00006	1.00121	0.00114576	0.001146
11	80	0.99891	1.00009	0.00118250	0.001183
12	48	0.99956	1.00079	0.00122652	0.001227
13	58	0.99890	1.00013	0.00123355	0.001234
14	22	0.99957	1.00082	0.00125487	0.001255
15	72	1.00003	1.00132	0.00129264	0.001293
16	27	0.99954	1.00088	0.00134685	0.001347
17	32	0.99921	1.00058	0.00137299	0.001373
18	20	0.99968	1.00108	0.00139763	0.001398
19	67	0.99931	1.00071	0.00140222	0.001402
20	8	0.99871	1.00012	0.00140488	0.001405
21	12	0.99877	1.00019	0.00141850	0.001418
22	35	0.99914	1.00058	0.00143623	0.001436
23	53	0.99863	1.00009	0.00145839	0.001458
24	10	0.99863	1.00010	0.00147377	0.001474
25	56	0.99916	1.00064	0.00147896	0.001479
26	100	0.99984	1.00132	0.00148015	0.001480
27	70	0.99858	1.00010	0.00151457	0.001515
28	45	0.99863	1.00019	0.00155542	0.001555
29	37	0.99883	1.00042	0.00158472	0.001585
30	16	0.99895	1.00055	0.00159983	0.001600
31	13	0.99940	1.00103	0.00162415	0.001624
32	83	0.99901	1.00063	0.00162423	0.001624
33	31	0.99961	1.00126	0.00164580	0.001646
34	66	0.99859	1.00025	0.00166257	0.001663
35	26	0.99922	1.00088	0.00166433	0.001664
36	25	0.99900	1.00069	0.00168873	0.001689
37	59	0.99927	1.00099	0.00172073	0.001721
38	51	0.99856	1.00031	0.00175037	0.001750
39	2	0.99904	1.00080	0.00175663	0.001757
40	61	0.99910	1.00089	0.00178752	0.001788
41	88	0.99927	1.00109	0.00181658	0.001817
42	87	0.99902	1.00085	0.00183206	0.001832
43	64	0.99948	1.00133	0.00184988	0.001850
44	63	0.99835	1.00021	0.00185980	0.001860
45	1	0.99850	1.00039	0.00188360	0.001884
46	60	0.99856	1.00046	0.00189288	0.001893
47	99	0.99924	1.00114	0.00189879	0.001899
48	96	0.99882	1.00074	0.00191386	0.001914
49	82	0.99900	1.00092	0.00191871	0.001919
50	42	0.99911	1.00106	0.00194818	0.001948

FIGURE 2.32
Data for simulated bearing-clearance experiment.

OBS	I	DS	DH	CLEAR	CLEARR
51	14	0.99933	1.00131	0.00198238	0.001982
52	69	0.99976	1.00175	0.00198985	0.001990
53	5	0.99886	1.00087	0.00200607	0.002006
54	97	0.99935	1.00136	0.00200660	0.002007
55	41	0.99842	1.00043	0.00200967	0.002010
56	91	0.998556	1.00059	0.00203534	0.002035
57	40	0.999214	1.00127	0.00206107	0.002061
58	85	0.998219	1.00029	0.00206777	0.002068
59	62	0.998677	1.00080	0.00211871	0.002119
60	36	0.998305	1.00043	0.00212039	0.002120
61	18	0.998381	1.00051	0.00213173	0.002132
62	39	0.999315	1.00156	0.00224081	0.002241
63	78	0.999243	1.00154	0.00229446	0.002294
64	15	0.999184	1.00148	0.00230003	0.002300
65	17	0.998789	1.00111	0.00232248	0.002322
66	52	0.998419	1.00076	0.00233807	0.002338
67	92	0.998837	1.00125	0.00241387	0.002414
68	33	0.999068	1.00149	0.00242243	0.002422
69	68	0.998738	1.00117	0.00242762	0.002428
70	46	0.998894	1.00134	0.00244692	0.002447
71	77	0.998773	1.00124	0.00246550	0.002466
72	24	0.998716	1.00119	0.00247453	0.002475
73	73	0.997726	1.00020	0.00247502	0.002475
74	19	0.998460	1.00094	0.00247740	0.002477
75	4	0.998848	1.00139	0.00254283	0.002543
76	93	0.998629	1.00123	0.00260304	0.002603
77	3	0.998413	1.00105	0.00264144	0.002641
78	95	0.999127	1.00178	0.00265321	0.002653
79	74	0.998667	1.00135	0.00267864	0.002679
80	90	0.999001	1.00169	0.00269164	0.002692
81	89	0.998062	1.00083	0.00277177	0.002772
82	9	0.998933	1.00172	0.00279072	0.002791
83	86	0.998791	1.00160	0.00280630	0.002806
84	79	0.998874	1.00169	0.00281787	0.002818
85	71	0.998802	1.00165	0.00284722	0.002847
86	49	0.998437	1.00133	0.00288965	0.002890
87	54	0.998951	1.00188	0.00293088	0.002931
88	6	0.998146	1.00112	0.00297113	0.002971
89	98	0.997649	1.00063	0.00297812	0.002978
90	81	0.998852	1.00183	0.00298056	0.002981
91	28	0.998532	1.00152	0.00298828	0.002988
92	29	0.998391	1.00140	0.00300552	0.003006
93	7	0.998748	1.00182	0.00306928	0.003069
94	75	0.998230	1.00141	0.00317548	0.003175
95	43	0.998071	1.00139	0.00331366	0.003314
96	38	0.998558	1.00189	0.00333455	0.003335
97	50	0.998317	1.00174	0.00342650	0.003426
98	57	0.998421	1.00188	0.00345471	0.003455
99	47	0.998218	1.00183	0.00360909	0.003609
100	30	0.998057	1.00199	0.00393338	0.003933

FIGURE 2.32
(Continued).

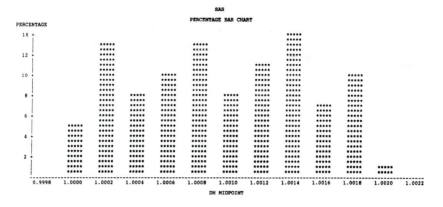

FIGURE 2.33
Histogram for sample of uniform distribution.

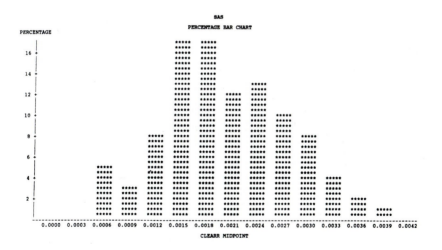

FIGURE 2.34
Distribution (histogram) of bearing clearance.

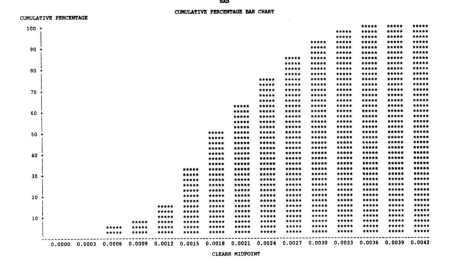

FIGURE 2.35
Cumulative distribution of bearing clearance.

Referring to the program listing (Fig. 2.40), we see that lines 8 to 10 may be thought of as simulating a physical experiment where we pick a shaft out of one bin, a bearing out of another, and assemble the two, measuring the diametral clearance of that actual pair of parts. The DO loop (lines 7 to 13) repeats this "experiment" 100 times. To summarize the clearance data in useful form, lines 14 to 16 rank-order the clearance values and print Fig. 2.32 which is the major useful result. From Fig. 2.32 we see that in a sample of 100, *no* pair of parts exhibited interference and the smallest clearance was 0.000463 cm. If the acceptable range of clearance were, say, 0.001 to 0.003, we see that 16 shaft-bearing pairs are unacceptable. Since this program is relatively inexpensive to run, it might be wise to rerun it with a sample size of, say, 500, to verify that the 100-item sample gives representative results.

While lines 1 to 16 alone would give us the practical results of main interest, we included some further SAS programming to show what can be done with software of this general class. Earlier we showed the utility of histograms for random variables. SAS will plot these as bar charts, using the PROC CHART statement, as in lines 17 to 19. PROC CHART

VARIABLE=CLEARR

MOMENTS

N	100	SUM WGTS	100
MEAN	0.00203305	SUM	0.203305
STD DEV	.000740992	VARIANCE	5.491E-07
SKEWNESS	0.156805	KURTOSIS	-0.373684
USS	.000467687	CSS	.000054358
CV	36.4473	STD MEAN	.000074099
T:MEAN=0	27.4369	PROB>|T|	0.0001
SGN RANK	2525	PROB>|S|	0.0001
NUM ¬= 0	100		

QUANTILES(DEF=4)

100% MAX	0.003933	99%	0.00392976
75% Q3	0.002588	95%	0.00333395
50% MED	0.001965	90%	0.0029873
25% Q1	0.00147925	10%	0.0011497
0% MIN	0.000463	5%	0.00073245
		1%	0.00046335
RANGE	0.00347		
Q3-Q1	0.00110875		
MODE	0.001624		

EXTREMES

LOWEST	HIGHEST
0.000463	0.003335
0.000498	0.003426
0.00057	0.003455
0.000677	0.003609
0.00073	0.003933

FIGURE 2.36
Results of univariate procedure on clearance.

allows several different types of graphs such as vertical bar graphs (VBAR), horizontal bar graphs (HBAR), and pie charts (PIE). Line 18 selects a vertical bar graph of the variable DH. After the slash (/) in this line, one needs to select the "type" of value to be plotted on the vertical axis of the bar graph. In this program we used two of the available types: PCT and CPCT. These are all that one would ever need, to graph histograms and cumulative distributions. If you want a histogram plotting percentage values on the vertical axis, use TYPE = PCT. For a cumulative plot use TYPE = CPCT. The only other input data needed is a listing of the midpoints of the bars in the graphs. You may choose these as you wish, to cover the range desired on the horizontal axis. In lines 18 and 19 we decided to use 13 equally spaced intervals. Examination of Fig. 2.33

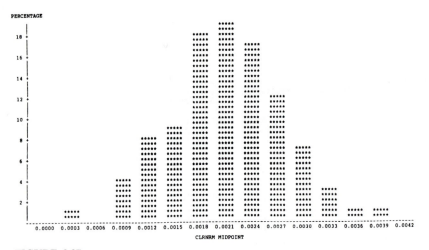

FIGURE 2.37
Histogram for a gaussian variable which has the same mean and standard deviation as the
actual clearance.

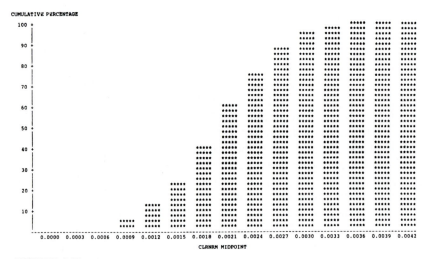

FIGURE 2.38
Cumulative distribution for data of Fig. 2.37.

VARIABLE=CLRNRM

MOMENTS

N	100	SUM WGTS	100
MEAN	0.00210973	SUM	0.210973
STD DEV	.000648424	VARIANCE	4.205E-07
SKEWNESS	0.143455	KURTOSIS	0.325368
USS	.000486721	CSS	.000041625
CV	30.7349	STD MEAN	.000064842
T:MEAN=0	32.5363	PROB>|T|	0.0001
SGN RANK	2525	PROB>|S|	0.0001
NUM ¬= 0	100		

QUANTILES(DEF=4)

100% MAX	0.004038		99%	0.00403277
75% Q3	0.00254125		95%	0.00334075
50% MED	0.002043		90%	0.002943
25% Q1	0.00169475		10%	0.0012572
0% MIN	0.000306		5%	0.00099355
			1%	0.00031213
RANGE	0.003732			
Q3-Q1	0.0008465			
MODE	0.001295			

EXTREMES

LOWEST	HIGHEST
0.000306	0.003353
0.000919	0.003364
0.000948	0.003435
0.000956	0.003515
0.000986	0.004038

FIGURE 2.39
Results of univariate procedure on data of Fig. 2.37.

shows how this PROC CHART plotting actually works. Other statistical packages provides similar capabilities.

Our reason for displaying a histogram of DH is to check whether its distribution really is uniform. As you can see, the histogram is not "level." However, nothing is wrong; this is just the usual behavior of finite samples. When we computer-generated a 100-item gaussian sample, it was not "perfect" either. (To establish confidence in random number generators, one needs to run sufficiently large samples. You might want to do this for the UNIFORM generator that is part of *your* software since the results are so easy to evaluate.) Lines 20 to 23 and 24 to 27 produce a histogram and a cumulative plot of the variable CLEARR, which is obtained as the difference between a gausssian and a uniform random variable, a combination whose theoretical distribution is *not* known, so

```
1.      // JOB ,
2.      // REGION=2048K,TIME=(0,24)
3.      /*JOBPARM LINES=2000,CARDS=0,DISKIO=1300,TAPEIO=0
4.      // EXEC SAS
5.      //SYSIN DD *
6.      DATA SAS01;
7.      DO I=1 TO 100;
8.      DS=.999+.0005*NORMAL(53615);
9.      DH=1.000+.002*UNIFORM(53615);
10.     CLEAR=DH-DS;
11.     CLEARR=ROUND(CLEAR,.000001);
12.     OUTPUT;
13.     END;
14.     PROC SORT;
15.     BY CLEARR;
16.     PROC PRINT;
17.     PROC CHART;
18.     VBAR DH / TYPE= PCT MIDPOINTS=.9998 1.0000 1.0002 1.0004
19.     1.0006 1.0008 1.0010 1.0012 1.0014 1.0016 1.0018 1.0020 1.0022;
20.     PROC CHART;
21.     VBAR CLEARR / TYPE=PCT MIDPOINTS=0.0 .0003 .0006 .0009
22.     .0012 .0015 .0018 .0021 .0024 .0027 .0030 .0033 .0036
23.     .0039 .0042;
24.     PROC CHART;
25.     VBAR CLEARR / TYPE=CPCT MIDPOINTS=0.0 .0003 .0006 .0009
26.     .0012 .0015 .0018 .0021 .0024 .0027 .0030 .0033 .0036
27.     .0039 .0042;
28.     PROC UNIVARIATE;
29.     VAR CLEARR;
30.     DATA SAS011;
31.     DO I=1 TO 100;
32.     CLRNRM1=.00203305+.000740992*NORMAL(53715);
33.     CLRNRM=ROUND(CLRNRM1,.000001);
34.     OUTPUT;
35.     END;
36.     PROC CHART;
37.     VBAR CLRNRM / TYPE=PCT MIDPOINTS=0.0 .0003 .0006 .0009
38.     .0012 .0015 .0018 .0021 .0024 .0027 .0030 .0033 .0036
39.     .0039 .0042;
40.     PROC CHART;
41.     VBAR CLRNRM / TYPE= CPCT MIDPOINTS=0.0  .0003 .0006 .0009
42.     .0012 .0015 .0018 .0021 .0024 .0027 .0030 .0033 .0036
43.     .0039 .0042;
44.     PROC UNIVARIATE;
45.     VAR CLRNRM;
46.     /*
```

FIGURE 2.40
SAS program for the bearing-clearance simulation.

we want to examine it graphically. Lines 28 and 29 employ a generally useful SAS procedure called UNIVARIATE. It accepts as input any set of observations and computes a list of useful statistical parameters, including the mean, standard deviation, and coefficient of variation. We used it here because we wanted to compare the empirical distribution just found for CLEARR with a gaussian distribution with the same numerical values of mean and standard deviation. We thus *first* ran only the first 29 lines of the program, to get the numbers (0.00203305 and 0.000740992) needed to complete lines 30 to 46. Once this has been done, we can then run the whole program as shown, since the "random" number generators repeat themselves precisely because we used fixed five-digit numeric seeds (53615 and 53715) rather than the zero (clock time) seed, which gives a *different* random sequence every time you run it.

In trying to decide whether the distribution of CLEARR is "close" to gaussian, the best comparison is probably that of the cumulative graphs, Figs. 2.35 and 2.38. Recalling that a 100-item sample of a perfect gaussian distribution is *not* perfect, we see that the two plots are really quite similar, so we may be justified in treating CLEARR as a gaussian variable.

2.13 OUTLIERS AND REJECTION OF OBSERVATIONS

Up to this point, when we have explained how to process a set of data to calculate some useful result, we have automatically and without question used *all* the observations in the data set in the calculations. In working with real-world data and practical problems, we should actually always perform some preliminary screening of our data before carrying out the desired processing. The purpose of such screening is to detect and discard any data points which we judge to be questionable, so that they do not improperly bias the calculation that we are about to do. Such decisions must be made prudently because apparently wild points (usually called *outliers*) may actually be valid data documenting unexpected but potentially important, behavior of our apparatus.

While one can build into the data processing software some rejection rules which automatically screen raw data for outliers, in most cases the judgment involves a complex consideration of many subtle factors which, in my opinion, should be reserved for careful human consideration. The simplest context in which such questions arise occurs when we have repeated a single measurement (such as the resistance of a given resistor) a number of times and find that one (or more) of the readings appears inconsistent with the rest. For a simple resistance measurement performed under laboratory conditions with a digital ohmmeter on a single resistor, the reasons for getting a wild value are limited mainly to mistakes of one kind or another:

Meter misread	Poor connection
Power-line surge	Meter malfunction
Resistor damaged in handling	Meter reading incorrectly copied

If we have read the resistance only twice, then we have no idea which reading is correct if they differ greatly, unless we have some additional knowledge about the expected value. If we take more and more readings, then it becomes easier to detect any outlier data that might occur. If we could estimate the true standard deviation of the data values; then any values beyond $\pm 3s$ from the average would reasonably be candidates for rejection. Often our sample size is so small that the estimate of s is quite uncertain. Also, in computing s, do we *include* the suspected wild values? Clearly the $\pm 3s$ criterion is a useful guideline, but we cannot avoid the responsibility of exercising our best engineering judgment after considering all the potential reasons for the questionable reading and the practical consequences of making a wrong decision. In any case, if readings *are* rejected, we should make note of this in our logbook and explain the basis of our judgment, also being careful to *retain* a record of the rejected data for future analysis, should this be necessary.

When the "repeated" readings, are *not* on a single test item but on a set of (supposedly identical) specimens, such as tensile testing of 10 steel bar samples, then the rejection of the outlier reading is less clear-cut than in the resistor example above. Now the "wild" reading of breaking stress may actually be the accurate behavior of that one specimen and should be included in calculations which will describe the behavior of the material being tested. On the other hand, the outlier could be the result of various possible mistakes, as in the resistor measurement and then should be excluded from further data processing. Again, the larger the sample size, the easier the decision. Rejection criteria more complex than the $\pm 3s$ rule given above are available,[1] but in my opinion they do not greatly improve on the quality of the decisions made.

A different context for outlier decisions is found when our experiment produces a graph of a dependent variable versus an independent variable or variables. In most such cases there are theoretical reasons (either known or unknown) for assuming that the curve or surface plotted should be "smooth." Any single point which deviates greatly from this smoothness may be suspect. In Fig. 2.41, points a and b are both such points; however, point a is more easily judged wild than point b. At the *ends* of a curve, a wild point could possibly be an accurate one which shows the beginning of a new regime of behavior rather than errors of some kind. Unless we have clear evidence of mistakes, such points should be tested by extending the range of the experiment so that these questionable points become interior rather than exterior points.

---

[1] Lipson and Sheth, *Statistical Design and Analysis of Engineering Experiments*, pp. 90–95.

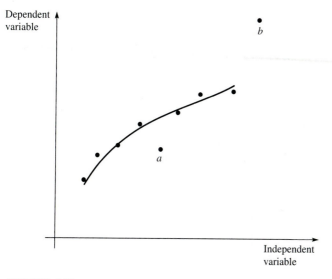

FIGURE 2.41
Outliers, interior and exterior.

PROBLEMS

2.1. A machine part has strength with mean value 16,000 psi and standard deviation 2000 psi. It is subjected to a stress with mean value 12,000 psi and standard deviation 3000 psi. Assume gaussian distributions and compute the probability of failure.

2.2. For a spring-mass system, $\omega_n = \sqrt{K_s/M}$. If

$$\bar{K}_s = 2000. \text{ N/cm} \qquad \sigma_k = 40. \text{ N/cm} \qquad \bar{M} = 0.8k_g \qquad \sigma_M = 0.02k_g$$

calculate $\bar{\omega}_n$ and σ_{ω_n}. For proper functioning, systems must have ω_n within ±10 percent of the mean value. If 10,000 systems are built, how many will fail this requirement?

2.3. A gaussian random variable has $\mu = 200$ and $\sigma = 25$. We wish to find a confidence interval for μ with a width of ±5 percent of μ. The confidence level is to be 98 percent. What is the smallest-size sample we could use to meet these requirements?

2.4. A quantity S is computed from measured values of three other quantities L, N, and M each of which has a mean value of 1.0, from $S = L^2NM^3$. We wish the standard deviation of S to be no more than 0.02 and want to find the maximum allowable standard deviations of L, N and M. Does this problem have a single answer? Why not? Make a reasonable assumption which *will* allow the problem to be solved, and then actually find σ_L, σ_N, and σ_M.

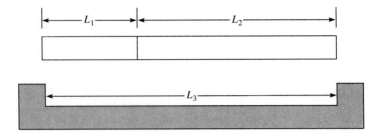

FIGURE P2.5

2.5. Parts 1 and 2 are butted together as shown in Fig. P2.5 and then pressed into part 3. For proper functioning, press fits with 0.001 to 0.003 cm of *interference* must be produced. Calculate the percentage of "good" assemblies if

$$\bar{x}_{L_1} = 1.000 \qquad \bar{x}_{L_2} = 2.002 \qquad \bar{x}_{L_3} = 3.000$$

$$\sigma_{L_1} = 0.0004 \qquad \sigma_{L_2} = 0.0006 \qquad \sigma_{L_3} = 0.0005$$

all in centimeters. If this percentage must be improved to 95 percent, to what values must the three σ's be reduced?

2.6. An unscrupulous engineer wishes to fake a set of resistor data so that it will appear to be gaussian with $\bar{R} = 1000\,\Omega$ and $\sigma_R = 100\,\Omega$, sample size = 100 resistors. A random-number-generating computer algorithm (like our SAS) is *not* available. Explain clearly how this could be done, and actually prepare the desired list of numerical values.

2.7. In a lab experiment on dc motor drives, theory gives the time constant τ as

$$\tau = \frac{(R_s + R_{sw} + R_m)J}{(R_s + R_{sw} + R_m)B + K_E K_T}$$

Suppose each of the quantities R_s, J, B, etc. exhibits uncertainty such that its mean value is 1.0 and its standard deviation is 0.10. Compute the mean value and standard deviation of τ. Then discuss why the theoretical and measured values of τ can easily be quite different.

2.8. The strength of a certain bolt is a gaussian random variable with $\bar{F} = 10,000\,\text{lb}_f$ and $\sigma_F = 500\,\text{lb}_f$.
a. Compute the probability of $F < 9000\,\text{lb}_f$.
b. Compute the probability of getting 5 bolts with $F < 9000$ in a sample of 10 bolts.

2.9. When planning an experiment whose purpose is to determine the numerical value of a mean value or standard deviation, we should decide on the size of confidence interval (the acceptable uncertainty) we want and then choose a

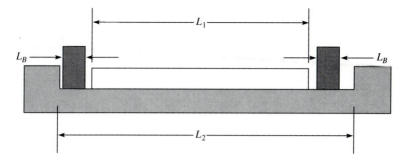

FIGURE P2.10

sample size n that will give this level of accuracy. This should be done *before* we run the experiment, but at this stage we have no numerical data on either $\bar{x}$ or s.

    a. For the case of finding a mean value, discuss the difficulty involved in choosing n. Then suggest some possible practical ways to deal with this difficulty.

    b. Repeat part *a* for the case where we are trying to estimate a standard deviation.

2.10. Part L_1 and parts L_B (brazing shims) are to be assembled into slot L_2 and then brazed. See Fig. P2.10. The parts should assemble without interference, but too large an air gap is also undesirable. Assume all dimensions are gaussian random variables with

$$\bar{L}_1 = 1.988 \qquad \bar{L}_2 = 2.000 \qquad \bar{L}_B = 0.004$$

$$\sigma_{L_1} = 0.001 \qquad \sigma_{L_2} = 0.001 \qquad \sigma_{L_B} = 0.0002$$

Compute the probability that parts won't assemble (interference) and also the probability of an air gap greater than 0.005.

2.11. We want to be 95 percent sure that the true value of a standard deviation computed from a sample of size n falls in a confidence interval of total width 30 percent around the calculated s value. What value of n should be used?

2.12. Shafts of diameter D are to be assembled into bearings of diameter d. Both D and d are gaussian random variables with $\bar{D} = 4.996$, $\sigma_D = 0.002$, $\bar{d} = 5.000$, and $\sigma_d = 0.002$, all in centimeters. Calculate the probability of a randomly chosen shaft and bearing *not* assembling, due to interference. See Fig. P2.12.

2.13. Now $\sqrt{x}$ is known to be gaussian with mean 5.0 and standard deviation 1.0. Find the probability that $x > 36$.

2.14. In the capillary tube example of Sec. 2.9, calculate the required numerical values of σ_l and σ_d to have the l and d variables each contribute an *equal*

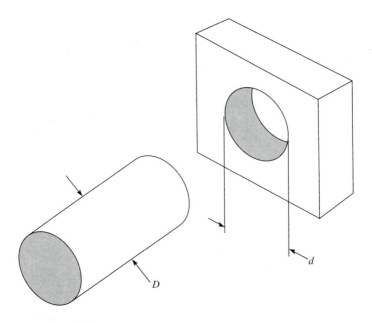

FIGURE P2.12

amount to the overall standard deviation $\sigma_{R_{fl}}$ and also meet the desired value for $\sigma_{R_{fl}}$ of 10 percent of R_{fl}.

2.15. If $y = xw^3$, $\bar{x} = 2.0$, $\sigma_x = 0.1$, and $\bar{w} = 5.0$, what is the maximum allowable σ_w if the maximum allowable σ_y is 10?

2.16. The numbers of cars per minute using a bridge is a random variable.

    a. What probability distribution function might make a good model for this situation? Why?

    b. How would you experimentally test the validity of your suggestion of part *a*? Include both data gathering and analysis.

    c. How would you deal with the obvious difference between traffic conditions at, say, noon and 2 o'clock in the morning?

2.17. We wish to estimate $\bar{x}$ with 99 percent confidence and limits $\bar{x} \pm h$. And σ is known to be 3.88. The cost of testing C increases with sample size n according to $C = 6 + 3n$ in dollars. The economic benefit B of knowing $\bar{x}$ with greater certainty (smaller h) is given by $B = 800/(h^2 + 5) + 10$ in dollars. Find the sample size n which maximizes our "profit."

2.18. Assume phone calls are processed through a central switching system where they could be monitored if desired. Considering only the time periods between 9 and 11 (mornings) and 2 and 4 (afternoons), which statistical model would you choose to represent the quantity "calls per minute"? Why? Now describe in detail the experiment you would set up to check your proposed distribution and how you would analyze the results.

2.19. We inspect a sample of 30 transistors and find 3 defective. In what range can we bracket the true percentage of defectives if we want to be (*a*) 90 percent sure, (*b*) 95 percent sure, and (*c*) 99 percent sure. Repeat these calculations for a sample of 100 which shows 10 defective.

2.20. A machine is supposed to produce 1000-Ω resistors. Its actual output is gaussian with mean 970 Ω and standard deviation 30 Ω. Customers want to buy resistors guaranteed to be 1000 $\Omega \pm 5$ percent. If we measure and sort our production to meet this goal, what percentage of "raw" production will be marketable as 1000 ± 5 percent? If we adjust the machine so that $\bar{R} = 1000 \Omega$ and σ_R is still 30 Ω, what percentage is now marketable? Some customers want 1000 $\Omega \pm 1$ percent and will buy 10,000 per month. How many resistors (raw production) must we make to meet this goal? How many ± 5 percent resistors will remain after we select the ± 1 percent?

2.21. Read Prob. 2.20. Customers who buy our "selected" (± 5 percent or ± 1 percent) resistors build them into circuits with other components and want to make statistical calculations about the circuits' behavior. They ask us whether they can treat resistors they purchase from us as gaussian. What should we tell them? Before you finalize your answer, think about and run the following "computer experiment." First calculate how large a sample it would take to get 100 resistors meeting the 1000 ± 5 percent criterion if $\bar{R} = 1000$ and $\sigma_R = 30 \Omega$. Then run a SAS (or other available software) program (see Sec. 2.6) to generate such data. Next run a SAS program to generate a 100-item sample of a gaussian model with $\bar{R} = 1000$ and $3\sigma = 50 \Omega$. Finally, use gaussian graph paper to plot and compare these two 100-item samples.

2.22. A manufacturing process produces steel shafts whose diameter approximately follows a gaussian distribution with mean value 2.350 in and standard deviation 0.004 in.

   a. What is the probability of getting a shaft smaller than 2.340 in? In a production lot of 10,000 shafts, how many smaller than 2.340 in would be expected?

   b. What is the probability of getting a shaft larger than 2.360 in?

   c. What percentage of the shafts will be outside the range of 2.335 to 2.365 in?

   d. If the interchangeability of this part requires that 95 percent of the diameters fall in the range of 2.345 to 2.355 in, an improved manufacturing process will be required. Irrespective of the mechanical details of this process, what standard deviation must it be capable of producing? (Assume its distribution will be gaussian with mean 2.350.)

   e. Assume the improved process of part *d* has been successfully implemented. One day a machine setting is inadvertently "bumped," changing the mean value to 2.353 but leaving the standard deviation at its improved value. What percentage of scrap will be manufactured?

2.23. We wish to compare two competitive manufacturing processes whose purpose is to produce a part whose nominal length is 2.5000 cm. We are interested in the following:

   a. How close to the goal of 2.5000 cm is the average length produced by each process? Is there a significant difference in the average lengths of the two processes?

b. How much *variability* does each process produce, and is there a *significant* difference in variability?

We need *reliable* numerical measures of these process characteristics to help us in choosing which process to use. Use confidence limits to study these questions, and calculate results for both the 95 and 50 percent levels of confidence.

Use SAS (or other available) programming to generate data simulating the two processes. Since we will thus *know* the correct answers, such a simulation is quite useful in developing some judgment in the use of these statistical tools. Let process 1 have mean 2.502 with standard deviation 0.003 while process 2 has 2.498 and 0.006. Generate samples with 36 items. Do the statistical tests, using only the first 6 items as your sample. Then repeat everything, using all 36 items.

2.24. Defective parts occur randomly in time in a manufacturing process, the long-term rate being 3.6 bad parts per hour. We wish to implement a shutdown sampling scheme such that there is only a 5 percent chance of shutting down when nothing is wrong. We also want the shutdown criterion to be the appearance of 2 or more defective items during the sampling time interval. What is the correct length of this sampling interval? If we use such a scheme, what is the *worst* rate of defective production that would *not* cause a shutdown? Do you think this would be acceptable?

Someone has suggested an alternative to the above approach. Why not just compute a running average, starting when the process is started and continuously updating the average as time goes by and defectives accumulate? Whenever this running average deviates significantly from 3.6, we shut down. Discuss the pros and cons of this scheme.

A final scheme uses our first method but augments it as follows. At the end of each sampling interval, where we check for 2 or more defectives within that period, we also check the number of defectives in the most recent 1-h period. If this number exceeds 6, we shut down. Discuss the advantages of this method, including any numerical calculations which might be useful.

Can you suggest any additional features, or totally different approaches, which might be useful for such practical problems?

2.25. a. For the flow resistance example of Section 2.9, employing available statistical software, use the gaussian random number generator to produce 20-item samples of diameter d and length l, using the numbers given there for the average values and standard deviations. (Should you use the *same* seed value in the random number generator for both d and l? Explain.) Using the 20 d and l values just generated, compute 20 values of R_{fl}/μ, the average value of R_{fl}/μ, and the standard deviation of R_{fl}/μ. Compare the ratio of standard deviation to average value found by simulation with the theoretical value of Equation 2.65. Discuss this comparison.

b. Repeat part a for a sample size of 100. Compare and discuss these results with those of part a.

   c Repeat parts *a* and *b* with the same average values but with standard deviations 5 times as large. Also compute the theoretical results. Why do we *expect* that the theory will now be less accurate than the simulation, for these larger standard deviations?

   d. If fluid viscosity μ were also a random variable, explain clearly how you would handle this situation, both theoretically and with simulation.

BIBLIOGRAPHY

1. Mary G. Natrella, *Experimental Statistics,* National Bureau of Standards Handbook 91, Government Printing Office, Washington, 1963.
2. C. Lipson and N. J. Sheth, *Statistical Design and Analysis of Engineering Experiments,* McGraw-Hill, New York, 1973.
3. G. E. P. Box, W. G. Hunter, and J. S. Hunter, *Statistics for Experimenters,* Wiley, New York, 1978.
4. E. L. Crow, F. A. Davis, and M. W. Maxfield, *Statistics Manual,* Dover, New York, 1960.
5. A. M. Neville and J. B. Kennedy, *Basic Statistical Methods for Engineers and Scientists,* International Textbook, Scranton, PA, 1964.

CHAPTER
3

MEASUREMENT SYSTEM DESIGN AND APPLICATION

The heart of the hardware of experimentation is the measurement system. Here the physical variables associated with the apparatus being tested are sensed and reduced to numerical values. If these numerical values are not sufficiently accurate, the experiment is meaningless. The proper design and application of the measurement system is thus vital to the success of any experimental study.

To design and apply measurement systems, we need two kinds of information. First, we must be familiar with accepted methods of specifying the accuracy of any measurement system. Second, we must be aware of the different devices available for measuring specific variables such as temperature, acceleration, pressure, and voltage, so that we can choose those most appropriate for our apparatus.

Pedagogically it would seem desirable that, in an engineering curriculum, a project lab would serve as a "capstone" course and would best be preceded by a course/lab devoted to measurement systems, using a text[1] focused on that subject. If this curricular plan were universally

---

[1] E. O. Doebelin, *Measurement Systems*: *Application and Design,* 4th ed., McGraw-Hill, New York, 1990.

used, there would be no real need for this chapter. Since curriculums are *not* uniform from department to department or school to school, it is necessary to include some material on measurement systems here. Of course, it cannot be presented in the breadth and depth of the specialized book referred to above, which has 960 pages, so a judicious choice must be made. I decided that it made the most sense to concentrate on *general principles* and not even try to address the details of specific devices for measuring specific variables. With a good grasp of the general concepts, you should be able to consult specialized texts for the details of, say, temperature measurement, if the experiment requires this. To keep the size of this book within reasonable limits, the treatment of general concepts must also be abbreviated, so you may need to consult the above reference (or others like it) again for details as the need arises in a particular experiment.

3.1 FUNCTIONAL ELEMENTS OF A MEASUREMENT SYSTEM

The operation of all measurement systems can be described by a suitable combination of only six basic functional elements. Not every system will require all six elements, and some systems may require several elements from one (or more) of the six types. That is, you will never have to "invent" a seventh basic function, but you will, for each example, have to select the proper mix of functions and interconnect them properly to represent that particular system. Figure 3.1 shows all six basic functional elements, arranged as the simplest possible system (a total of six elements). More complex examples might have, say, 10 elements, which of course implies that this particular system required some of the six basic functions to be *repeated* at different locations in the instrument.

In explaining Fig. 3.1, note first that the measurement system's purpose is to extract information from some *measured medium* about some *measured quantity* (measurand) and communicate this data to a human observer. The measurement system thus has two interfaces with the "outside world," one where it receives input from the measured medium and one where it outputs information to the observer. Between

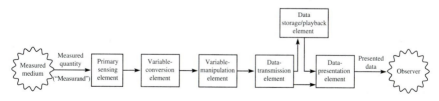

FIGURE 3.1
Functional block diagram of measurement system (or instrument).

these two interfaces the measurement system must provide the functions necessary to "bridge the gap." In general, the functions need not be in the *sequence* shown in Fig. 3.1, except that, by definition, *primary sensing* will always be the first and *data presentation* will always be the last in the left-to-right chain. The primary sensing element is that which first receives energy from the measured medium and produces an output depending in some way on the measured quantity. This energy transfer from measured medium to measurement system means that *the act of measurement always disturbs the measured quantity to some degree, making perfect measurement an unreachable goal.* In designing and applying measurement systems, this fundamental limitation must be always remembered, and measures must be taken to minimize, or compensate for, its effects on measurement accuracy.

The next of the six basic functions is *variable conversion,* shown second in Fig. 3.1, whereas in general it might occur elsewhere in the chain or not at all. Some instruments require conversion of the information signal from one physical form to another. For example, we may need to convert temperature to a related voltage if we wish to transmit temperature information over an electric cable. A *variable-manipulation* function is useful if our need is not to convert to a different physical variable but rather to perform some kind of mathematical operation on the data. A simple example is an electronic amplifier which multiplies a small voltage by a known constant (say 1000 V/V) to produce a larger and more useful voltage. For mechanical signals a simple lever or gear train could provide similar functions. For digital signals a micro-computer provides extremely versatile manipulation functions.

When the functional elements of an instrument or measurement system are actually physically separated, it becomes necessary to transmit the data from one to the other. An element performing this function is called a *data-transmission element.* It may be as simple as a shaft-and-bearing assembly or as complicated as a telemetry system for transmitting signals from satellites to ground equipment by radio. If the information about the measured quantity is to be communicated to a human being for monitoring, control, or analysis purposes, it must be put into a form recognizable by one of the human senses, sight being the most common. An element performing this "translation" function is called a *data-presentation element,* and it includes the simple *indication* of a pointer moving over a scale, the *analog recording* of a pen writing on a chart, or *digital recording* as on a computer printer. While paper records are often adequate, some applications require a *data-storage/playback* function which can retrieve the stored data upon command. The magnetic tape recorder-reproducer is the classical example here; however, many recent instruments digitize the electric signals and store them in a computerlike memory for later playback.

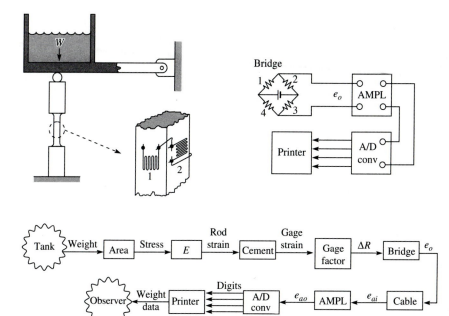

FIGURE 3.2
Tank-weighing system.

We have now listed and explained the six basic functional elements of measurement systems and instruments. When you are trying to understand the operation of an instrument new to you, or if you are designing a new instrument for a particular purpose, organizing your thinking in terms of these basic functions will be very useful. Let's go now through a specific example to make the generic descriptions given above more concrete for you. In Fig. 3.2 a strain gage force transducer senses the weight W of a liquid in a process tank, using strain gages in a Wheatstone bridge to produce full-scale output voltage e_o of 50 mV. This voltage is amplified by 200 V/V to bring the full-scale input to the analog-to-digital (A/D) converter to a standard value of 10 V. The printer responds to the A/D converter's bit signals by printing the weight in engineering units, such as newtons or pounds force.

3.2 INPUT/OUTPUT CONFIGURATION OF INSTRUMENTS AND MEASUREMENT SYSTEMS

In Fig. 3.1 we show the signal flow from the quantity we wish to measure to the data presented to the human observer. Unfortunately, all

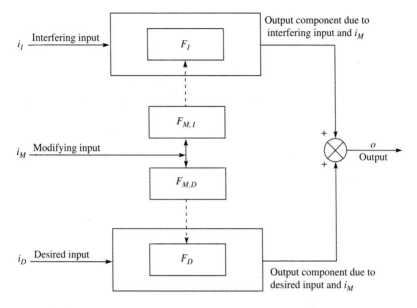

FIGURE 3.3
Input/output configuration of measurement systems.

measurement systems are sensitive not only to this desired input but also to various spurious inputs which contaminate the presented data with errors. To show this aspect of measurement more clearly, we present Fig. 3.3. Here, inputs are classified into three basic types: desired, modifying, and interfering. *Desired* inputs are those physical variables that we intend to measure. Spurious (undesired) inputs are classified into two basic subclasses, interfering and modifying. *Interfering* inputs contribute an error component to the output in a rather direct manner, similar to the desired input. In the example of Fig. 3.2, if temperature expansion causes a lengthening of the pivoted arm supporting the tank, then even though the liquid weight W has not changed, the vertical force felt by the strain-gaged column will be different, giving an error in the recorded weight. Here, temperature would be thought of as an interfering input. *Modifying* inputs have a more subtle effect, causing errors by modifying the input/output relation between the output and desired and/or interfering inputs. In our tank-weighing system, a temperature change in the strain-gaged element itself can cause a change in its elastic modulus E, giving a different strain (and thus an error) even though the weight W has not changed at all. In fact, this same temperature has several spurious effects, some interfering and others modifying. If the strain gages and the underlying metal have different thermal expansion coefficients, their differential expansion is an interfering input. The strain gages are also electric resistors whose resistance changes directly with temperature, whether they are strained or not, giving yet another interfering input. Finally, the gage factor, which relates resistance change to strain change, may be temperature-sensitive, producing a modifying input. We could list many other spurious inputs, some important, others insignificant, for this system. It should be clear that an analysis of this type is very important in studying the accuracy of *any* measurement system.

The letter symbols $(F_D,\ F_I,$ etc.) in Fig. 3.3 represent an input/output relation between the input and output of the particular block. For steady signals, F_D might, for example, be a simple constant like 1.3672 V/lb$_f$, relating a voltage output to a force input. For dynamic signals, we will shortly define the operational transfer function. We can use the letter symbols to represent this, or other useful relations, depending on the needs of the particular study. Also note that Fig. 3.3 was shown for the purpose of defining the three types of inputs and is not directly usable for any specific example. That is, if a system has one desired, three interfering, and two modifying inputs, you must draw a block diagram showing the proper types and interconnections of blocks for that particular configuration, taking Fig. 3.3 as a guideline for the basic types of signals and blocks to use.

Once we recognize the existence of spurious inputs and accept Fig. 3.3 as a reasonable way to represent their influence in measurement systems, the next thought that occurs is, How can we fight against these bad effects? Methods of correction for interfering and modifying inputs have been classified[1] into five major categories:

1. Method of inherent insensitivity

2. Method of high-gain feedback

3. Method of calculated output corrections

4. Method of signal filtering

5. Method of opposing inputs

The reference gives an extensive discussion of these techniques, including many examples. Some of these corrective methods are applicable at the design stage of the instrument or measurement system while others can be applied to existing apparatus. If a serious problem relating to spurious

---

[1] Doebelin, ibid., pp. 23–35.

inputs is encountered, you may want to consult the reference for some helpful hints.

3.3 STATIC CALIBRATION AND STATIC PERFORMANCE SPECIFICATIONS

We are now ready to start explaining methods for specifying the accuracy of measurement systems.[1] When the quantity being measured is steady or varying only slowly, accuracy is defined in terms of the *static performance characteristics*. Rapidly varying quantities require some additional specifications, the *dynamic performance characteristics*. We begin with the static characteristics.

Numerical values for all the static performance characteristics are found by a process called *static calibration*. In measurement systems it is logical to define error as the difference between the reading of the instrument and the true value of the measured quantity. We need immediately to make clear that this "true value" can never be known with perfect accuracy since calibration involves comparing the instrument's reading with that of a *calibration standard*. These standards are themselves physical devices and thus are imperfect, so they are fundamentally incapable of telling us the "true" value. The best we can do, in actual practice, is to procure a standard that is known to be sufficiently more accurate than the device to be calibrated so that the calibration will be useful. A rule of thumb often used is that the standard should, if possible, be about 10 times more accurate than the instrument being calibrated. Of course, before a new instrument is calibrated, we don't yet know its accuracy, so a 10-times-better rule cannot be precisely applied. However, we often have at least a rough idea of the ranges we are working in, so this is rarely a real problem. Also, it is clear that if we are working with instruments that are pushing the state of the art, a standard which is 10 times better may not even exist, so we must do the best we can.

What are these "standards" that we have been referring to? Many times they are simply measuring instruments whose design and use have been highly refined so that their readings are closer to the true value than any other known devices. In pressure measurement, for example, the gas-operated piston gage is a standard used to calibrate other instruments in the range of 1.4 kPa to 17 MPa. Its uncertainty is about ±57 parts per million (ppm). A pressure calibration apparatus essentially consists of a

---

[1] Ibid., chap. 3.

gas tank to which both the standard and the gage to be calibrated are connected, so that both "feel" the same pressure. The difference between the reading of the gage and that of the standard gives a correction which can be used to improve the accuracy of the gage when it is put into service after calibration. Standards are sometimes *reference materials.* The temperature standard, for example, uses the freezing point of silver (961.93°C) as a precision "fixed point" on the temperature scale. Clearly we can't use just any piece of silver for such precise work; we need a carefully prepared pure sample, a "reference material."

Standards technology exhibits a hierarchy. That is, we have primary, secondary, tertiary, etc., standards. Primary standards are the most accurate, and most industrially developed nations maintain national laboratories to provide calibration services to industry at the primary (and lower) levels of accuracy. In the United States this function is performed by the National Institute of Standards and Technology (NIST) [formerly the National Bureau of Standards (NBS)]. Large, high-technology companies will also often maintain their own calibration facilities, at an accuracy level appropriate to their needs, sometimes even exceeding (temporarily) the capabilities of the federal labs.

A comprehensive static calibration documents the response of the instrument to both its desired input and any significant spurious inputs. The main calibration, of course, is with respect to the desired input. Here, we must first identify the significant spurious inputs and, as best we can, hold them constant at known levels while we exercise the desired input over the range of values desired. We must, of course, have procured a suitable standard and arranged an apparatus such that the standard and the instrument to be calibrated experience the same value of the measured quantity. If calibration of one or more spurious inputs is judged necessary, a similar process is followed, except that the accuracy of standards used to determine the numerical values of spurious inputs often can be modest, since the effects of spurious inputs will be small in most cases.

3.3.1 Precision, Accuracy, and Bias

Having given a general overview of the calibration process, we now want to explain the details of how it is used to get numerical values for describing measurement system accuracy. Let's use a pressure gage calibration[1] as our example. Suppose the only spurious input worthy of

---

[1] Ibid., p. 54.

consideration is ambient temperature, so we hold this at $20 \pm 1°C$ while we do our pressure calibration. If our desired range of calibration is, say, 0 to 10.0 kPa, a typical procedure would be to start at 0 kPa and take readings at about 10 discrete pressure values covering the desired range. Points would be taken as we increased the pressure from 0 to 10 and also as it was decreased from 10 to 0, giving a total of about 20 points.

Figure 3.4a gives a table and graph of the pressure calibration data: in this case 22 total points were used. This data exhibits features typical of calibration data in general. When a true value is repeated (say, the up-going and down-coming points at 6 kPa), the instrument does *not* give the same reading. In fact, if we repeated this (or any other) point many times, the instrument readings would scatter according to some statistical distribution, perhaps close to gaussian. We thus recognize the presence of random error as part of the total inaccuracy. Next we note that the overall trend of the calibration curve seems to be linear (a straight-line relation between true pressure and gage reading). Most practical instruments are carefully designed to give this linearity because it maximizes convenience of use. When a calibration is completed, it is used to convert an instrument reading of an unknown pressure into our best estimate of the true value. The conventional way of doing this is to fit a curve (or straight line) to the calibration data points and use this curve (or its mathematical equation) to convert readings to true values. Note that, for this example (and in general), the line shown in Fig. 3.4a does not pass through (0, 0), (1, 1), (2, 2), etc. That is, in addition to the random error already noted, in general there will be also a systematic (reproducible) component of error which we can discover and correct by calibration.

The long-accepted method for fitting the "best" straight line to a set of scattered points is the least-squares scheme. Here we set up a standard calculus minimization problem to find the slope m and the intercept b of the line which makes the sum of the squares of all the vertical deviations between the data points and the line a minimum. The formulas for m and b are well known[1] and in fact most readers will have a hand calculator with a canned program for this computation. Let's now give a specific formula for the best-fit calibration line:

$$q_o = mq_i + b \tag{3.1}$$

where $q_o \triangleq$ output quantity (dependent variable)

$q_i \triangleq$ input quantity (independent variable)

In our present example we find that $m = 1.08$ and $b = -0.85$; thus if we

---

[1] Ibid., p. 55.

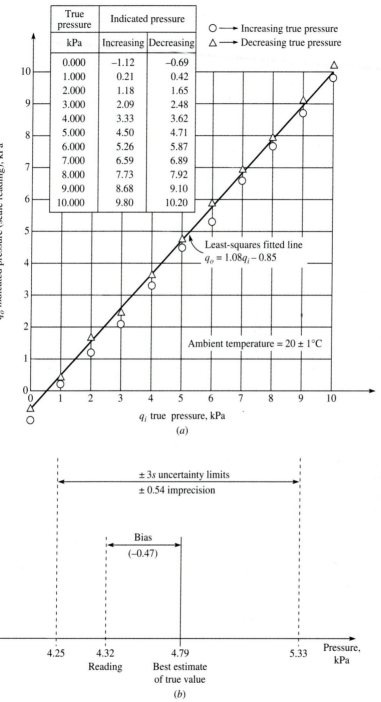

True pressure	Indicated pressure	
kPa	Increasing	Decreasing
0.000	−1.12	−0.69
1.000	0.21	0.42
2.000	1.18	1.65
3.000	2.09	2.48
4.000	3.33	3.62
5.000	4.50	4.71
6.000	5.26	5.87
7.000	6.59	6.89
8.000	7.73	7.92
9.000	8.68	9.10
10.000	9.80	10.20

O ⟶ Increasing true pressure
△ ⟶ Decreasing true pressure

Least-squares fitted line
$q_o = 1.08q_i - 0.85$

Ambient temperature = $20 \pm 1°C$

q_o indicated pressure (scale reading), kPa

q_i true pressure, kPa

(a)

± 3s uncertainty limits
± 0.54 imprecision

Bias
(−0.47)

4.25   4.32   4.79   5.33   Pressure, kPa
        Reading Best estimate
                 of true value

(b)

FIGURE 3.4
Static calibration to define bias and uncertainty.

wish to convert instrument readings (q_o) to true values (q_i), we can use the formula

$$q_i = \frac{q_o + 0.85}{1.08} \tag{3.2}$$

While Eq. (3.2) gives us the best estimate of the true value by correcting for the bias portion of the total inaccuracy, it does not consider the random error, so we must now provide for this. To get the most statistically reliable estimate of the random error in q_i, we assume that the nature and numerical value of the statistical scatter in the data are the same over the entire range of calibration. This is not exactly true (and actually quite difficult and expensive to document), but it allows us to "pool" all the data points, giving a larger sample. It can be shown that the standard deviation of the true value is given by

$$s_{q_i}^2 = \frac{1}{N} \sum \left(\frac{q_o - b}{m} - q_i \right)^2 \tag{3.3}$$

Carrying out this calculation for our present example data gives the standard deviation of q_i as 0.18 kPa. This value is used as follows in giving a complete statement of the accuracy of a given pressure reading.

Suppose that we use the gage to read an unknown pressure and the reading is 4.32 kPa. Using Eq. (3.2) to compute the best estimate of the true value and then attaching an uncertainty range of ±3 standard deviations, we finally state the pressure as 4.79 ± 0.54 kPa. *Before* this instrument was calibrated, had we taken this same reading (4.32 kPa), our information would have been incorrect and incomplete since we would have believed the pressure to be 4.32 kPa and would have had no idea that even this value had a random uncertainty. Calibration corrects for the systematic error (also called the *bias*) and puts numerical limits (±3 standard deviations) on the random error (also called *imprecision* or *uncertainty*). Figure 3.4b gives a graphical interpretation. In specifying the random component of the error, one could use measures other than 3 standard deviations, although this choice seems to be the most popular. Since this use is *not*, however, universal, you must be careful to always *state* the meaning you intend. When 3 standard deviations are used, we can say, using the above numbers to illustrate the general situation, "Our best estimate of the pressure is 4.79 kPa, and we are 99.7 percent sure (gaussian distribution is assumed) that the true value is between 4.25 and 5.33." If we had used, say, ±2 standard deviations, then we would be 95 percent sure. Some authors use the *probable error*, ±0.674 standard deviation. In this case we are 50 percent sure. (In fact, the use of ±2 standard deviations and 95 percent probability has recently been recommended as a standard practice for most applications by several standards agencies.[1,2] The main reason for preferring $\pm 2s$ over $\pm 3s$ is that the 99.7 percent probability associated with $\pm 3s$ limits implies *accurate* knowledge of the extreme tails of the probability density function. Such knowledge is rarely available in practical measurement situations. Thus, we cannot be confident that a $\pm 3s$ limit *really* guarantees 99.7 percent probability (that we will be wrong only 3 times out of a thousand). Being wrong 1 out of 20 times ($\pm 2s$, 95%) seems to be a more justifiable description of the uncertainty.)

Thus, we see that the total error of any measured value has two components: a systematic error which can be corrected by proper calibration and a random uncertainty which can be numerically specified but which *cannot* be removed. The inclusion in published experimental results of the random error (imprecision) adds greatly to the utility of such results, yet a recent study[3] showed that this was done in only about 25 percent of the papers in high-quality journals, so there is considerable room for improvement.

3.3.2 Static Sensitivity

While accuracy, as we have just defined it, is the prime consideration in describing measured data, other static characteristics are useful for specific purposes. The *static sensitivity* of an instrument tells us how much output signal the device produces for each unit of input. For our earlier pressure gage example, note that the actual physical output variable of the gage is the angular rotation of the pointer over a scale marked in units of kilopascals, so now we need to know how many angular degrees correspond to 1 kPa. If this number is, say, 5.0°/kPa, then we can use the slope of the calibration curve in Fig. 3.4 to compute the gage's static sensitivity as $(5.0)(1.08) = 5.40°/\text{kPa}$. In comparing instruments, the device with the larger sensitivity produces a larger change in its output signal for a given change in the measured quantity. Those who have little experience with measurement often confuse sensitivity and accuracy, so

---

[1] B. N. Taylor and C. E. Kuyatt, Guidelines for Evaluating and Expressing the Uncertainty of NIST Measurement Results, NIST Tech. Note 1297, NIST, Gaithersburg, MD 20899, Jan. 1993, sec. 6.5.

[2] Guide to the Expression of Uncertainty in Measurement, 1st ed., International Organization for Standardization (ISO), 1993, sec. G.1.2, American National Standards Institute, 105–111 S. State St., Hackensack, NJ 07601, phone (201) 642-4900.

[3] H. T. McClelland and M. L. Birkland, "A Survey of the Experimental Determination of Precision in Materials Research," *Factors That Affect the Precision of Mechanical Tests,* ASTM STP1025, American Society for Testing and Materials, Philadelphia, 1989, pp. 240–242.

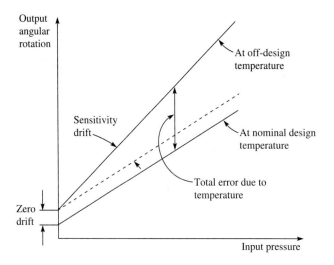

FIGURE 3.5
Zero drift and sensitivity drift.

try not to fall into this trap. The two characteristics are entirely different. A measurement system can have a very good (high) sensitivity and simultaneously have terrible accuracy.

3.3.3   Zero Drift and Sensitivity Drift

The spurious inputs described earlier can cause a calibration curve, such as Fig. 3.4*a,* to shift its position, causing errors. For our pressure gage example, temperature can cause a thermal expansion (interfering input) which would shift the entire curve in the vertical direction. Simultaneously, temperature might change the elastic modulus of the gage's spring element, giving a different sensitivity, which again results in error (see Fig. 3.5). These effects are called, respectively, *zero drift* and *sensitivity drift.* If we have calibrated these effects, we may be able to show them to be negligible in a particular application. If not, we can arrange to measure the temperature and correct the pressure data accordingly. Some instruments ("smart transducers") include built-in sensors for spurious inputs and a microcomputer to do the correction calculations on a continuous basis.

3.3.4   Linearity

A straight-line calibration curve has a number of advantages. If the instrument's output signal is being displayed versus time on a video screen or recorder chart, the immediate visual interpretation of its

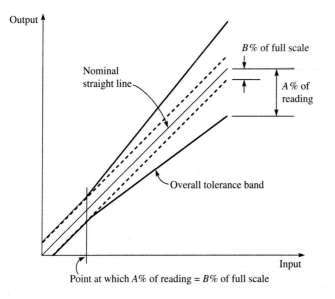

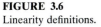

Point at which $A\%$ of reading = $B\%$ of full scale

FIGURE 3.6
Linearity definitions.

meaning is more correct; if one peak on the chart *appears* twice as high as another, the measured variable *really is* twice as large. Furthermore, any data processing to be performed on the signal is simpler when the instrument is linear rather than nonlinear. It is thus natural that instrument specifications relating to conformance to linearity would be common. While several definitions of linearity exist, the most common is the *independent linearity*. Here the reference straight line is the least-squares fitted line of Fig. 3.4, and linearity refers to the deviation of the calibration points from that line. Since bias error is easily corrected once calibration is complete, it is conventional to shift the line so that it goes through $(0, 0)$ before considering linearity (see Fig. 3.6). Nonlinearity can be expressed as a percentage of the actual reading or as a percentage of the full-scale reading. A *combination* of these two is widely used and leads to the following type of specification:

Independent nonlinearity
$$= \pm A\% \text{ of reading or } \pm B\% \text{ of full scale, whichever is greater} \quad (3.4)$$

The first part ($\pm A$ percent of reading) recognizes the desirability of a constant-percentage nonlinearity, while the second ($\pm B$ percent of full scale) admits the impossibility of testing for extremely small deviations near zero. That is, if a fixed percentage of reading is specified, the absolute deviations approach zero as the reading approaches zero, requiring absolute perfection, which is impossible in a real-world instrument. Once we have decided on the desired numerical values of A and B, we can draw the overall tolerance band of Fig. 3.6. If all the calibration points fall within this band, the linearity specification is met.

3.3.5   Resolution, Threshold, Hysteresis, and Dead Band

In the calibration curve of Fig. 3.4, the points are quite widely spread, about 10 over the entire range. To determine the resolution of a measurement system, we go to any point of interest and then make *very small* changes in the true value. In any real instrument, tiny changes away from any point result in *no* change at all in the instrument output; the instrument is unable to resolve these minute changes. We gradually increase the input until the first noticeable output change occurs. The input change that was needed to produce this first perceptible output change is called the *resolution* of the instrument in that neighborhood. The resolution can be numerically different in different regions of the calibration curve. Resolution is clearly an important characteristic since it sets a limit on the smallest changes of the measured variable that we are able to detect. When resolution is measured at the origin (true value = 0) of the calibration graph, it is called the *threshold* of the instrument. Thus the threshold is the smallest detectable value of the measured quantity while resolution is the smallest perceptible change.

While the 22 points of the calibration graph in Fig. 3.4 (and calibration graphs in general) could be taken in any sequence (perhaps in a random sequence), a monotonic increasing change followed by a monotonic decrease back to the initial point is more common. This scheme is often preferred since it exposes instrument defects caused by mechanical friction, free play in mechanisms, magnetic and electrical hysteresis effects, etc., most clearly and maximizes their contribution to the total error. That is, we usually want to be *conservative* in specifying instrument accuracy, so we use calibration methods which allow the worst case to occur. In Fig. 3.7, we illustrate these concepts, showing also the definition of hysteresis (also called *dead space* or *dead band*).

3.3.6   Scale Readability

Since the majority of instruments that have analog rather than digital output are read by a human observer who notes the position of a pointer on a calibrated scale, it is desirable for data takers to state their opinions

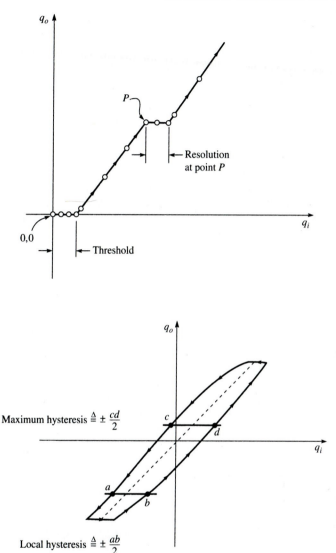

FIGURE 3.7
Threshold, resolution, and hysteresis.

as to how closely they believe they can read this scale. This characteristic, which depends on both the instrument and the observer, is called the *scale readability*. While this characteristic logically should be *implied* by the number of significant figures recorded in the data, it is probably good practice for the observer to stop and *think* about this before taking data and to then *record* the scale readability. This may also be a good place to suggest that all data, including scale readabilities, be given in decimal rather than fractional form. Since some instrument scales are marked off in quarters and halves, this requires data takers to convert to decimal form *before* recording the data. That is, the person most expert in deciding whether pressure data read from a gage as $21\frac{1}{4}$ psi should be recorded as 21.250 or 21.3 is the person actually reading the gage, not the data analyst who reads and uses the data sheet several days or weeks later. When we use a digitial data acquisition system, of course, there is no concern for scale readability; however, we still need to decide how many digits are *really* significant when our system prints out a pressure as 21.2386 psig. That is, just because we can resolve 1 part in 212,386 tells us nothing about how accurate this value is!

3.3.7 Span

The range of variable that an instrument is designed to measure is sometimes called the *span.* Thus a thermometer used for body temperature measurements might have a span of 12°F from 94 to 106. Sometimes we wish to express the span in terms of a ratio of the highest to the lowest value. Then the term *dynamic range* is often used, and it is given in decibels (dB), where dB $\triangleq 20 \log N$, with N being the ratio of high to low value. Thus a dynamic range of 60 dB means that the instrument can handle a range of 1000 to 1.

3.3.8 Generalized Input Impedance

In Fig. 3.1 we pointed out that measurements are always imperfect because the measured variable is disturbed by the act of measurement, which always extracts some energy from the measured medium. The magnitude of this type of error depends on how much energy is withdrawn; some instruments withdraw much less energy than others and thus are preferable in this respect. The static performance characteristic which addresses this question is called the *generalized input impedance.*[1] Here we give a brief treatment; the reference is available for more details.

The general concept is best introduced through a specific example,

---

[1] E. O. Doebelin, *Measurement Systems,* 3d ed., McGraw-Hill, New York, 1983, pp. 87–97.

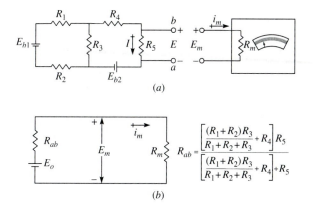

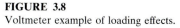

FIGURE 3.8
Voltmeter example of loading effects.

the voltmeter being the most familiar to most readers. In Fig. 3.8 we wish to measure the voltage E_{ab} with an ordinary voltmeter. The problem is that as soon as we connect the voltmeter to the terminals *ab,* the voltage *changes* because the voltmeter draws a current from the circuit, changing the current going through resistance R_5 and thus the voltage across R_5. A *perfect* voltmeter would have an infinite input resistance R_m and would draw no current at all, leaving the current I undisturbed and giving a perfect reading. A careful analysis in the reference shows that the error in a real voltmeter can be kept small by making the input resistance of the voltmeter sufficiently large *relative* to the output resistance of the circuit at the terminals being measured. That is, a *perfect* voltmeter requires an infinite input resistance, but any degree of accuracy short of perfection can be achieved in practical work by maintaining a proper ratio of meter resistance to circuit resistance.

The circuit output resistance R_{ab} is defined as that between the terminals *ab* when all voltage sources (such as the two batteries) are replaced by their internal resistance, often taken as zero. Figure 3.8 gives a formula for R_{ab} when the battery resistance is taken as zero. Analysis also shows that the relation between the undisturbed voltage E and its measured value E_m is given by

$$E_m = \frac{R_m}{R_{ab} + R_m} E \tag{3.5}$$

Note that if R_m is infinite (perfect voltmeter), the measured value is identical to the actual value. If R_m is, say, 9 times R_{ab}, then the measured value is only 90 percent of the true value, a 10 percent error. Of course, if we *know* both R_m and R_{ab}, we could calculate a suitable correction;

however, it is obviously more convenient to have R_m so large that no correction is needed.

Our voltmeter example can be generalized in two ways. To extend the concept to dynamic electrical measurements, we can replace electrical resistance with its generalization, electrical impedance. To encompass all physical variables, rather than being restricted to voltage, we must define a generalized impedance. Both these extensions have been made in the above-quoted reference. We find that whenever we "attach" an instrument to some physical system, the error caused by this connection (called a *loading effect*) can be minimized by using an instrument whose generalized input impedance is sufficiently large *relative* to the generalized output impedance of the measured system. For example, force transducers must be sufficiently stiff relative to the stiffness of the mechanical system into which they are inserted. (Actually, it turns out that there is a class of applications where the instrument should have a *low* impedance relative to that of the measured system. This is the case for electrical *ammeters,* as the reader may already know.)

3.3.9   Conclusion

While our list of static performance characteristics does not include all those that have ever been defined, we have touched on the most commonly useful ones at this point. Next we go on to discuss the dynamic characteristics.

3.4   DYNAMIC PERFORMANCE CHARACTERISTICS OF MEASUREMENT SYSTEMS

When physical variables are rapidly changing, measurement is usually more difficult than in the static case, and we need to define new ways of specifying accuracy. The mathematical model most often used here is the same one that is widely used for vibration analysis, circuit analysis, the study of automatic control systems, etc. This model is the linear ordinary differential equation with constant coefficients, or a simultaneous set of such equations. If the reader has had a course in system dynamics,[1] much of the upcoming material will be familiar since most such courses

---

[1] E. O. Doebelin, *System Dynamics,* Merrill, Columbus, OH, 1972; *System Modeling and Response,* Wiley, New York, 1980 (most recent printing of both available from Long's Bookstore, 1836 N. High St., Columbus, OH 43201.)

concentrate on systems modeled with equations of the above type. Some of the static characteristics discussed in Sec. 3.3 contribute nonlinear or random effects to the dynamic behavior. Usually these effects are small enough relative to dynamic errors to neglect, and it is conventional to do so. In any case, these effects make the equations analytically unsolvable, so we don't have much choice. [When these effects are *not* negligible, the numerical simulation methods (ACSL, SIMULINK, etc.) mentioned earlier are useful.] We will begin by defining a general model, suitable for linear instruments of arbitrary complexity, but then we will concentrate on a few special cases which include a large percentage of practical applications. In our model we will call the physical variable to be measured q_i and the response or output signal of the measurement system q_o. For most systems we can make simplifying assumptions that allow the relation between input and output to be described by an equation of the form

$$a_n \frac{d^n q_o}{dt^n} + a_{n-1} \frac{d^{n-1} q_o}{dt^{n-1}} + \cdots + a_1 \frac{dq_o}{dt} + a_0 q_o$$

$$= b_m \frac{d^m q_i}{dt^m} + b_{m-1} \frac{d^{m-1} q_i}{dt^{m-1}} + \cdots + b_1 \frac{dq_i}{dt} + b_0 q_i \quad (3.6)$$

where $q_o \triangleq$ output quantity

$q_i \triangleq$ input quantity

a's, b's $\triangleq$ combinations of system physical parameters, assumed constant

$t \triangleq$ time

To derive an equation of this type for any specific measurement system, one must apply the appropriate physical laws and make proper simplifying assumptions. Once the equation has been derived, it is used as follows. To check the dynamic accuracy in measuring any specific time-varying input quantity q_i, we must express this time variation as a mathematical function and insert it into the right-hand side of Eq. (3.6). Then we must solve the differential equation to find the instrument response q_o as a known function of time. By overlaying a graph of q_o on that of q_i we can easily see whether measurement is accurate at all times. One advantage of the mathematical model assumed above is that an analytical method for solving such equations is available no matter how complex the equation becomes (see Doebelin reference above for details). Today, however, most engineers would instead use a simulation package (such as the CSMP we used earlier) to quickly and easily get a numerical solution and the desired overlaid graphs.

While the general model of Eq. (3.6) allows treatment of measurement systems of any complexity, most practical problems can be adequately studied by using one or several of three simple special cases.

These are the zero-order, first-order, and second-order instruments, which we now discuss.

3.4.1 The Zero-Order Instrument

Not only is the zero-order instrument the simplest instrument model, but also it has absolutely perfect dynamic performance and thus serves as a standard of comparison for all other models. To collapse Eq. (3.6) into the zero-order model, we take all the a's and b's to be zero, except for a_0 and b_0. This gives us a simple *algebraic* (rather than differential) equation

$$a_0 q_o = b_0 q_i \qquad (3.7)$$

At first it appears that it requires two numbers (a_0 and b_0) to completely specify a zero-order instrument. However, we can clearly always divide through by a_0 to get

$$q_o = \frac{b_0}{a_0} q_i = K q_i \qquad (3.8)$$

$$K \triangleq \frac{b_0}{a_0} \triangleq \text{static sensitivity} \qquad (3.9)$$

We see now that it really only requires *one* number K, called the *static sensitivity*, to completely define any specific zero-order instrument. Note also that since q_o is exactly proportional to q_i at every instant of time, the instrument output signal is a perfect copy of the measured variable.

We now define two standard inputs widely used to evaluate the quality of dynamic measurement for all kinds of measurement systems. These are the *step input* and the *sinusoidal input.* For a step input, the measured variable is assumed to instantly change from one value (say zero) to another value, which we call q_{is}. For a zero-order instrument, the response to the step input is, of course, easily found, and we show it in Fig. 3.9. For a sinusoidal input we set

$$q_i = A_i \sin \omega t \qquad (3.10)$$

where $A_i \triangleq$ amplitude of sinusoidal oscillation

$\omega \triangleq$ frequency of sinusoidal oscillation

The response of any linear system [such as the general measurement system of Eq. (3.6)] to a sinusoidal input is called the system's *frequency response.* In terms of the differential equation solution, we define the sinusoidal response as *only* the "steady-state" part (particular solution) of the equation. The transient portion (complementary solution) is com-

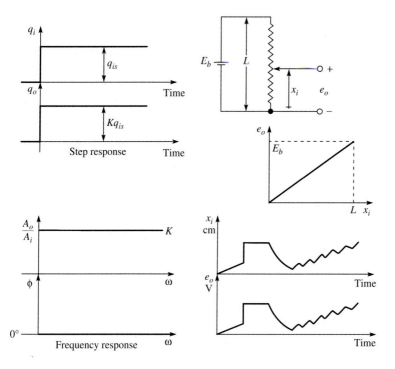

FIGURE 3.9
Example and behavior of zero-order instrument.

pletely ignored. For the general case, when we make the input signal sinusoidal and wait for transients (complementary solution) to die out, we always find that the output signal in this "sinusoidal steady state" is also a perfect sine wave of exactly the same frequency as our input sine wave. However, the *amplitude* of the output sine wave will differ from that of the input, and the output may have a *phase-angle shift* with respect to the input. These changes in amplitude and phase will be different at different input frequencies, and we want to compute and graph this variation with frequency. We will plot the ratio of the output amplitude to the input amplitude (called the *amplitude ratio* from now on) and the phase angle between output and input (defined as positive if output leads input) both against the same frequency scale. These two curves (always shown together) are called the *frequency-response curves,* and they are very useful in understanding the dynamic response of all kinds of systems, not just measurement systems.

Because its equation is algebraic rather than differential, the zero-order instrument's frequency-response graphs are very easily found. Clearly, for every frequency from zero to infinity, the amplitude ratio will be numerically equal to K, the instrument's static sensitivity (also called the *steady-state gain*), and there is no phase shift between input and output, giving zero phase angle (see Fig. 3.9). In Fig. 3.9 we use A_o as the amplitude of the output signal and ϕ as the phase angle. This frequency response (flat amplitude ratio and zero phase shift for all frequencies) is absolutely perfect, and we will use it as a standard of comparison for less perfect models. Note that, no matter how rapidly the measured variable changes (how high the frequency), the output signal faithfully reproduces the input.

While no real-world device ever perfectly conforms to *any* mathematical model, there are a few instruments which are adequately modeled as zero-order for their usual applications. Figure 3.9 shows one of these—the potentiometer displacement transducer. Here a strip of resistance material is excited with a voltage and provided with a sliding contact, which is mechanically connected to the object whose motion we wish to measure. If the resistance element is carefully manufactured to have a uniform resistance per unit length, we may write

$$e_o = \frac{x_i}{L} E_b = K x_i \tag{3.11}$$

where $K \triangleq E_b/L$, V/cm. Note that for the "arbitrary" x_i time variation shown in Fig. 3.9, the instrument's output voltage is a perfect reproduction.

We now want to introduce two concepts used in system dynamics that are also useful in studies of measurement systems: the transfer function and the block diagram. For the time being (we will shortly extend this definition), we define the *transfer function* of a linear system as the ratio of the output variable to the input variable. Thus, to compute the output variable, we simply multiply the transfer function by the input variable. For our simple example of the zero-order instrument, we see that its transfer function is just the static sensitivity K. Once we have defined the concept of transfer function, we can draw useful block diagrams by showing the input signal going into a block containing the transfer function, with the output signal leaving the block, as in Fig. 3.10.

3.4.2   The First-Order Instrument

While a few instruments are sufficiently close to perfection that the zero-order model is appropriate, most display dynamic errors which cannot be ignored. We then need to include more terms in our general

FIGURE 3.10
Block diagram of zero-order instrument.

model, the next step up in complexity being the *first-order* instrument, defined by

$$a_1 \frac{dq_o}{dt} + a_0 q_o = b_0 q_i \tag{3.12}$$

We can immediately simplify this to

$$\tau \frac{dq_o}{dt} + q_o = K q_i \tag{3.13}$$

where

$$K \triangleq \frac{b_0}{a_0} \triangleq \text{static sensitivity}$$

$$\tag{3.14}$$

$$\tau \triangleq \frac{a_1}{a_0} \triangleq \text{time constant}$$

We see that it takes two numerical values, K and τ, to completely specify a first-order instrument. The static sensitivity K always (i.e., for *any* order instrument) has the same meaning. For the steady state after a step input, the output will always be K times the input. For a first-order instrument, the time constant τ determines the speed of response; the smaller its value, the faster the response to any kind of input.

Solution and graphing of Eq. (3.13) for a step input produce

$$q_o = K q_{is}(1 - e^{-t/\tau}) \tag{3.15}$$

and Fig. 3.11. When $t = \tau$, 63.2 percent of the total response has been achieved; however, it takes an *infinite* length of time to get to the final value because of the asymptotic behavior. A useful measure of speed for any instrument is the 5 *percent settling time*, the time for the output signal to get *and stay* within 5 percent of the final value. For any first-order instrument, this is equal to 3 time constants. When we do a physical analysis of any specific first-order instrument, we obtain a formula relating τ to the physical parameters of the device. This formula allows us to design and/or use the instrument so as to minimize τ and thus obtain the fastest possible response. For a liquid-in-glass thermometer,[1] for example, we get the result

$$\tau \triangleq \frac{\rho C V_b}{U A_b} \quad \text{seconds} \tag{3.16}$$

---

[1] Doebelin, *Measurement Systems,* 4th ed., pp. 105–114.

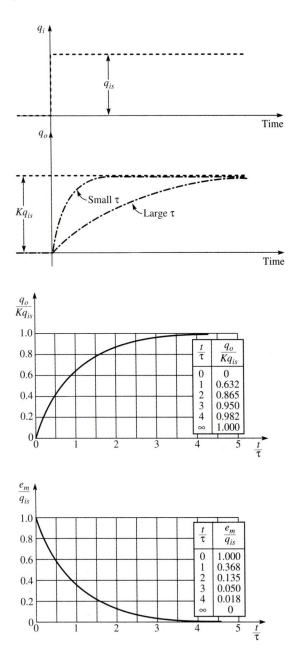

FIGURE 3.11
Step response of first-order instrument.

where $\rho \triangleq$ density of thermometer-filling fluid

$C \triangleq$ specific heat of thermometer-filling fluid

$V_b \triangleq$ volume of thermometer bulb

$U \triangleq$ overall heat transfer coefficient across bulb wall

$A_b \triangleq$ surface area of thermometer bulb

We see that in this case, to minimize the time constant and thus speed up the response, we need to minimize the product $\rho C V_b$ and maximize $U A_b$ as much as possible. For an existing thermometer, all the parameters except U are fixed; however, U varies with the kind of fluid in which the thermometer is immersed and how fast the fluid might be flowing. We see thus that we cannot assign a fixed number as the time constant in this case; its numerical value will vary depending on how the thermometer is used.

Now we need to extend the concept of transfer function beyond the simple zero-order case. To do this, first we define the differential operator D as follows:

$$D \triangleq \frac{d}{dt} \triangleq \text{time derivative} \qquad (3.17)$$

Then we can rewrite Eq. (3.6) as

$$(a_n D^n + a_{n-1} D^{n-1} + \cdots + a_1 D + a_0)q_o$$
$$= (b_m D^m + b_{m-1} D^{m-1} + \cdots + b_1 D + b_0)q_i \quad (3.18)$$

We now treat this equation *as if* it were algebraic and form the ratio of output to input as follows:

$$\frac{q_o}{q_i}(D) \triangleq \frac{b_m D^m + b_{m-1} D^{m-1} + \cdots + b_1 D + b_0}{a_n D^n + a_{n-1} D^{n-1} + \cdots + a_1 D + a_0}$$
$$\triangleq \text{operational transfer function} \qquad (3.19)$$

[If the reader is familiar with Laplace transform methods, the Laplace transfer function $(q_o/q_i)(s)$ is identical in form to Eq. (3.19), except that D is everywhere replaced by s.] Note that since the D operator does *not* take on numerical values, the transfer function is *not* the numerical ratio q_o/q_i that it was for the special case of the zero-order instrument. Rather, the transfer function is just a shorthand way of specifying the differential equation relating the two quantities, and it allows us to draw useful block diagrams. To remind ourselves of this special meaning, we *always* write transfer functions as $(q_o/q_i)(D)$, not just q_o/q_i. Applying these concepts to the first-order instrument, we get its operational transfer function as

$$\frac{q_o}{q_i}(D) = \frac{K}{\tau D + 1} \qquad (3.20)$$

and its block diagram is shown in Fig. 3.12.

FIGURE 3.12
Block diagram of first-order instrument.

We could find the frequency response of the first-order instrument by solving the differential equation (using one of several available methods) with q_i a sine wave. There is, however, a much quicker way, which we now give without any proof.[1] It uses the *sinusoidal transfer function,* which is easily obtained from the operational transfer function by substituting $i\omega$ for D wherever it appears in Eq. (3.19). For the first-order instrument we get

$$\frac{q_o}{q_i}(i\omega) \overset{\Delta}{=} \frac{K}{i\omega\tau + 1} = \frac{K}{\sqrt{(\omega\tau)^2 + 1}}\angle\tan^{-1}(-\omega\tau) \qquad (3.21)$$

where $i \overset{\Delta}{=} \sqrt{-1}$

  $\omega \overset{\Delta}{=}$ sinusoidal frequency, rad/sec

Note that, in general, sinusoidal transfer functions are complex numbers, and we always want to convert them to *polar form,* i.e., express them as a magnitude and an angle. When we do this, it turns out that the magnitude is the amplitude ratio for q_o and q_i and the angle is the phase angle between them. Thus we can compute the frequency response of any linear system very easily.

We now give without proof[2] a criterion for specifying the quality of dynamic measurement in terms of the frequency-response curves of any linear measurement system. For accurate measurement of sine waves from zero frequency (static measurement) to an upper frequency ω_{max}, we require that the amplitude ratio curve be *flat* within some stated tolerance (often ±5 percent) and that the phase-angle curve be a *straight line* through the origin $(0, 0)$, also within some stated tolerance. Comparing this with the perfect zero-order system, we see that the amplitude-ratio requirement seems reasonable but the phase-angle rule is puzzling. The perfect (zero-order) phase angle stays at zero for all frequencies, while our rule allows the phase angle to get as large as it pleases, so long as its variation with frequency is close to linear. It can be shown[3] that the

[1] Ibid., pp. 98–102.

[2] Ibid., pp. 182–187.

[3] Ibid., pp. 182–187.

only price paid for allowing this type of phase-angle variation is that the instrument output signal is *delayed in time* relative to the input signal. It is otherwise accurate; its size and shape duplicate closely those of the measured variable. In many (but not all) applications, such a time delay is of little consequence, and thus the above criteria are regularly used in practice. Two situations where we *should* take this time delay into account occur when

1. The instrument is embedded in a feedback control loop.
2. We are making *multichannel* measurements.

In a feedback control system, phase lag causes stability problems and cannot be ignored. With multichannel measurements, if the time delay for all sensor channels is not the same, there will be a time skew between channels and the recorded numerical values will *not* be truly simultaneous, causing errors in any calculations using data from several channels. If, however, we *know* the delay in each channel (Fig. 3.13 shows how to calculate it), then we can *correct* for these effects in our data processing computer programs.

Having given frequency-response accuracy criteria for the *general* case, we can now apply them to any special case, such as the first-order instrument. It should be obvious that smaller time constants will allow accurate measurement to higher frequencies; however, some numerical examples will be helpful. In Fig. 3.14 we see that $\tau = 0.001$ sec meets the 5 percent amplitude-ratio criterion to about 77. Hz while $\tau = 0.01$ is accurate only to 7.7 Hz. It is also true (though not obvious from Fig. 3.14) that the 5 percent amplitude-ratio criterion overrides a 5 percent phase-angle criterion for first-order systems. That is, if we choose the time constant to meet the amplitude requirement, the phase requirement is more than met.

We have spent considerable time studying instrument response to sine waves, and you may be wondering whether sine waves are really that important. Do we often encounter in practice the need to measure sinusoidal variables? Actually, we *don't ever* see perfect sine waves, just as we never see perfect step inputs; however, both are useful models of those signals that we *do* see. For the case of sine waves, an even stronger argument for their importance can be made. It can be shown[1] that signals of *any* waveform (periodic, transient, random, etc.) can be converted to the *frequency-domain* form where calculation of system response requires *only* a knowledge of the system's frequency-response curves. To apply

---

[1] Ibid., pp. 141–178.

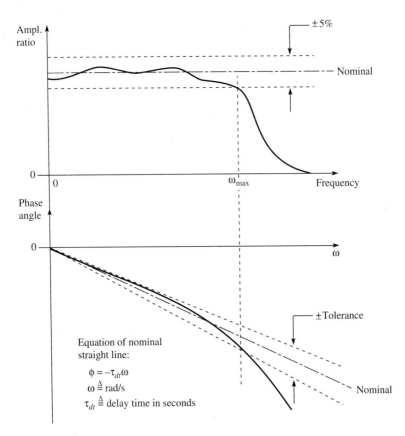

FIGURE 3.13
Dynamic accuracy of general measurement system, frequency-domain interpretation.

these methods, we need some mathematical tools beyond the scope of this text (Fourier series, Fourier transforms, power spectral density, etc.), so we merely make you aware of the existence of these methods, leaving exposition of the details to the reference.

While response to steps and sine waves is of major importance, certain other standard inputs are useful in revealing particular features of instrument behavior. Figure 3.15 displays the *ramp response* of the first-order system. Here the measured variable varies linearly with time ("ramplike"), starting at zero at time zero. The differential equation is easily solved for this case; we show only the results. Note that in comparing instrument output with the input, we choose to plot not q_o but rather q_o/K. This is useful not only here but in *every* case, since a perfect response always has $q_o/K = q_i$, allowing easy visual determination of measurement errors. Note in Fig. 3.15 that, unlike the step response,

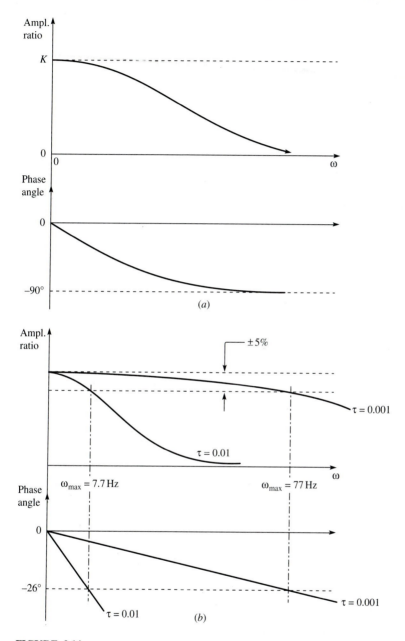

FIGURE 3.14
Frequency-response of "fast" and "slow" first-order instruments.

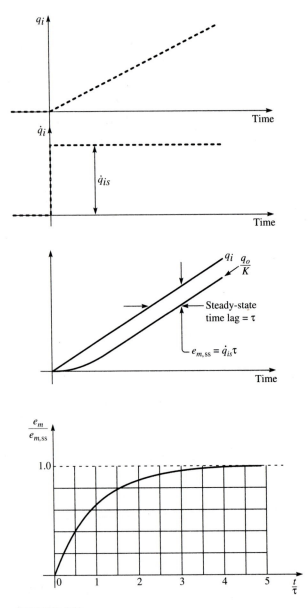

FIGURE 3.15
Ramp response of first-order instruments.

waiting a "long time" does *not* get us as close as we wish to the correct value. If we call the instantaneous measurement error e_m, then there is a *steady-state* error $e_{m,ss}$ which persists no matter how long we wait. This steady-state error is directly proportional to τ and to the input rate of change dq_i/dt. We also see another useful interpretation of the time constant as the time delay (once we have reached the ramp steady state) between the occurrence of the true value and its appearance at the instrument output.

We have now presented the main useful facts about the dynamic response of first-order instruments. The speed of response for any type of input is governed entirely by the numerical value of the time constant; the static sensitivity K has no effect on the speed at all. If we can cut the time constant in half, e.g., we can improve the speed by a factor of 2. Theoretical analysis of any instrument will show whether a first-order model might be appropriate, but experimental testing is necessary to confirm whether this model is sufficiently accurate. Such testing,[1] which we did not describe, usually takes the form of step input or sinusoidal experiments.

3.4.3 The Second-Order Instrument

When a first-order model does not predict well the observed behavior of an instrument, particularly if the instrument's step response exhibits oscillations, we may wish to try the second-order model, defined by

$$a_2 \frac{d^2q_o}{dt^2} + a_1 \frac{dq_o}{dt} + a_0 q_o = b_0 q_i \tag{3.22}$$

The standard form of this model is

$$\left(\frac{D^2}{\omega_n^2} + \frac{2\zeta D}{\omega_n} + 1 \right) q_o = K q_i \tag{3.23}$$

where

$$\omega_n \triangleq \sqrt{\frac{a_0}{a_2}} \triangleq \text{undamped natural frequency, rad/sec} \tag{3.24}$$

$$K \triangleq \frac{b_0}{a_0} \triangleq \text{static sensitivity} \tag{3.25}$$

$$\zeta \triangleq \frac{a_1}{2\sqrt{a_0 a_2}} \triangleq \text{damping ratio, dimensionless} \tag{3.26}$$

---

[1] Ibid., pp. 188–194.

We see that it takes three numerical values to specify any second-order instrument. As always, K has no effect whatever on the speed of response. For most applications, the damping ratio ζ has an optimum value of about 0.65, based on our earlier frequency-response criterion of flat amplitude ratio and straight-line phase angle. This leaves ω_n as the main parameter affecting the speed. It turns out that the speed is directly proportional to the undamped natural frequency; if we can, say, double it, we increase the speed by a factor of 2.

A simple example of a second-order instrument is the force-measuring spring scale of Fig. 3.16. The input force f_i deflects the spring (stiffness K_s N/m), giving an output displacement x_o. We assume the moving mass M to be rigid and the sliding friction between the piston and the cylinder to be viscous with coefficient B N/(m/sec). Newton's law gives

$$\sum \text{forces on mass} = (\text{mass})(\text{acceleration})$$

$$f_i - B\frac{dx_o}{dt} - K_s x_o = M\frac{d^2 x_o}{dt^2} \tag{3.27}$$

This clearly fits our definition of a second-order instrument, and we can immediately define its standard parameters:

$$K \triangleq \frac{1}{K_s} \quad \text{m/N} \tag{3.28}$$

$$\omega_n \triangleq \sqrt{\frac{K_s}{M}} \quad \text{rad/sec} \tag{3.29}$$

$$\zeta \triangleq \frac{B}{2\sqrt{K_s M}} \tag{3.30}$$

This instrument exhibits a *tradeoff* between speed of response and static sensitivity, a feature not uncommon in many instruments. By tradeoff we mean that design changes intended to improve (increase) static sensitivity result in a *decrease* in the speed of response. In this example we can increase sensitivity only by using a softer spring, but unfortunately this reduces the natural frequency and thus the speed.

Figure 3.17a shows the step response for any second-order instrument. Note that we now require a *family* of curves if we wish to display the effect of the damping ratio. This family does, however, work for any value of K and any value of ω_n since a normalization is possible by plotting $q_o/(Kq_{is})$ against $\omega_n t$. Instruments with $\zeta = 1.0$ are called *critically damped* and are the fastest possible (for a given ω_n) without any overshooting and oscillating. This is *not*, however, the usually preferred

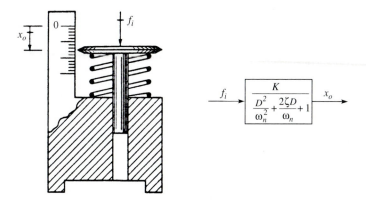

FIGURE 3.16
Spring scale example of second-order instrument.

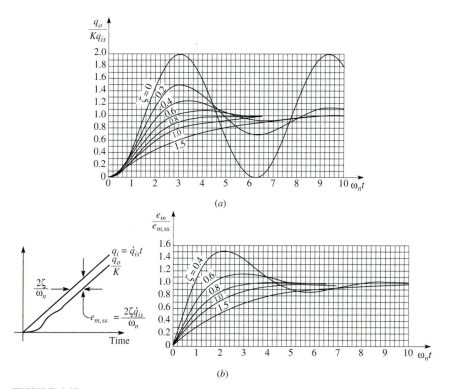

(a)

(b)

FIGURE 3.17
Step and ramp responses of second-order instruments.

situation. Rather, the optimum value of 0.65 quoted earlier is more often used. It gives the best frequency response, and while its step response has a slight overshoot, it is somewhat faster than critically damped. Instruments with $\zeta < 1.0$ are called *underdamped* while those with $\zeta > 1.0$ are called *overdamped*.

Figure 3.17b shows the ramp response, where we again note the presence of a *steady-state* error, which decreases for smaller ζ and larger ω_n. The frequency response of Fig. 3.18b confirms visually our choice of $\zeta = 0.65$ as optimum. It appears to give the flattest amplitude ratio and

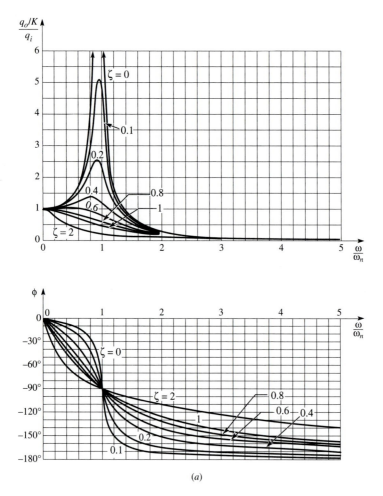

(a)

FIGURE 3.18
Frequency response of second-order instruments.

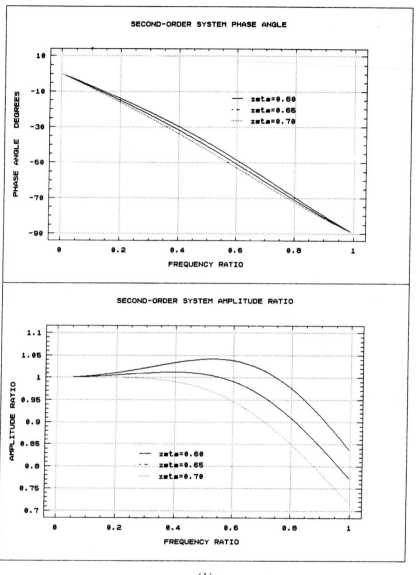

(b)

FIGURE 3.18
(Continued).

straightest phase angle over the widest frequency range. We also see that this optimum is rather broad, so we need not get *exactly* 0.65, anything close is OK. While, for the step response, $\zeta < 1.0$ is the dividing line between systems that overshoot and those that don't, a peak of amplitude ratio in the frequency response requires $\zeta < 0.707$. Also, the oscillation frequency observed in the step response of underdamped systems is given by $\omega_{nd} \triangleq \omega_n(1 - \zeta^2)^{0.5}$ while the frequency of the peak in the frequency response is $\omega_{np} \triangleq \omega_n(1 - 2\zeta^2)^{0.5}$.

Summarizing the dynamic behavior of second-order instruments, we can say that we will usually strive to make the damping ratio close to 0.65 and choose a natural frequency high enough to meet our accuracy requirements. Sometimes the choice of a suitable value for ω_n can be made by using the step, ramp, and frequency-response information given above. If the signal to be measured is not a simple waveform, we may have to run a mathematical study of its specific behavior. This is most efficiently done by using a simulation language, as we show in the next section.

3.4.4 Using Simulation to Study Dynamic Accuracy

Often when we are designing a measurement system to do an experimental study of some device or process, we have previously done some sort of theoretical modeling of the device or process. Even if this theoretical analysis is only approximate, it can be used to estimate the dynamic behavior of the signals that we wish to measure. Knowing the general nature of these signals, we can choose measuring instruments which are sufficiently fast to capture them accurately. Since these theoretical models may be rather crude in some cases, we should be conservative, if we can, and choose instruments to be somewhat faster than appears necessary from this analysis. Since the simulation languages mentioned earlier can easily model in a single run both the process under experimental test and the instruments used to make the measurements, this method of designing the measurement system is very efficient.

To illustrate the procedure, we now show a simple example. (The simulation packages actually allow study of *much* more complex situations with relatively little effort.) Our example relates to the testing of automotive air bags. The model of air-bag dynamic behavior which we will use is much simplified but should serve our educational purpose nicely. Our experiment involves activating the air bag and measuring the gas pressure inside it as the bag inflates and deflates in a fraction of a second. We need to choose a pressure transducer which will measure this pressure transient with acceptable accuracy. At time zero, activation of the air bag causes the filling gas to flow into the bag at a mass flow rate which we assume is constant until it stops at 0.025 seconds. As the pressure inside the bag rises, a vent valve is opened when the pressure first exceeds 4.0 psig, and this valve remains open thereafter, eventually deflating the bag. The vent valve passes a mass flow rate which is instantaneously proportional to the pressure difference between the inside of the bag and the surrounding atmosphere, assumed to be at 0.0 psig. We will not carry out a careful physical analysis, but rather we just assume some simple relations and numerical values, since our purpose here is not to teach you about air bags but rather to illustrate a design method for measurement systems.

With the CSMP simulation package we employed in an earlier study, a program might go as follows:

```
 1 TITLE TRANSDUCER SELECTION STUDY        Title of program
 2 METHOD RKSFX      Select a fixed-step integrator.
 3 * MODEL OF SYSTEM TO BE MEASURED      Comment
 4 * AUTOMOTIVE AIR BAG                        Comment
 5 MIDOT = 20.*(STEP(0.0) – STEP(.025))     Mass inflow rate to air bag
 6 NOSORT      Start some FORTRAN which must be kept in sequence
 7 VALVE = 0.0     Begin selection of outlet valve condition
 8 IF(P.GT.4.0) VALVE = 1.0     Continue valve open/close logic
 9 IF (P.LT.4.0.AND.PDOT.LT.0.0) VALVE = 1.0    Complete valve logic.
10 SORT     End of special FORTRAN section for valve logic
11 MODOT = 50.*(P – 0.0)*VALVE     Air-bag mass outflow rate
12 PDOT = 10.*(MIDOT – MODOT)     Rate of change of air-bag pressure
13 P = INTGRL(0.0, PDOT)     Numerical integration of PDOT to get P
14 * END OF AIRBAG MODEL
15 * MODEL OF FIRST-ORDER PRESSURE TRANSDUCER
                                         Comment
16 PM1 = REALPL(0.0, .005, P)     First-order system, TAU = .005
17 * MODEL OF SECOND-ORDER PRESSURE TRANSDUCER
                                         Comment
18 PM2A = CMPXPL(0.0, 0.0, .65, 1000., P)     Second-order system
19 PM2 = PM2A*1.E6
20 * MODEL OF THIRD-ORDER TRANSDUCER     Comment
21 P3DOT3 = (P – 7.5E – 6*P3DOT2 – .0063*P3DOT – PM3)/5.E-9
                                         Third derivative of P
22 P3DOT2 = INTGRL(0.0, P3DOT3)     Numerical integration of
                                         P3DOT3
23 P3DOT = INTGRL(0.0, P3DOT2)     Numerical integration of
                                         P3DOT2
24 PM3 = INTGRL(0.0, P3DOT)     Numerical integration of P3DOT
25 TIMER FINTIM = .15     DELT = .0001     OUTDEL = .003
26 OUTPUT P PM1 PM2 PM3     Plot four graphs.
27 PAGE GROUP     Use same scale for all four graphs.
28 END
```

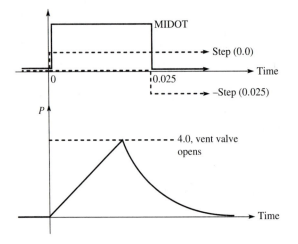

FIGURE 3.19
Simulation of air-bag pressure.

Line 5 uses CSMP's step functions (which can be delayed by any amount of time) to model the turning on and off of the air-bag inflow (see Fig. 3.19). Lines 6 through 10 are a short FORTRAN section used to model the vent valve actuation. We want the valve to be closed until the rising pressure reaches 4 psi, whereupon the valve should stay open thereafter, irrespective of what P does. This logic could be implemented in several ways. The method used here requires knowledge of the slope of the pressure-time curve. This slope is the first derivative of pressure P and occurs naturally, as PDOT, in line 12. The rest of the program is devoted to modeling various pressure sensors (transducers). Each of these will have as its input the predicted air-bag pressure P. Since first- and second-order physical systems are so common in dynamic analysis, CSMP provides the modules REALPL and CMPXPL to model them with single statements. In REALPL the first entry inside the parentheses is the initial condition, in our case zero, while the second is the time constant, which we set at 0.005 second. The last entry is the input to the first-order system, pressure P in our case. CMPXPL has a similar format except that two initial conditions (here both zero) are needed, and we must enter the damping ratio (0.65) and natural frequency (1000. rad/sec). Since CMPXPL has a steady-state gain of $1/\omega_n^2$ and we want a gain of 1.0, line 19 multiplies by 1.E6 to get the measured value PM2.

Our final sensor model is a third-order type, not previously discussed and used here to show how easily *any* order instrument can be studied with simulation. Its differential equation is taken as

$$5.E - 9\frac{d^3 q_o}{dt^3} + 7.5E - 6\frac{d^2 q_o}{dt^2} + 0.0063\frac{dq_o}{dt} + q_o = q_i \qquad (3.31)$$

The numerical values of the coefficients determine the dynamic response, but we have no easy-to-use criteria such as τ, ζ, or ω_n to help us in choosing trial values. It is, of course, clear that the response will approach the ideal zero-order as the coefficients of the derivative terms approach zero. Also, the simulation method allows us to try many combinations quickly as we search for acceptable values that also correspond to commercially available instruments. Line 21 in the program is simply a FORTRAN statement for the highest derivative, which is then successively integrated in lines 22 to 24 to finally get the instrument output PM3.

Figure 3.20 shows the results of a run made with the numerical values used in the above program. Here we have used the CSMP printer plot rather than the continuous-line electrostatic plotter used in our earlier CSMP example. While the electrostatic plotter produces smooth curves, the printer plot has the advantage of showing accurate tabular values along the edge of the graph. We can now see, in the curve labeled *P,* what the theoretically predicted pressure waveform looks like and how well (or poorly), the various sensors measure it. Our initial program provided only a *visual* display of measurement error as the graphical difference between the "true" pressure P and a sensor response. We could easily modify the program to compute the instantaneous error for any of the sensors and plot a graph and/or print a table of this error. The computed error can be calculated as a percentage of the full-scale value of P, thus we could quickly scan such a table and tell whether, say, a 5 percent error criterion had been exceeded. As mentioned earlier, CSMP is no longer actively marketed, so new users of simulation would select some currently supported package, such as ACSL or SIMULINK.

3.5 ACCURACY (UNCERTAINTY) ANALYSIS OF COMPLETE SYSTEMS

In this section we want to show how to deal with the following types of situations:

1. In analyzing and/or designing an instrument composed of several components, how does the inaccuracy of each component propagate into the overall instrument error? Which components contribute most to the total error and should thus receive the most development effort for improvements?

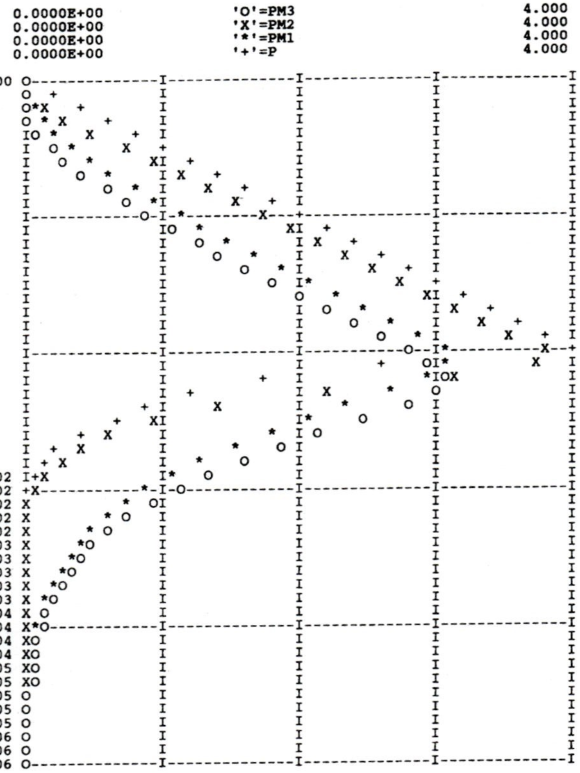

Plot header / axis labels:

```
          0.0000E+00                      4.000
          0.0000E+00                      4.000
          0.0000E+00                      4.000
          0.0000E+00
```

Legend:

```
'O'=PM3
'X'=PM2
'*'=PM1
'+'=P
```

TIME	P	PM1	PM2	PM3
0.00000E+00	0.00000E+00	0.00000E+00	0.00000E+00	0.00000E+00
1.00000E-03	0.20000	7.83799E-02	0.13250	1.22222E-03
2.00000E-03	0.40000	0.14881	0.31151	1.41490E-02
3.00000E-03	0.60000	0.24933	0.52048	5.16263E-02
4.00000E-03	0.80000	0.36788	0.73299	0.11811
5.00000E-03	1.0000	0.50119	0.94001	0.21100
6.00000E-03	1.2000	0.64659	1.1419	0.32511
7.00000E-03	1.4000	0.80189	1.3414	0.45553
8.00000E-03	1.6000	0.96529	1.5406	0.59866
9.00000E-03	1.8000	1.1353	1.7400	0.75195
1.00000E-02	2.0000	1.3108	1.9398	0.91357
1.10000E-02	2.2000	1.4907	2.1399	1.0821
1.20000E-02	2.4000	1.6742	2.3399	1.2563
1.30000E-02	2.6000	1.8608	2.5400	1.4352
1.40000E-02	2.8000	2.0497	2.7400	1.6179
1.50000E-02	2.9999	2.2407	2.9400	1.8038
1.60000E-02	3.1999	2.4333	3.1399	1.9922
1.70000E-02	3.3999	2.6273	3.3399	2.1827
1.80000E-02	3.5999	2.8223	3.5399	2.3750
1.90000E-02	3.7999	3.0182	3.7399	2.5686
2.00000E-02	3.9999	2.8876	3.7374	2.7634
2.10000E-02	2.6037	2.6258	3.1138	2.9487
2.20000E-02	2.1207	2.3370	2.2105	3.0412
2.30000E-02	1.7366	2.0554	1.4347	2.9703
2.40000E-02	1.2107	1.7816	0.93471	2.7561
2.50000E-02	0.89170	1.5186	0.65049	2.4656
2.60000E-02	0.59823	1.2796	0.44864	2.1594
2.70000E-02	0.42350	1.0697	0.29028	1.8663
2.80000E-02	0.25687	0.88917	0.17507	1.5938
2.90000E-02	0.15580	0.73610	0.10048	1.3461
3.00000E-02	9.44953E-02	0.60759	5.69005E-02	1.1263
3.10000E-02	5.73141E-02	0.50043	3.30043E-02	0.93587
3.20000E-02	3.47628E-02	0.41153	1.99797E-02	0.77408
3.30000E-02	2.10847E-02	0.33803	1.25000E-02	0.63841
3.40000E-02	1.27885E-02	0.27742	7.88105E-03	0.52555
3.50000E-02	7.75661E-03	0.22753	4.90264E-03	0.43289
3.60000E-02	4.70461E-03	0.18653	2.98476E-03	0.35126
3.70000E-02	2.85349E-03	0.15287	1.78540E-03	0.29126
3.80000E-02	1.73073E-03	0.12525	1.06541E-03	0.23888
3.90000E-02	1.04974E-03	0.10269	6.35718E-04	0.19583
4.00000E-02	6.36699E-04	8.40343E-02	3.37118E-04	0.16048
4.10000E-02	3.86177E-04	6.88213E-02	2.38339E-04	0.13148
4.20000E-02	2.34227E-04	5.61495E-02	1.43419E-04	0.10770
4.30000E-02	1.42066E-04	4.61495E-02	8.77598E-05	8.82102E-02
4.40000E-02	8.61674E-05	3.77885E-02	5.33799E-05	7.22405E-02
4.50000E-02	5.22632E-05	3.09413E-02	3.22805E-05	5.91578E-02
4.60000E-02	3.16992E-05	2.53423E-02	1.94652E-05	4.84188E-02
4.70000E-02	1.92265E-05	2.07429E-02	1.17462E-05	3.96653E-02
4.80000E-02	1.16614E-05	1.69834E-02	7.11009E-06	3.24780E-02
4.90000E-02	7.07301E-06	1.39234E-02	4.31797E-06	2.65924E-02
5.00000E-02	4.29000E-06	1.13926E-02	2.60201E-06	2.17730E-02

FIGURE 3.20

Response of three different transducers to air-bag pressure.

2. If an instrument has several spurious inputs, each of which contributes its own error, how do we combine these individual errors to assess the instrument total error? Which spurious inputs introduce the largest errors and thus require the greatest attention to control?

3. If we are going to calculate from a known formula a result based on measured values from several instruments, each of which contributes its own error, how do we combine these errors to assess the overall accuracy of the computed result? Which instruments contribute most to the total error and are thus candidates for improvement or replacement? If we *require* a certain accuracy in the computed result, how do we choose instruments to meet this goal?

All these important questions can be dealt with by using the same general approach, a method developed in Sec. 2.9. There we saw that, for small percentage changes in the independent variables, the standard deviation of the dependent variable could be closely approximated by using a *root-sum-square* (rss) method. The relative importance of the individual contributions is given by taking the partial derivative of the dependent variable with respect to the independent variable being considered. When this partial derivative is large relative to that of another independent variable, it means that the dependent variable is especially sensitive to that independent variable.

When using the method of Sec. 2.9 for measurement system error analysis, we should note that it applies to only the *random* portion of the total error. If we should wish to combine *systematic* errors of components into an overall error, then the appropriate method sums the absolute values of the component errors, *not* the square root of the sum of the squares. Given a set of numerical values, if we treat them as random rather than systematic errors, then the total error (computed by the rss scheme) is always less than that computed by the absolute-value method. This seems reasonable since the largest values of two independent random quantities are *not* likely to occur simultaneously; thus they shouldn't just be added to get the combined effect. Systematic errors, on the other hand, are considered fixed (rather than randomly varying), and thus they do add directly.

Studies of this sort are often called *uncertainty analyses* and have been treated at some length in the literature.[1,2] Figure 3.21 shows

[1] R. B. Abernethy, *Measurement Uncertainty Handbook,* Instrument Society of America, Research Triangle Park, NC, 1980.

[2] Tech. Note # 1927, NIST, Gaithersburg, MD, 1993.

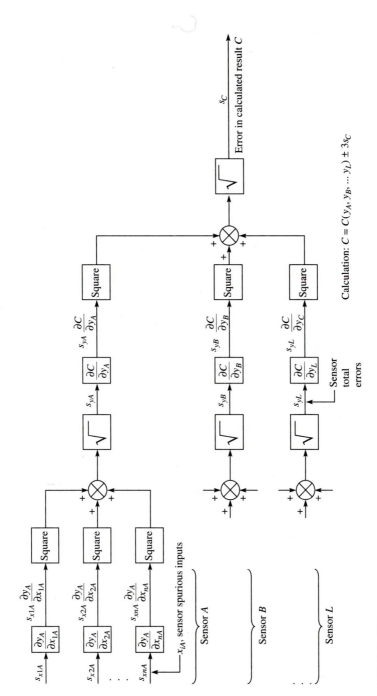

FIGURE 3.21
Combination of individual errors to get total error.

graphically how errors combine for each individual instrument in a measurement system and then how these total errors for each instrument combine to create the total error in a result calculated from the set of measurements. Let's now do an example to clarify the details. Suppose we wish to calculate from measurements the efficiency of a heat exchanger whose heat input is electric power W_i watts and whose heat output is a stream of liquid whose temperature is raised as the stream goes through the exchanger. The *thermal power W_o* watts, which is the heat-exchanger output, is the product of the liquid mass flow rate and the temperature rise across the exchanger; so we need to measure these quantities and then do the calculation (see Fig. 3.22).

Suppose that the inlet and outlet temperature sensors provide the following data, where the plus-and-minus tolerances are taken to be ± 3 standard deviations (random errors):

$$T_i = 34. \pm 1.°C \qquad T_o = 87. \pm 1.°C$$

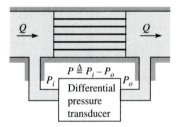

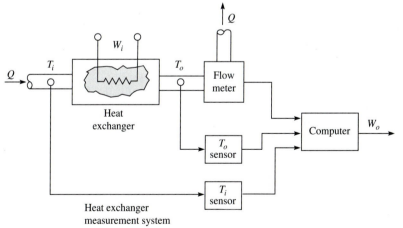

FIGURE 3.22
Example for error calculations.

The flowmeter is a laminar type[1] which uses the pressure drop p across a bundle of N capillary tubes to infer the volume flow rate Q from the theoretical relation

$$Q = \frac{\pi D^4}{128\mu L} pN \qquad (3.32)$$

where $\mu \triangleq$ fluid viscosity

$\quad D \triangleq$ tube diameter

$\quad L \triangleq$ tube length

In actual practice, the flowmeter might have received an overall calibration relating Q directly to measured p values and giving also the $\pm 3s$ uncertainty. For our example, let's first show how uncertainties in flowmeter parameters μ, L, and D would influence the uncertainty in Q. We need to first compute the "sensitivity" of Q to each of the three parameters, i.e., the partial derivatives of Q with respect to each:

$$\frac{\partial Q}{\partial D} = \frac{4\pi D^3 pN}{128\mu L} \qquad \frac{\partial Q}{\partial L} = -\frac{\pi D^4 pN}{128\mu L^2} \qquad \frac{\partial Q}{\partial \mu} = -\frac{\pi D^4 pN}{128\mu^2 L} \qquad (3.33)$$

Let's see what the partial of Q with respect to D can tell us. The tube inside diameter D could deviate from nominal size for a variety of reasons, such as expansion due to temperature changes or gradual reduction caused by deposition of solid contaminants in the flowing liquid. Note that, because D appears *cubed*, a, say, 5 percent change in D results in a 16 percent change in Q. Changes in μ and L, which appear squared in the denominator, each contribute about 10 percent to the Q error for 5 percent individual changes. If the pressure measurement p is incorrect by 5 percent, this causes a 5 percent error in Q. [Another way of looking at this is to solve Eq. (3.32) for p, rather than Q, and compute partial derivatives of p with respect to the parameters, assuming Q is fixed. This would tell us how much p would change (for a given Q) when a parameter varied. Either viewpoint gives equivalent information.] Clearly, this type of analysis is very helpful for both designers and users of instruments.

For our heat-exchanger experiment let's assume that the flowmeter had been given an overall calibration at a fluid temperature of 87, with the following results:

$$Q = 0.874p \pm 0.013 \qquad \text{cm}^3/\text{sec}$$
$$p \triangleq \text{pressure difference} \qquad \text{N/cm}^2 \qquad (3.34)$$

---

[1] Doebelin, *Measurement Systems*, 4th ed., p. 573.

where the plus-or-minus tolerance is 3 standard deviations. Suppose our calibrated differential pressure transducer gives

$$p = 3.34 \pm 0.006 \qquad N/cm^2$$

and thus

$$Q = (0.874)(3.34) \pm [(0.874)(0.006)^2 + (0.013)^2]^{0.5}$$

$$q = 2.919 \pm 0.014 \qquad cm^3/sec$$

(3.35)

The formula for calculating the output thermal power is

$$W_o = \rho Q c (T_o - T_i) \tag{3.36}$$

where $\rho \triangleq$ fluid density

$\quad c \triangleq$ fluid specific heat

To compute a value for W_o and its associated uncertainty, we need the partial derivatives.

$$\frac{\partial W_o}{\partial \rho} = Qc(T_o - T_i) \qquad \frac{\partial W_o}{\partial Q} = \rho c(T_o - T_i) \tag{3.37}$$

$$\frac{\partial W_o}{\partial c} = \rho Q(T_o - T_i) \qquad \frac{\partial W_o}{\partial T_o} = \rho Qc \qquad \frac{\partial W_o}{\partial T_i} = -\rho Qc \tag{3.38}$$

In completing our numerical example, let's assume that the fluid density and specific heat are known so accurately that we treat them as exact values. (What would you do if this were *not* true?) Using $c = 0.007143 \ J/(kg \cdot C°)$ and $\rho = 0.002543 \ kg/cm^3$, we get

$$W_o = \rho Qc(T_o - T_i) = (0.002543)(2.919)(0.007143)(87 - 34)$$

$$= 0.002811 \ W \tag{3.39}$$

for our best estimate of the output thermal power. We will now use the rss method to calculate the uncertainty associated with this value.

$$s_{W_o} = \sqrt{\left(\frac{\partial W_o}{\partial Q} s_Q\right)^2 + \left(\frac{\partial W_o}{\partial T_i} s_{T_i}\right)^2 + \left(\frac{\partial W_o}{\partial T_o} s_{T_o}\right)^2} \tag{3.40}$$

Substituting the given numerical values gives

$$W_o = 0.002811 \pm 0.000076 \ W \tag{3.41}$$

where, as usual, the uncertainty is quoted as $\pm 3s$.

When we are designing a complete measurement system, we should have in mind at the outset the level of accuracy *required* in the final calculated results to meet the goals of the study. We can then choose our instruments and methods so that this level of accuracy will actually be achieved, assuming, of course, that our needs are not beyond the state of the art. As in most other *design* (rather than analysis) situations, this type of problem has many possible solutions, not just a single unique one. That is, if we require, say, 5 percent accuracy in our final calculations, there are an *infinite number* of combinations of individual sensor errors that can meet this goal. One way available to deal with this apparent dilemma is the *method of equal effects*. Here, to get started on our design, we choose to give each sensor an *equal* contribution to the total system error. That is, using Eq. (3.40) as an example, we choose s_Q, s_{T_i}, and s_{T_o} such that the squared terms under the square root sign are all equal. In general, this approach leads to the result (using the terminology of Fig. 3.21)

$$s_{y_i} = \frac{s_C}{\sqrt{L}(\partial C/\partial y_i)} \tag{3.42}$$

As soon as we decide on the allowable overall error s_C, knowing the formula for computing C, we can compute the various partial derivatives and then get the allowable individual errors s_{y_i} from Eq. (3.42), where L is the number of sensors involved in the calculation. Once we have thus obtained tentative values for the individual sensor-error requirements, we can compare these with the capabilities of available commercial sensors. If we can meet all the requirements at acceptable cost, our initial design is feasible and we can implement it. If, on the other hand, some of the sensor requirements *cannot* be met, then we must try to "overdesign" some of the other sensors so that we can relax the requirements that were originally not feasible. This trial-and-error process is continued until all requirements are met or until we find that the originally assumed overall error is not technically or economically feasible and must itself be relaxed.

At this point we have presented the bare essentials needed to consider the design of a measurement system with respect to accuracy requirements. Recall that the choice of specific sensors for specific physical variables was *not* attempted here, but rather was left for specialist texts referenced earlier. Also, our discussion of accuracy was abbreviated, so that you may again have to consult references when specific needs arise. Within these limitations, however, those who previously had little background in measurement system design should now be ready to tackle practical problems in this area with some confidence.

PROBLEMS

3.1. Using your own experience or by reference to instrumentation texts, make and explain functional block diagrams like Fig. 3.1 for instruments used to measure:

    a. Temperature   *b.* Velocity
    c. Humidity   *d.* Pressure
    e. Acceleration   *f.* Force
    g. Flow rate   *h.* Voltage

3.2. Repeat Prob. 3.1 for the spurious-input considerations of Fig. 3.3.

3.3. In Fig. 3.2, identify each of the blocks shown with one of the standard functions.

3.4. Discuss the statement "It is impossible to determine the true length of a piece of wire."

3.5. For the situation of Fig. 3.4, discuss what kind of experiment and data analysis would be needed to determine whether the statistical scatter was numerically the same for the entire range of calibration.

3.6. Discuss the statement "An instrument can have high precision and low accuracy."

3.7. Sometimes calibrations like that of Fig. 3.4 exhibit moderate scatter around a *curve* rather than a straight line. Discuss how you might modify the methods of the text to deal with this situation.

3.8. For the situation in Fig. 3.8, what is the minimum value of R_m if the error in voltage measurement for circuits with $1000 < R_{ab} < 10,000$ is to be no more than 1 percent?

3.9. Analyze the circuit of Fig. 3.8 when the instrument is an ammeter rather than a voltmeter, showing that now the instrument should have a *low* resistance.

3.10. For the displacement-measuring potentiometer of Fig. 3.9, if the output terminals are connected to an oscilloscope, give some reasons why the response is not *perfectly* instantaneous. (Even with the oscilloscope connected, in practice we would treat the system *as if it were* instantaneous.)

3.11. Go through the steps of differential equation solution to obtain Eq. (3.15).

3.12. Analysis of the thermometer of Eq. (3.16) shows that increasing the bulb volume will increase the static sensitivity. Intuitively, why do you think this is so? If the instrument designer decides to increase sensitivity in this way, what price is paid?

3.13. Verify Eq. (3.21), using standard methods of differential equation solution.

3.14. If a first-order thermometer is to follow a temperature changing steadily at $20°$ per minute, what is the largest allowable time constant if the steady-state tracking error is to be no more than $1°$.

3.15. A quantity varying periodically with time has frequency components (Fourier-series harmonics) whose highest important frequencies are at or below 100 Hz. Choose the time constant for a first-order instrument such that the amplitude-ratio error is no worse than 5 percent for the most critical frequency.

3.16. Use the sinusoidal transfer function to find the amplitude ratio and phase shift of second-order instruments.

3.17. In the example of Fig. 3.22, if the fluid specific heat and density each have $\pm 3s$ uncertainties of 2 percent of their mean value, recompute the result of Eq. (3.41).

3.18. In Eq. (3.16), assume that each variable on the right-hand side has the value 1.00 ± 0.01, where 0.01 is $3s$. Compute the $\pm 3s$ limits for τ.

3.19. Repeat Prob. 3.18 for Eq. (3.28).

3.20. Repeat Prob. 3.18 for Eq. (3.29).

3.21. Repeat Prob. 3.18 for Eq. (3.30).

3.22. Three sensors in a multichannel data system each meet the frequency-response requirements for flat amplitude ratio and straight-line phase angle; however, the three phase-angle graphs all have different slopes. These slopes are -1.2, -2.1, and $-3.1°/Hz$. Make calculations and explain how you would correct for time skew in this situation.

BIBLIOGRAPHY

1. E. O. Doebelin, *Measurement Systems,* 4th ed., McGraw-Hill, New York, 1990.
2. F. S. Tse and I. E. Morse, *Measurement and Instrumentation in Engineering,* Marcel Dekker, New York, 1989.
3. R. B. Abernethy, *Measurement Uncertainty Handbook,* Instrument Society of America, Research Triangle Park, NC, 1980.
4. H. H. Ku (ed.), *Precision Measurement and Calibration: Statistical Concepts and Procedures,* Special Publication 300, vol. 1, National Bureau of Standards (now National Institute for Standards and Technology), Washington, 1969.
5. H. W. Coleman and W. G. Steele, Jr., *Experimentation and Uncertainty Analysis for Engineers,* Wiley, New York, 1989.
6. Tech. Note # 1927, NIST, Gaithersburg, MD, 1993.

CHAPTER
4

EXPERIMENT PLANS

While each experiment or test will naturally have its own peculiarities, we saw in Chap. 1 that we can identify certain *classes* of experiments which share important common features. In this chapter we will present overall plans for laying out experiments which fall into several common and important classes. These plans will address the general procedure to be followed but will not go into other important details of planning the entire experimental project. Filling in these further details will be left for Chap. 5.

Perhaps the simplest class of experiment which we listed in Chap. 1 involved the determination of a numerical value for a material property, dimension, or parameter, of a component or system. Many times our goal is to establish a good estimate of the *mean value* of such a quantity. Unless we have some previous experience with the situation, an experiment which obtains only a *single* observation of the quantity under study is somewhat unsatisfactory statistically; we get no inkling of the possible variability in the physical phenomenon being examined. If we require an estimate of the possible range of the numerical value, then we must make *several* observations. This situation always involves a tradeoff since more extensive experimentation costs more time and money. While the actual cost of repeating an observation can usually be estimated fairly accurately, the monetary *benefit* of knowing a parameter more precisely is often difficult to evaluate; thus the tradeoff decision is rarely clear-cut and requires use of judgment and experience in addition to statistical calculations.

A second common class of experiment focuses not on the mean value of a quantity but on its variability. For example, in many manufacturing processes we are able to rather easily adjust the process to produce the average value that we wish, but process variability often depends on certain intrinsic factors that are unknown and/or difficult to adjust. In evaluating process capabilities, we thus are interested mainly in variability. Here of course we cannot use samples of size 1; variability is revealed only with larger samples. However, we still need to decide *how large* a sample can be economically justified.

Our third class of experiment is just an extension of the first class described above and deals with situations where we wish to *compare* mean values for two alternative materials, treatments, processes, etc. For example, does heat treatment *A* or *B* give a higher yield point for a particular steel, or is there no significant difference?

Similarly, the fourth class of experiment is an extension of the second above in that it is concerned with comparisons of the *variability* of alternative materials, methods, etc.

When we are interested in finding out what are the significant factors (independent variables) that determine the final quality or performance of a product or process, we encounter the fifth class of experiment. As mentioned earlier, this class of test has recently become extremely important in the improvement of all kinds of manufacturing processes and products. Two standard statistical approaches to such problems have been known for a long time. They also had been widely applied in agricultural, social science, and other fields where models based on physical laws are difficult or impossible to apply. Until recently, however, their routine application to industrial manufacturing had been quite limited. These two methods are the *analysis of variance* (also called ANOVA) and *multiple regression*. When the independent variables are "qualitative" (not known as numerical values), the tendency has been to use analysis of variance. For example, if we are studying the effect on automobile tire wear of different brands of tires, different road surfaces, and climate, none of these factors are measurable numerical values. However, when the factors *are* measurable as numbers, then multiple regression may be the preferred approach. Again for a tire-testing example, if we are doing laboratory wear testing on a rotating drum machine and we want to study the effect on wear of different amounts of various constituents going into the rubber, these independent variables *can* be assigned numerical values. These two methods (ANOVA and multiple regression) are actually closely related, and in the interest of keeping the number of new concepts to be learned to a minimum, we will emphasize the multiple regression approach and show how it can be used for *both* of the two situations just mentioned. We will also discuss a popular Japanese version of these techniques, the Taguchi method.

The class of experiments called 5B, where all dependent and independent variables are quantitative, can also be implemented in a *theoretical* rather than an experimental manner, that is, as a computer experiment. Here, no laboratory data gathering whatever takes place, but rather we generate our raw data from a computer model of the physical system. The most common version of this approach uses widely available finite-element software as the analysis tool which generates the "experimental" data to be used as input to multiple-regression software that fits an assumed model to the data.[1] Recall that finite-element methods produce only specific *numerical* answers, *not* general letter formulas into which we can insert any parameter values we please, and in this sense are analogous to laboratory experiments, which also produce no general-formula results. The finite-element software could, of course, be used to explore (with many individual runs) the effect of system parameters on the dependent variable of interest, say, the stress at a critical point in a machine part. This approach, however, becomes too expensive in many cases. We instead make only *a few* finite-element runs, at parameter combinations judiciously chosen according to the guidelines of statistical experiment design and then fit, using multiple regression, the simplest acceptable model to this data. If a reasonably accurate model can be found, it can then be efficiently used in many ways to understand and improve our design. Since finite-element analyses are never perfect and *can* have large errors, at least a few *actual* (not computer) experiments should at some point be run to validate the model being used.

The sixth and last class of experiment, *accelerated testing,* can actually be applied to many experimental situations and may be thought of as a *version* of other classes, such as the five we just described above. Because of its practical importance, however, we discuss it as a separate class. As the name implies, the goal here is to quickly obtain information on product quality, life, etc., by using a test environment which speeds up the failure process relative to the normal service environment. This can both save money directly and shorten product development time, which is vital in today's competitive markets. The design of accelerated tests can sometimes be simple and obvious. If a refrigerator door latch should last 10 years at 20 operations per day in normal use, we can simply run the test rig at 10 operations per minute to compress 20 years of real time into

---

[1] Anthony Rizzo, Quality Engineering with FEA and DOE, *Mechanical Engineering,* May 1994, pp. 76–78.

about 5 days of test time. Other applications of accelerated testing can be extremely subtle and complex. The ASTM moist sulfur dioxide test (ASTM G87) simulates the corrosive effects of heavy-industry environments on painted structures and equipment, using cycles of heat and cold, condensation and drying, and injection of SO_2 gas into a test chamber. The U.S. Navy is also an extensive user of such tests since ship funnel exhaust gases mingle with moist and salty sea air to produce intense corrosive effects. The chemical and physical effects which determine the nature and speed of corrosion in either the test chamber or the service environment are extremely complicated, making correlation of test results with service failure prediction very difficult.

Having given a brief overview of some major classes of experiment, we next pursue each of them in more detail in the rest of this chapter. Figure 4.1 summarizes the classes to be discussed.

4.1 ESTIMATION OF PARAMETER MEAN VALUES

While it would be theoretically desirable to always estimate numerical values from samples larger than one item, we need to admit at the outset that, in practice, this is not always done, because of time and/or cost penalties. Also, past experience with a similar situation may tell us that the variability is not large enough to worry about. In some cases, only one test item may *exist,* making multiple measurements impossible.

Some important classes of experiments	
1	Estimation of parameter mean value
2	Estimation of parameter variability
3	Comparison of mean values
4	Comparison of variabilities
5A	Modeling the dependence of a dependent variable on several qualitative independent variables
5B	Modeling the dependence of a dependent variable on several quantitative independent variables
6	Accelerated testing

FIGURE 4.1
Types of experiments.

When the above-listed considerations do *not* apply, we should design our experiment (choose the sample size) based on statistical guidelines so as to achieve the level of precision felt appropriate for the particular application. This can be done by using confidence intervals as discussed in Sec. 2.10. There we saw that the confidence interval shrank as the sample size n increased, according to $1/n^{0.5}$; however, the confidence interval also depends on the standard deviation s, which would generally be unknown before we had run any tests. Fortunately, we often have some past experience to help us in estimating a tentative value for s. For example, most metals exhibit a coefficient of variation (standard deviation divided by mean value) of about 2 to 7 percent for their ultimate strength and yield point.[1] If we are running tests on a new steel, the first specimen tested will allow us to estimate the nominal-strength value. We can then use the 2 to 7 percent values to estimate a standard deviation, allowing then some predictions about confidence limits as a function of sample size.

To get a little practice with problems of this sort, let's now run a "computer experiment" (simulation) to generate some numerical values. We have already used this concept in Chap. 2, and we will make several more applications in this chapter. The great advantage of such simulations in learning how to use various statistical tools and interpret their results is that we always know in advance the "correct" answer and thus can easily judge the quality of computed predictions. In Chap. 2 we used the statistical package called SAS. I will now generate the needed data with another widely used item of software called STATGRAPHICS.[2] There are 5 or 10 statistical packages of a comprehensive nature on the software market. They all have a similar *scope,* but each has certain features which might make it preferable to a certain analyst or for a particular problem type. It is often quite difficult to choose between them, unless one has actually used them all, not just superficially but in depth. This involves considerable time and expense which an engineer who uses statistics occasionally can hardly justify. My earlier use of SAS in this book was on an IBM mainframe using batch processing. While writing the book, I purchased STATGRAPHICS for use with my personal computer at home, and thus I found it convenient to use it from this point on in the book.

I first used STATGRAPHICS to generate a gaussian random sample of 100 yield stresses with a mean value of 87,400 psi and a

[1] E. B. Haugen, *Probabilistic Mechanical Design,* Wiley, New York, 1980, pp. 596–604.

[2] Manugistics, Inc., 2115 E. Jefferson St., Rockville, MD 20852, phone 1-800-592-0050.

standard deviation of 1500 psi (see Fig. 4.2*a*). We will treat these data as if we were actually running a tensile test machine and testing one specimen after another. Note that in a *real* test the random variation in these numbers would be partly due to the scatter in the actual strength of the material and partly due to stress and strain measurement errors. If proper data on the measuring devices and techniques were available, we might be able to estimate the contribution of measurement to the total scatter and thus isolate the effects of material variability alone. To avoid these complicating details, let's just assume that measurement errors are negligible, so that our data reflect the actual material behavior.

We see from Fig. 4.2*a* that the first specimen had a yield point stress of 87,300 psi (the random number generator actually gives six digits, 87,283.3, but rounding is appropriate in the light of measurement accuracy rarely approaching 1 part in 873). Once we have this one number in hand, we can use Eq. (2.67) to get some idea of the sample size needed to achieve various levels of accuracy in the estimate of the true mean value of the yield strength. We also need to estimate the standard deviation from the "historical" data on the coefficient of variation quoted above. Suppose we do some calculations for assumed standard deviations of 1000 and 2000 psi and a 95 percent confidence interval on the mean value, leading to the table of Fig. 4.3. The smallest sample that allows such a calculation is a sample of 2, which we see results in confidence intervals of about 10 or 20 percent, respectively, for assumed σ values of 1000 and 2000. We generally need to know yield strengths more accurately than this, so we need a larger sample. If σ actually were 1000, we see that a sample of 10 would give a confidence interval a little better than ±1 percent, which might often be adequate. These results can be generalized by rewriting Eq. (2.67) as

$$\text{Confidence ratio} \triangleq \frac{\text{confidence interval}}{\text{mean value}} = \pm \frac{t_{\alpha/2,\,n-1}}{\sqrt{n}} \frac{s}{\bar{x}}$$

$$\tag{4.1}$$

$$\frac{\text{Confidence ratio}}{\text{Coefficient of variation}} = \frac{t_{\alpha/2,\,n-1}}{\sqrt{n}}$$

As soon as we can estimate the mean value and standard deviation, we can use Eq. (4.1) and the second column of Fig. 4.3*a* to choose tentative sample sizes. For example, if $s/\bar{x}$ is 0.02 and we want a confidence interval which is 5 percent of the mean value, then according to Eq. (4.1) we need in column 2 the number 2.5, which we see corresponds roughly to $n = 3$. We can, of course, use the comprehensive t table of Fig. 2.23 to augment the sparse values of Fig. 4.1 when needed for such calculations. Figure 4.3*b*, based on Eq. (4.1) and the t table, allows quick estimates.

All the above calculations were made using only data from the first

(1) 873	(18) 878	(35) 877	(52) 865	(69) 860	(86) 873
(2) 898	(19) 842	(36) 889	(53) 887	(70) 863	(87) 866
(3) 897	(20) 854	(37) 901	(54) 857	(71) 858	(88) 880
(4) 858	(21) 891	(38) 878	(55) 872	(72) 858	(89) 853
(5) 890	(22) 870	(39) 899	(56) 863	(73) 845	(90) 895
(6) 884	(23) 883	(40) 897	(57) 882	(74) 888	(91) 878
(7) 883	(24) 876	(41) 863	(58) 860	(75) 876	(92) 879
(8) 881	(25) 893	(42) 876	(59) 861	(76) 867	(93) 864
(9) 889	(26) 843	(43) 891	(60) 873	(77) 910	(94) 891
(10) 878	(27) 876	(44) 878	(61) 861	(78) 870	(95) 846
(11) 905	(28) 872	(45) 860	(62) 881	(79) 880	(96) 871
(12) 852	(29) 868	(46) 858	(63) 905	(80) 889	(97) 886
(13) 868	(30) 867	(47) 900	(64) 880	(81) 854	(98) 865
(14) 859	(31) 876	(48) 874	(65) 849	(82) 862	(99) 879
(15) 898	(32) 868	(49) 886	(66) 844	(83) 871	(100) 898
(16) 874	(33) 875	(50) 873	(67) 883	(84) 865	
(17) 887	(34) 873	(51) 912	(68) 874	(85) 888	

(a)

n	Mean strs. 100 psi	95% conf. int. on mean stress	Std. deviation 100 psi	95% conf. int. on std. dev.
1	873			
2	885	728–1045	17.8	7.9–567
3	889	854–924	14.2	7.4–89.2
4	881	850–912	19.5	11.1–72.9
5	883	862–905	17.4	10.4–50.0
6	883	867–900	15.6	9.7–38.2
7	883	870–896	14.2	9.1–31.3
8	883	872–894	13.2	8.7–26.9
9	884	874–893	12.5	8.5–24.0
10	883	875–892	11.9	8.2–21.8
20	877	869–885	17.4	13.3–25.5
40	878	873–883	15.5	12.7–19.9
100	875	872–878	15.7	13.8–18.3

(b)

FIGURE 4.2
Yield stress data set, gaussian distribution.

n	$\dfrac{t_{.025,\,n-1}}{\sqrt{n}}$	95% conf. int. std. dev. = 1000	95% conf. int. std. dev. = 2000	Conf. intervals % of 87,300 psi
2	8.99	± 8990 psi	± 17980 psi	± 10, ± 20
3	2.48	± 2480 psi	± 4960 psi	± 2.8, ±5.6
4	1.59	± 1590 psi	± 3180 psi	± 1.8, ± 3.6
5	1.24	± 1240 psi	± 2480 psi	± 1.4, ± 2.8
10	0.715	± 715 psi	± 1430 psi	± 0.8, ± 1.6
100	0.199	± 199 psi	± 398 psi	± 0.2, ± 0.4

(a)

FIGURE 4.3
Calculations to estimate necessary sample size.

specimen tested. Let's now see what develops as we test more specimens, as recorded in Fig. 4.2b. For a sample of 2, the confidence interval is very wide, but it does include the "true value" of 87,400. For 10 specimens we see that the confidence interval includes the true value, but the true value is located very near the lower limit of 87,500. However, the size of the confidence interval as calculated from the actual data *is* about ±1 percent, as predicted in Fig. 4.3, so our method of estimating sample sizes seems to work quite well.

Since many statistical calculations assume a gaussian distribution and we know that real data are *never* perfectly gaussian and sometimes are quite far from gaussian, our next simulation uses a set of yield strength values generated by a random number generator in STAT-GRAPHICS that produces a *triangular* distribution. We have not yet discussed the triangular distribution; its probability density function has a triangular shape, and it requires specification of three numerical values for its definition. These are the lower limit, the midpoint, and the upper limit. To get a triangular distribution that is comparable to our earlier gaussian data, I decided to make the lower limit 3 standard deviations below 87,400, the midpoint equal to 87,400, and the upper limit 3 standard deviations above 87,400. This is not the only possible interpretation, but it seems to be somewhat logical in that the gaussian curve has 99.7 percent of its area within ±3s while the triangular curve has 100 percent between its lower and upper limits. Also, the two average values are exactly the same.

Figure 4.4 shows the data sample generated and some calculations based on this strength data. Again, we are pretending that we are running a tensile strength testing machine and producing a new reading of yield strength as we test each new specimen. The results shown for confidence limits on the mean value and standard deviation were computed by using the same standard formulas as used in Fig. 4.2. That is, in the real-world

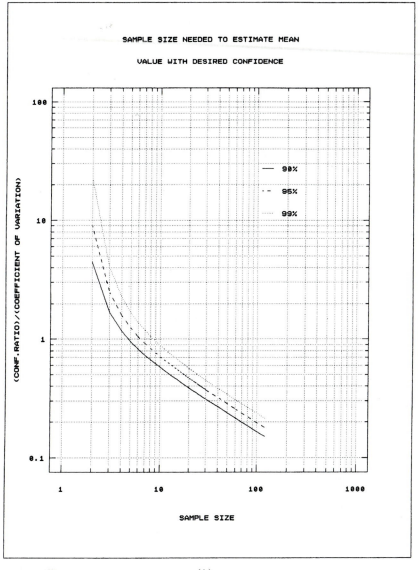

FIGURE 4.3
(Continued).

(b)

experiment we *wouldn't know* that the data was not gaussian, so we proceed as if it were. The main point to be gleaned from this simulation is that when we get in the neighborhood of 10 specimens, the confidence interval is about the same width (as a percentage of the mean value) as it was with gaussian data, and it *does* contain the true value 87,400 within it. Our calculation methods appear to be quite tolerant of deviations from

n	Mean strs. 100 psi	95% conf. int. on mean stress	Std. deviation 100 psi	95% conf. int. on std. dev.
1	842			
2	851	734–967	13.0	5.8–413
3	853	828–879	10.2	5.3–64.2
4	856	840–871	9.7	5.5–36.1
5	860	845–875	12.3	7.4–35.4
6	863	849–877	13.7	8.5–33.5
7	866	852–880	14.9	9.6–32.7
8	868	856–880	14.3	9.5–29.2
9	868	858–878	13.4	9.1–25.8
10	872	859–884	17.7	12.2–32.2
20	867	859–876	18.7	14.2–27.3
40	871	864–877	19.4	15.9–24.9
100	871	867–875	18.5	16.2–21.5

(1) 842	(18) 906	(35) 880	(52) 864	(69) 883	(86) 862
(2) 860	(19) 845	(36) 879	(53) 855	(70) 882	(87) 864
(3) 859	(20) 864	(37) 855	(54) 880	(71) 868	(88) 850
(4) 863	(21) 864	(38) 898	(55) 862	(72) 876	(89) 867
(5) 876	(22) 876	(39) 849	(56) 869	(73) 872	(90) 852
(6) 880	(23) 851	(40) 850	(57) 852	(74) 837	(91) 891
(7) 884	(24) 881	(41) 887	(58) 871	(75) 860	(92) 846
(8) 878	(25) 871	(42) 902	(59) 867	(76) 885	(93) 866
(9) 870	(26) 850	(43) 871	(60) 864	(77) 835	(94) 841
(10) 907	(27) 844	(44) 874	(61) 865	(78) 877	(95) 877
(11) 847	(28) 905	(45) 850	(62) 891	(79) 897	(96) 885
(12) 865	(29) 894	(46) 876	(63) 908	(80) 902	(97) 856
(13) 850	(30) 877	(47) 855	(64) 891	(81) 873	(98) 874
(14) 877	(31) 869	(48) 894	(65) 868	(82) 877	(99) 856
(15) 846	(32) 882	(49) 883	(66) 874	(83) 872	(100) 914
(16) 852	(33) 911	(50) 862	(67) 854	(84) 875	
(17) 877	(34) 892	(51) 915	(68) 861	(85) 842	

FIGURE 4.4
Yield stress data set, triangular distribution

the gaussian assumptions, and thus we can safely use them on most real-world data. For $n = 10$ the mean value is actually better centered in the confidence interval than it was with the gaussian data. Don't attach too much significance to this last statement! Remember that if we *reran* the gaussian data with a new seed in the random number generator, this new gaussian data could give sample values quite different from our first set of gaussian data. To remind you that it takes very large samples to make a set of perfect data from a theoretical random number generator behave like the underlying distribution, Fig. 4.5 compares our 100-item triangular data with the perfect curves for both the density function and the cumulative function. The deviation from ideal is quite obvious. Since STATGRAPHICS makes it very easy to "play around" with situations like this, I tried a sample size of 1000 which, while still not perfect, gave a histogram that was quite close to the desired triangle.

Although we have done only one example in this section, it is typical of most situations where you are trying to estimate a mean value; so you should now be prepared to deal with most other experiments of this class when you encounter them.

4.2 ESTIMATION OF PARAMETER VARIABILITY

As mentioned earlier in this chapter, variability in a product or process is often a prime consideration in the design and operation of manufacturing processes of all kinds. Again we can return to Chap. 2 for the needed analytical tool—the confidence limit on the standard deviation. Although a single data point allows a mean-value estimate, we need at least two points for even the crudest variability calculation, and we want to get a little practice in choosing proper sample sizes for this kind of situation.

Our example here will involve the manufacture of an electric low-pass filter, as in Fig. 4.6. This type of filter is widely used in many kinds of electronic equipment to allow low-frequency components in the input signal voltage e_i to pass through to e_o while blocking the passage of high-frequency components. This discrimination based on frequency content is a result of the filter's first-order-system dynamic behavior, specifically the amplitude-ratio curve of Fig. 4.6. Note that we use here for the first time a *logarithmic* plotting scheme quite common in system dynamics studies. Amplitude ratios are expressed in decibels (dB) according to the following definition:

$$dB \triangleq \text{decibel value of amplitude ratio} \triangleq 20 \log \text{(actual amplitude ratio)}$$

$$(4.2)$$

When frequency is plotted on a logarithmic scale, of course we can never show a point at zero frequency, but we can get as close as we please. A

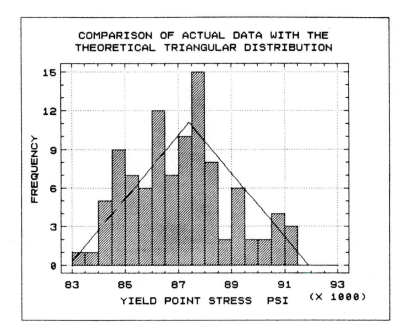

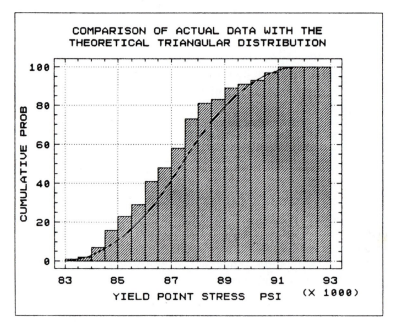

FIGURE 4.5
Deviation of 100-item sample from ideal behavior.

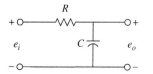

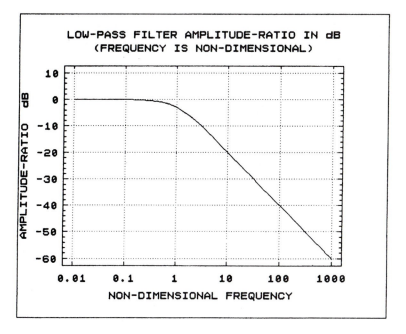

FIGURE 4.6
Low-pass filter and its frequency response.

10-to-1 frequency ratio is called a *decade* while an *octave* is a 2-to-1 ratio, and an actual amplitude ratio of 1.0 corresponds to 0 dB. To make this curve *universal* for all first-order systems, we have also made the frequency axis nondimensional. That is, the numbers are values of $\omega\tau$, not ω, where ω must be given in radians per unit time and τ is the system time constant.

We plan to control the quality of the filters by requiring that the amplitude ratio at one specific frequency be kept within specified tolerance limits. For first-order systems, a frequency often used for specifying performance is the *break-point frequency*. This frequency is the reciprocal of the system time constant τ. At this frequency, every first-order system has an actual amplitude ratio of 0.707 (-3 dB) and a phase angle of $-45°$. Since the resistors and capacitors used in filter construction each exhibit random variability, the amplitude ratio at a specified frequency will also vary. We want this $\pm 3s$ variation to not exceed ± 10 percent of 0.707 at a frequency of 10 Hz (62.8 rad/s). Note in Fig. 4.6 that an actual frequency of 62.8 rad/s corresponds to a non-dimensional frequency of 1.0 and that $\tau = 1/62.8$ sec. The nominal design values of R and C have been fixed at 100,000. Ω and 0.1592 μF, and capacitors with variability $\pm 3s = \pm 0.0054$ μF are available. Now we need to calculate the allowable variation in the resistors and set up an experiment to check resistors as they come in from the vendor to see whether they meet our specifications.

First we use the methods of Sec. 2.9 to calculate the allowable resistor variation. The actual (not decibel) amplitude ratio R_{amp} is given by

$$R_{\mathrm{amp}} = \frac{1}{\sqrt{(\omega RC)^2 + 1}} \tag{4.3}$$

We need also the partial derivatives

$$\frac{\partial R_{\mathrm{amp}}}{\partial C} = -\frac{1}{2} \frac{2(\omega RC)\omega R}{[(\omega RC)^2 + 1]^{1.5}} = 2.22 \times 10^6 \frac{1}{\mathrm{F}} \tag{4.4}$$

$$\frac{\partial R_{\mathrm{amp}}}{\partial R} = -\frac{1}{2} \frac{2(\omega RC)\omega C}{[(\omega RC)^2 + 1]^{1.5}} = 3.59 \times 10^{-6} \frac{1}{\Omega} \tag{4.5}$$

Using Eq. (2.61), we get

$$(s_{R_{\mathrm{amp}}})^2 = \left(\frac{\partial R_{\mathrm{amp}}}{\partial C} s_C \right)^2 + \left(\frac{\partial R_{\mathrm{amp}}}{\partial R} s_R \right)^2 \tag{4.6}$$

Substituting numerical values gives us finally

$$s_R = 6473 \ \Omega \tag{4.7}$$

which makes our specification for the resistors

$$R = 100,000 \pm 19,419 \ \Omega \tag{4.8}$$

We now know what is required of the resistors so that the allowed variability in the filter amplitude ratio will be achieved. Using the methods of Eq. (2.71) and/or Fig. 2.26, we can now begin to plan an experiment for testing purchased resistors to see whether they meet our needs. This consists mainly in choosing a proper sample size to locate the measured standard deviation of resistance within an acceptable confidence interval. Examination of the graph of Fig. 4.7 reveals that if we accept confidence intervals that are a given *percentage* of the standard deviation, we don't need to know the standard deviation before testing commences. Since such percentage criteria actually *are* what we usually want, this aspect of choosing the sample size appears to be more convenient than it was for checking mean values in Sec. 4.1, where we needed an estimate of standard deviation *before* we could start testing.

Confidence intervals for standard deviation									
Sample size	**50% confidence**			**90% confidence**			**95% confidence**		
	Lo	**Hi**	**± %**	**Lo**	**Hi**	**± %**	**Lo**	**Hi**	**± %**
6	0.867s	1.36s	25.	0.672s	2.09s	71.	0.624s	2.45s	91.
11	0.893s	1.22s	16.	0.739s	1.59s	43.	0.699s	1.76s	53.
16	0.907s	1.16s	13.	0.775s	1.44s	33.	0.739s	1.55s	40.
21	0.916s	1.14s	11.	0.798s	1.36s	28.	0.765s	1.44s	34.
41	0.936s	1.09s	7.7	0.847s	1.23s	19.	0.821s	1.28s	23.
61	0.946s	1.07s	6.2	0.871s	1.18s	15.	0.849s	1.22s	18.
101	0.957s	1.05s	4.8	0.897s	1.13s	12.	0.879s	1.16s	14.

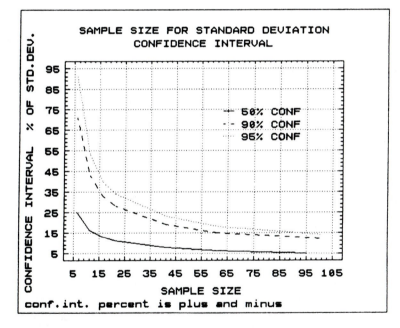

FIGURE 4.7
Choosing sample size for confidence interval on standard deviation.

While the *procedure* is apparently more direct, the numbers in Fig. 4.7 quickly make us aware that the needed sample sizes for estimating standard deviations are *much* larger than we saw earlier for mean values. For example, at the 95 percent confidence level, a sample size of 11 gives a confidence interval that has a width of ±53 percent. (Don't be misled by the plus-or-minus notation. Recall that these confidence intervals are *not* symmetric about the nominal value since they are based on the chi-squared distribution, itself nonsymmetric. Thus, ±53 percent really means 30 percent below and 76 percent above, as seen from the table values of 0.699*s* to 1.76*s*. While the *intervals* are not symmetric, the *probabilities* are. That is, of the 95 percent confidence, we are 47.5 percent sure that the true value is within the subinterval 0.699*s* to 1.0*s* and 47.5 percent sure it is within 1.0*s* to 1.76*s*.)

We still need to make a decision on the sample size to use in checking incoming resistors for acceptance, even though we are now aware that insisting on a small confidence interval will require very large sample sizes compared to those needed for estimating mean values. It is really impossible to make such decisions rationally without having detailed *economic* data on the costs and benefits of different sample sizes. The situation is further complicated by the fact that the *final* result (what percentage of filters fails to meet our specification) depends on several nebulous factors, some of which are as follows:

1. What are the *true* distribution functions for our resistors and capacitors?
2. Both the mean values and the standard deviations of the actual populations will drift with time. That is, these random variables are not "stationary."
3. An alternative manufacturing philosophy to the random assembly we have been assuming is *selective assembly*. Here we measure each *R* and *C* and then carefully pair them to get the desired filter behavior. What are the economics of this approach compared with those of random assembly?

These factors and others not listed here all conspire to make sample-size decisions less "scientific" and more dependent on past experience, judgment, company policy, etc., than we might have hoped. One possible approach is to select, by using Fig. 4.7 for guidance, a tentative sample size which seems reasonable in terms of company experience and practice and then to run some computer simulations to check the consequences of that choice of sample size. For example, if a sample size of about 100 is considered, we see that (with 95 percent confidence) the standard deviation predicted from such a sample could be incorrect by as much as +16 or −12 percent. If this "worst case" actually occurred, what effect would it have on the quality of our filters?

Simulation could be used in several different ways to study this kind of situation. Let's postulate the following scenario. Suppose that we have measured 100 resistors out of a batch just shipped to us, and we find the standard deviation to be just barely acceptable, i.e., right at the limit given by Eq. (4.7), or 6473 Ω. We are concerned, of course, that with this sample size, the *actual* standard deviation might be as much as 16 percent higher, and thus *not* acceptable. We can use STATGRAPHICS (or other similar software) to simulate the "assembly" of resistors and capacitors into filters and the "measurement" of the filters' amplitude ratio at the test frequency, giving the resistor and capacitor populations whatever statistical properties we wish. We first generate gaussian samples (we could, of course, try many other distributions) of R's and C's with chosen mean values and standard deviations. We will generate random samples of 100 R's and C's and then compute the amplitude ratio at 10 Hz according to Eq. (4.3), for 100 pairs of components, just as if we were selecting them one pair at a time from bins. Let's start with the ideal situation, where the mean and standard deviations of R and C are exactly what we specify. Recall that our specified allowable range for the amplitude ratio was 0.707 ± 10 percent, which rounds to 0.63 to 0.78.

Figure 4.8*a* shows the results for the first 100 filters, sorted into rank order. We see that none of these filters has an amplitude ratio outside the specified bounds. Let's next allow *both* the mean values and the standard deviations for the R's and C's to deviate from specification—a situation certain to occur in actual practice. We first let the mean values of both R and C increase by 1 percent and also let their standard deviations both be 10 percent larger than specified. A STATGRAPHICS simulation of this situation gives the table of Fig. 4.8*b*. Again, none of the 100 filters is outside the allowed bounds, but we can see (even without doing the calculation) that the mean value has shifted to a slightly lower value. This is obviously due to the larger mean values of R and C, which appear in the denominator of the amplitude ratio. Finally, we keep the mean values at their +1 percent values, but we set both standard deviations at +30 percent, which gives Fig. 4.8*c*. Now 2 of the 100 filters are out-of-specification, at the low end (the two 0.62 values).

We see that analysis using a combination of theoretical formulas and computer simulation can be very effective in choosing a proper sample size for acceptance testing of our resistors. As mentioned earlier, however, economic factors would have to be carefully integrated with these statistical considerations before a final decision could be made.

4.3 COMPARISON OF MEAN VALUES

We often need to compare one material, heat treatment, source of supply, etc., with an alternative and then make a choice of the "better" of the two. Sometimes better is defined in terms of the mean value of some measurable parameter; the better aluminum alloy is the one with the higher fatigue strength, the better brakelining material is the one with

Variable: AAERNFIL.fltr_r_amp   (Length = 100)

(1) 0.64	(18) 0.68	(35) 0.7	(52) 0.71	(69) 0.72	(86) 0.73
(2) 0.65	(19) 0.68	(36) 0.7	(53) 0.71	(70) 0.72	(87) 0.73
(3) 0.65	(20) 0.68	(37) 0.7	(54) 0.71	(71) 0.72	(88) 0.73
(4) 0.66	(21) 0.68	(38) 0.7	(55) 0.71	(72) 0.72	(89) 0.73
(5) 0.67	(22) 0.69	(39) 0.7	(56) 0.71	(73) 0.72	(90) 0.74
(6) 0.67	(23) 0.69	(40) 0.7	(57) 0.71	(74) 0.72	(91) 0.74
(7) 0.67	(24) 0.69	(41) 0.7	(58) 0.71	(75) 0.72	(92) 0.74
(8) 0.67	(25) 0.69	(42) 0.7	(59) 0.71	(76) 0.72	(93) 0.74
(9) 0.67	(26) 0.69	(43) 0.7	(60) 0.71	(77) 0.72	(94) 0.74
(10) 0.67	(27) 0.69	(44) 0.7	(61) 0.71	(78) 0.72	(95) 0.75
(11) 0.67	(28) 0.69	(45) 0.7	(62) 0.71	(79) 0.72	(96) 0.75
(12) 0.68	(29) 0.69	(46) 0.7	(63) 0.71	(80) 0.72	(97) 0.75
(13) 0.68	(30) 0.7	(47) 0.7	(64) 0.72	(81) 0.73	(98) 0.75
(14) 0.68	(31) 0.7	(48) 0.7	(65) 0.72	(82) 0.73	(99) 0.75
(15) 0.68	(32) 0.7	(49) 0.71	(66) 0.72	(83) 0.73	(100) 0.75
(16) 0.68	(33) 0.7	(50) 0.71	(67) 0.72	(84) 0.73	
(17) 0.68	(34) 0.7	(51) 0.71	(68) 0.72	(85) 0.73	

(a)

Variable: AAERNFIL._r_amp1_10   (Length = 100)

(1) 0.63	(18) 0.67	(35) 0.69	(52) 0.7	(69) 0.71	(86) 0.73
(2) 0.63	(19) 0.67	(36) 0.69	(53) 0.7	(70) 0.71	(87) 0.73
(3) 0.64	(20) 0.67	(37) 0.69	(54) 0.7	(71) 0.72	(88) 0.73
(4) 0.65	(21) 0.68	(38) 0.69	(55) 0.7	(72) 0.72	(89) 0.73
(5) 0.65	(22) 0.68	(39) 0.69	(56) 0.7	(73) 0.72	(90) 0.73
(6) 0.66	(23) 0.68	(40) 0.69	(57) 0.7	(74) 0.72	(91) 0.73
(7) 0.66	(24) 0.68	(41) 0.69	(58) 0.7	(75) 0.72	(92) 0.73
(8) 0.66	(25) 0.68	(42) 0.69	(59) 0.7	(76) 0.72	(93) 0.74
(9) 0.66	(26) 0.68	(43) 0.7	(60) 0.7	(77) 0.72	(94) 0.74
(10) 0.66	(27) 0.68	(44) 0.7	(61) 0.7	(78) 0.72	(95) 0.74
(11) 0.66	(28) 0.68	(45) 0.7	(62) 0.71	(79) 0.72	(96) 0.74
(12) 0.67	(29) 0.69	(46) 0.7	(63) 0.71	(80) 0.72	(97) 0.75
(13) 0.67	(30) 0.69	(47) 0.7	(64) 0.71	(81) 0.72	(98) 0.75
(14) 0.67	(31) 0.69	(48) 0.7	(65) 0.71	(82) 0.72	(99) 0.75
(15) 0.67	(32) 0.69	(49) 0.7	(66) 0.71	(83) 0.72	(100) 0.75
(16) 0.67	(33) 0.69	(50) 0.7	(67) 0.71	(84) 0.72	
(17) 0.67	(34) 0.69	(51) 0.7	(68) 0.71	(85) 0.72	

(b)

FIGURE 4.8
Simulation results for three different situations.

the lower wear rate, etc. We saw in Sec. 2.10 that such comparisons are best made by using appropriate confidence intervals and that we could implement this concept by using individual intervals and the overlap graphs of Fig. 2.25 or a single combined interval, as in Eq. (2.75). The combined interval is probably the choice of most statisticians, and we

Variable: AAERNFIL.r_amp   (Length = 100)

(1) 0.62	(18) 0.67	(35) 0.69	(52) 0.7	(69) 0.72	(86) 0.73
(2) 0.62	(19) 0.67	(36) 0.69	(53) 0.7	(70) 0.72	(87) 0.73
(3) 0.63	(20) 0.67	(37) 0.69	(54) 0.7	(71) 0.72	(88) 0.73
(4) 0.64	(21) 0.67	(38) 0.69	(55) 0.7	(72) 0.72	(89) 0.74
(5) 0.65	(22) 0.67	(39) 0.69	(56) 0.7	(73) 0.72	(90) 0.74
(6) 0.65	(23) 0.67	(40) 0.69	(57) 0.7	(74) 0.72	(91) 0.74
(7) 0.65	(24) 0.67	(41) 0.69	(58) 0.7	(75) 0.72	(92) 0.74
(8) 0.65	(25) 0.67	(42) 0.69	(59) 0.7	(76) 0.72	(93) 0.74
(9) 0.65	(26) 0.67	(43) 0.69	(60) 0.7	(77) 0.72	(94) 0.75
(10) 0.65	(27) 0.68	(44) 0.69	(61) 0.7	(78) 0.72	(95) 0.75
(11) 0.66	(28) 0.68	(45) 0.7	(62) 0.71	(79) 0.72	(96) 0.75
(12) 0.66	(29) 0.68	(46) 0.7	(63) 0.71	(80) 0.72	(97) 0.75
(13) 0.66	(30) 0.69	(47) 0.7	(64) 0.71	(81) 0.72	(98) 0.75
(14) 0.66	(31) 0.69	(48) 0.7	(65) 0.71	(82) 0.73	(99) 0.76
(15) 0.66	(32) 0.69	(49) 0.7	(66) 0.71	(83) 0.73	(100) 0.76
(16) 0.66	(33) 0.69	(50) 0.7	(67) 0.71	(84) 0.73	
(17) 0.67	(34) 0.69	(51) 0.7	(68) 0.72	(85) 0.73	

(c)

FIGURE 4.8
(Continued).

emphasize it here; however, the overlap graphs are sometimes a useful adjunct, especially when one is explaining results to coworkers who have little experience in statistics.

In Eq. (2.74) we can get some useful approximate results by letting both the sample sizes and standard deviations of the two samples be equal. We also let n get "very large"; this makes $f \to \infty$ and the t distribution approach the gaussian. Then the t value to be used in Eq. (2.75) becomes fixed and known when we choose the confidence level ($t = 1.96$ for 95 percent confidence, e.g.). From Fig. 4.9 we can see that if

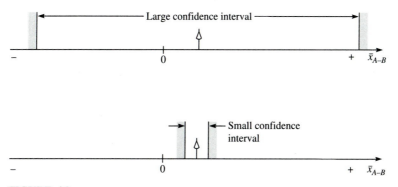

FIGURE 4.9
Small confidence interval needed to compare two mean values.

we want a clear indication as to whether the two mean values are *really* different, we need a confidence interval that is small *relative* to $\bar{x}_{A-B}$, say, 10 percent of $\bar{x}_{A-B}$. With the assumptions above, the 95 percent confidence interval in Eq. (2.75) becomes $\pm 2.77 s(n)^{-0.5}$, and if we set this equal to $0.10\bar{x}_{A-B}$, we can estimate the needed sample size n from

$$n = 767\left(\frac{s}{\bar{x}_{A-B}}\right)^2 \tag{4.9}$$

As usual, the problem with using such formulas is that, before running the experiment, we don't have accurate numbers for s and $\bar{x}_{A-B}$ and may not even have rough estimates. If numbers *are* available, of course we use them to estimate the sample size. If not, we must simply proceed with testing and use test data from the first few specimens to estimate s and $\bar{x}_{A-B}$ if we want to predict approximately how large a sample will be needed.

Since the assumptions leading to Eq. (4.9) were invoked only to achieve a result useful for rough estimating, we might want to use computer simulation again to get a little experience with actual behavior. Let's pretend that we are in the following situation. We manufacture miniature thermoelectric coolers and are developing a "new and improved" model which is intended to have a better overall efficiency. The current model has been subjected to quite thorough testing, so we have efficiency data for it based on a sample of size 30, which gave a mean efficiency of 10.3 percent and a standard deviation of 0.18 percent. We want to test our new model to see whether it really does have a significantly improved efficiency; theoretical studies predict that it should be about 11.6 percent. Using Eq. (4.9) to do some experiment planning, we assume the new model to exhibit a standard deviation similar to the old one, but we opt for a conservative (high) value of 0.20 for s in the equation. Using the theoretically predicted efficiency improvement of 1.3 percent as $\bar{x}_{A-B}$, we can estimate the sample size n as about 18.

We will now use STATGRAPHICS to simulate the actual experiment, enjoying the usual simulation benefit of knowing from the outset what the "correct" answers really are. First we employ the gaussian random number generator to produce our "experimental" data for both the old and new coolers. (Recall that it is not difficult to simulate nongaussian data if we should so choose.) Our "old cooler" data is made of sample size 30 with a mean of 10.3 percent and a standard deviation of 0.18 percent, giving the list of "observations" called COOLEFFA in Fig. 4.10. We have not bothered to round these data to reflect a reasonable measurement accuracy; such rounding can be left for the *final results*. In simulating data for the new cooler, we decide to use a mean of 11.5 percent and a standard deviation of 0.22 percent, different from what was

Variable: AAERNFIL.COOLEFFC

(1)	10.2244
(2)	10.6737
(3)	10.4285
(4)	10.2172
(5)	10.2937
(6)	10.0475
(7)	10.7309
(8)	10.6477
(9)	10.4604
(10)	10.3756
(11)	10.5099
(12)	10.7808
(13)	10.3457
(14)	10.686
(15)	10.4036

Variable: AAERNFIL.COOLEFFA

(1)	10.5009	(19)	10.222
(2)	10.2158	(20)	10.1898
(3)	10.0536	(21)	10.3609
(4)	10.313	(22)	10.4691
(5)	10.082	(23)	10.1769
(6)	10.5963	(24)	10.3772
(7)	10.4076	(25)	10.1063
(8)	10.5293	(26)	10.4223
(9)	10.309	(27)	10.0859
(10)	10.2596	(28)	10.3458
(11)	10.0482	(29)	10.6432
(12)	10.2725	(30)	10.0842
(13)	10.2214		
(14)	10.4831		
(15)	10.2076		
(16)	10.1343		
(17)	10.0803		
(18)	10.3958		

Variable: AAERNFIL.COOLEFFB

(1)	11.0042
(2)	11.129
(3)	11.6312
(4)	11.4313
(5)	11.2996
(6)	11.575
(7)	11.0827
(8)	11.4999
(9)	11.4081
(10)	11.8422
(11)	11.5893
(12)	11.2477
(13)	11.7225
(14)	11.4751
(15)	11.5561

FIGURE 4.10
Data sets for thermoelectric cooler experiments.

expected, since a *real* experiment would exhibit such unpredictability. With these input values the gaussian generator produces the data called COOLEFFB in Fig. 4.10.

In addition to simulating our experimental observations, STAT-GRAPHICS provides convenient tools for analyzing data, whether it is real or simulated data. A routine called TWO-SAMPLE ANALYSIS implements our Eqs. (2.74) and (2.75) to compare the mean values of two sets of observations. [It also implements our Eq. (2.76) for comparing standard deviations. We will use this feature in Sec. 4.4.] While Fig. 4.10 shows 15 observations of the new cooler's efficiency, we will consider them one at a time, just as if we were carrying out the real experiment, one test at a time. This is made simple in STATGRAPHICS by the use of a command called N TAKE. If we enter, say, 5 TAKE COOLEFFB, this creates a new data set which is the first five items in COOLEFFB, which we may then analyze as we please.

We see from Fig. 4.10 that the first test of the new coolers gives an efficiency of 11.0 percent, but Eq. (2.74) allows no calculation of the confidence interval since at least two observations are needed to estimate the new standard deviation. Since 11.0 is larger than 10.3 (the efficiency of the old coolers), we are, of course, encouraged; but we need to do some more testing before we conclude that our new design is a definite improvement. The next test gives us a sample size of 2 and allows STATGRAPHICS to compute the results of Fig. 4.11a. We see that the difference between the mean values is now -0.78 (we know the true value is -1.2), and the 95 percent confidence interval extends from -1.67 to $+0.11$, so the true difference in mean values is still quite uncertain. Two graphical displays of these results are available in STATGRAPHICS, a comparison of histograms and a box-and-whisker plot. Figure 4.11a shows the histograms of the two data sets with the old cooler histogram plotted upward at the left and the new cooler histogram plotted downward at the right, giving a good visual comparison of the actual data but perhaps being somewhat misleading with respect to the validity of the apparent difference in mean values. In the box-and-whisker plot of Fig. 4.11, the box encloses 50 percent of the observations, the line in the box locates the median, and the whiskers (vertical lines extending from the box) extend to the highest and lowest observations in the data set. (When there are only two observations, the whiskers are not used.)

We could, of course, repeat the above work for $n = 3, 4, 5$, etc., but we choose to examine only 5, 10, and 15. Figure 4.11b is for $n = 5$, and we see that the difference in means is now estimated as -1.01, with a confidence interval from -1.31 to -0.71. For $n = 10$, in Fig. 4.11c, we get -1.10 and -1.29 to -0.91. Finally in Fig. 4.11d ($n = 15$) we have -1.14 and -1.29 to -1.00, and we are now in a range of sample sizes where the difference in mean values can be defined quite clearly. We can also see that our early estimate of 18 as a reasonable sample size was not far off the mark.

When the standard deviations are comparable in size to the true difference of mean values, it becomes more difficult to detect this difference reliably. To illustrate this numerically, I made up a new data set COOLEFFC (see Fig. 4.10) with a mean value of 10.5 and a standard deviation of 0.22, and I compared this with our original data set COOLEFFA (mean, 10.3; standard deviation, 0.18). Figure 4.12a to d shows the results of this study and should be self-explanatory.

Before we leave this section, note that a *practical* interpretation of

```
                      Two-Sample Analysis Results
--------------------------------------------------------------
                                    COOLEFFA      Sample 2
Sample Statistics: Number of Obs.   30            2
                   Average          10.2865       11.0666
                   Variance         0.0288865     7.7837E-3
                   Std. Deviation   0.16996       0.0882253
                   Median           10.266        11.0666

Difference between Means = -0.780114
Conf. Interval For Diff. in Means:       95      Percent
  (Unequal Vars.)  Sample 1 - Sample 2   -1.66543 0.105203     1.6 D.F.
```

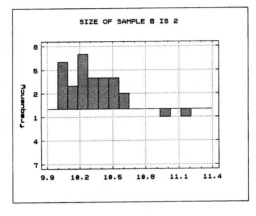

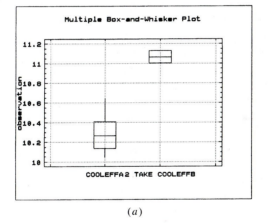

(*a*)

FIGURE 4.11
Effect of sample size on comparison of mean values.

experimental results and statistical calculations such as those above must always be undertaken. If we have just proved that our new design of thermoelectric coolers has, on average, a better (higher) efficiency than our earlier model, what is the practical economic significance to our company, our country, or the world? At first glance, an increase in

Two-Sample Analysis Results

```
-----------------------------------------------------------------
                                COOLEFFA        Sample 2
Sample Statistics:  Number of Obs.      30          5
                    Average             10.2865     11.299
                    Variance            0.0288865   0.0609238
                    Std. Deviation      0.16996     0.246828
                    Median              10.266      11.2996

Difference between Means = -1.01258
Conf. Interval For Diff. in Means:       95      Percent

  (Unequal Vars.)  Sample 1 - Sample 2   -1.31416 -0.710998    4.7 D.F.
```

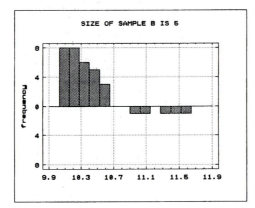

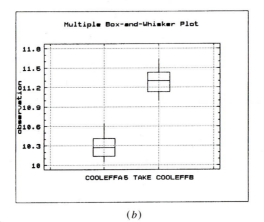

(*b*)

FIGURE 4.11
(Continued).

average efficiency seems beneficial in that less energy will be wasted in carrying out the cooling tasks performed by the coolers. We need to explore further, however, since the new design may require *more* energy (or scarce materials) in its manufacture. Also, if customers purchase our coolers for incorporation into a larger system of their design, and if they

Two-Sample Analysis Results

		COOLEFFA	Sample 2
Sample Statistics:	Number of Obs.	30	10
	Average	10.2865	11.3903
	Variance	0.0288865	0.0700697
	Std. Deviation	0.16996	0.264707
	Median	10.266	11.4197

Difference between Means = -1.10385
Conf. Interval For Diff. in Means: 95   Percent

  (Unequal Vars.) Sample 1 - Sample 2   -1.29921 -0.908498   11.6 D.F.

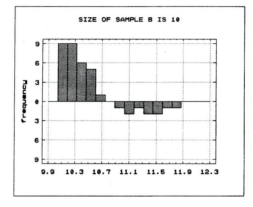

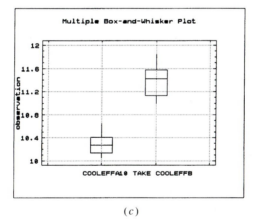

(c)

FIGURE 4.11
(Continued).

use the average efficiency numbers we supply them in their design
calculations, then their larger system may not meet its specifications (or
may even fail catastrophically) when one of our units happens to exhibit a
lower than average value. Thus it may be necessary to consider both the
average efficiency and its variability.

Two-Sample Analysis Results

		COOLEFFA	COOLEFFB
Sample Statistics:	Number of Obs.	30	15
	Average	10.2865	11.4329
	Variance	0.0288865	0.0577377
	Std. Deviation	0.16996	0.240287
	Median	10.266	11.4751

Difference between Means = -1.14646
Conf. Interval For Diff. in Means: 95   Percent
 (Unequal Vars.) Sample 1 - Sample 2   -1.29066 -1.00227   21.2 D.F.

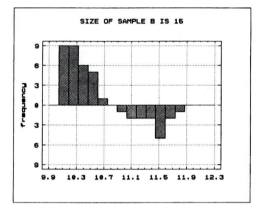

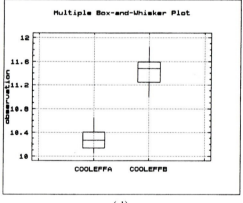

(*d*)

FIGURE 4.11
(Continued).

4.4 COMPARISON OF STANDARD DEVIATIONS

In the example of Sec. 4.3 the performance criterion under study (efficiency) is determined by a complex interaction of many design

Two-Sample Analysis Results

```
-------------------------------------------------------------------
                                    COOLEFFA     Sample 2
Sample Statistics: Number of Obs.   30           2
                   Average          10.2865      10.449
                   Variance         0.0288865    0.100928
                   Std. Deviation   0.16996      0.317692
                   Median           10.266       10.449

Difference between Means = -0.162558
Conf. Interval For Diff. in Means:   95      Percent

  (Unequal Vars.)  Sample 1 - Sample 2   -3.04401 2.71889    1.0 D.F.
```

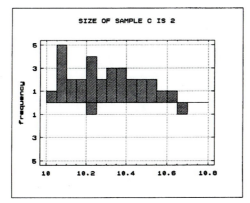

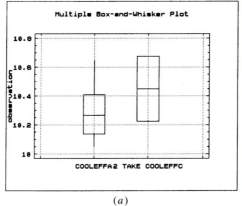

(a)

FIGURE 4.12
Large scatter makes comparison of means difficult.

factors and is not subject to easy adjustment if we wish to raise or lower it. On the other hand, in manufacturing processes (such as turning, milling, and grinding) which remove material from an initially oversize workpiece, we can rather easily make machine adjustments to get the average dimension we desire in the part. However, the variability about

Two-Sample Analysis Results

		COOLEFFA	Sample 2
Sample Statistics:	Number of Obs.	30	5
	Average	10.2865	10.3675
	Variance	0.0288865	0.0364957
	Std. Deviation	0.16996	0.191039
	Median	10.266	10.2937

Difference between Means = -0.0810088
Conf. Interval For Diff. in Means: 95   Percent

(Unequal Vars.) Sample 1 - Sample 2   -0.313198 0.15118   5.1 D.F.

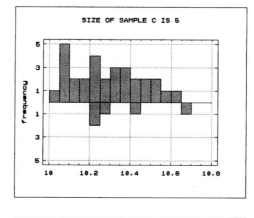

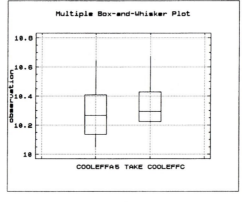

(*b*)

FIGURE 4.12
(Continued).

this average value is usually *not* easy to change since it depends in a complex way on an interaction of factors, some of which are intrinsic to the machine tool being used. It is quite common thus to be faced with the need to compare competitive machine tools and/or processes in order to determine which has the least variability.

```
                          Two-Sample Analysis Results
----------------------------------------------------------------------------
                                       COOLEFFA        Sample 2
Sample Statistics: Number of Obs.      30              10
                   Average             10.2865         10.4099
                   Variance            0.0288865       0.0499532
                   Std. Deviation      0.16996         0.223502
                   Median              10.266          10.402
Difference between Means = -0.123473
Conf. Interval For Diff. in Means:     95      Percent
  (Unequal Vars.)  Sample 1 - Sample 2   -0.290732 0.0437868   12.7 D.F.
```

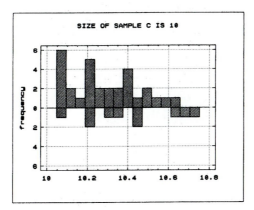

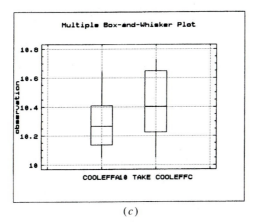

(c)

FIGURE 4.12
(Continued).

We can use the accepted method of Eq. (2.76) in Sec. 2.10 to compare the standard deviations of two samples of gaussian data. Here we used the F distribution to find the ratio of the variances (squares of standard deviations) for our two samples. Let's again use STATGRAPHICS to both generate our data sets and analyze them. Suppose we are designing laser beam machines for cutting parts out of

Two-Sample Analysis Results

		COOLEFFA	COOLEFFC
Sample Statistics:	Number of Obs.	30	15
	Average	10.2865	10.455
	Variance	0.0288865	0.0462135
	Std. Deviation	0.16996	0.214973
	Median	10.266	10.4285

Difference between Means = -0.168564
Conf. Interval For Diff. in Means:　　　95　　Percent
(Unequal Vars.)　Sample 1 - Sample 2　　-0.300132 -0.036995　　23.0 D.F

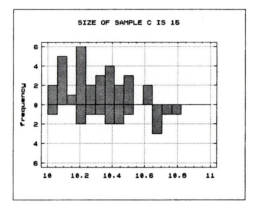

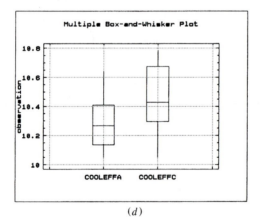

(d)

FIGURE 4.12
(Continued).

sheet metal. We have developed, to the prototype stage, two different design concepts, and we wish to evaluate them for variation in the dimensions of the parts produced. In one machine the laser beam is stationary, and the part is moved beneath the beam on a servo-driven *XY* table. In the competitive design, the part is stationary while the laser beam is steered in the desired path by a set of servo-driven mirrors. We have decided on a part shape which will efficiently "exercise" the two machines and a measurement scheme to determine the critical dimensions produced, so we can proceed with an experiment to compare the two machines.

To simulate this experiment, first we will generate two data sets which have the same standard deviation and see what size sample is required to reliably detect that there is no significant difference between them. From the table of Fig. 2.27 (90 percent confidence level), if we try a sample size of, say, 100, then the ratio of standard deviations will be in a confidence interval from 0.85 to 1.18. Going to a sample of 500 reduces this interval to 0.93 to 1.07. From these preliminary estimates we can see that unless the experimental cost per observation is very low, we will usually have to be satisfied with a fairly wide confidence interval, since it takes such large samples to narrow the interval. Let's take the mean value of the dimension produced by machine A to be 10.00 mm with a standard deviation of 1.00 mm, and we let STATGRAPHICS generate a gaussian data set with 100 observations. In Fig. 4.13a this data set is called LSRDIMNA. For the alternative machine we take the mean value at 10.50 with standard deviation also 1.00, giving the data set LSRDIMNB in Fig. 4.13b. (The two machines should be adjusted to give the *same* mean value, but we take them slightly different since this is more realistic.)

We now pretend that we produce one part at a time on each machine, measure each part, and use STATGRAPHICS to analyze the data so produced in order to decide whether the two machines have the same, or different, variabilities. For the smallest sample size (2) that allows calculation, we find that $s_A = 1.18$ and $s_B = 0.15$, and the 90 percent confidence interval for their ratio goes from 0.6 to 98.1. Clearly, no reliable conclusion can be reached from these sparse data. While we could easily repeat these calculations for n equal to 3, 4, 5, etc., we choose to display in Fig. 4.14 only the results for $n = 5$, 10, 20, 50, and 100. As we predicted theoretically above, even for $n = 100$, there is still a fair amount of uncertainty about the true relative values of the two standard deviations. With this simulation available for guidance, however, we can face real-world problems of this type, knowing what to expect, and thus make the best decisions possible under difficult conditions.

To get a little more background in this area, I made up a third data set (LSRDIMNC in Fig. 4.13c) with a mean value of 10.50 and a standard deviation of 2.00, and I compared this with LSRDIMNA, giving the results of Fig. 4.15. Again we see that sample sizes of about 100 are needed before we can be reasonably sure that the ratio of standard deviations is about 0.5.

Variable: AAERNFIL.LSRDIMNA   (Length = 100)

(1) 9.92221	(19) 7.84495	(37) 11.8147	(55) 9.84203
(2) 11.5979	(20) 8.64352	(38) 10.2981	(56) 9.26004
(3) 11.5237	(21) 11.1141	(39) 11.6407	(57) 10.5639
(4) 8.9183	(22) 9.76082	(40) 11.5605	(58) 9.05319
(5) 11.0937	(23) 10.5875	(41) 9.27076	(59) 9.15047
(6) 10.6478	(24) 10.1001	(42) 10.1635	(60) 9.9091
(7) 10.6165	(25) 11.2686	(43) 11.1152	(61) 9.15842
(8) 10.4427	(26) 7.93408	(44) 10.2629	(62) 10.4979
(9) 11.0124	(27) 10.1145	(45) 9.04601	(63) 12.0745
(10) 10.2551	(28) 9.89236	(46) 8.93448	(64) 10.3681
(11) 12.0652	(29) 9.58472	(47) 11.7375	(65) 8.3452
(12) 8.56272	(30) 9.53326	(48) 9.99651	(66) 8.0007
(13) 9.61414	(31) 10.1076	(49) 10.7951	(67) 10.5874
(14) 9.00901	(32) 9.58739	(50) 9.9259	(68) 9.98206
(15) 11.5937	(33) 10.0521	(51) 12.5456	(69) 9.05218
(16) 9.98073	(34) 9.90116	(52) 9.41928	(70) 9.26112
(17) 10.8829	(35) 10.2322	(53) 10.8626	(71) 8.92096
(18) 10.2822	(36) 11.0063	(54) 8.84661	(72) 8.93942

(73) 8.04201	(91) 10.2376
(74) 10.9215	(92) 10.357
(75) 10.1507	(93) 9.33106
(76) 9.51891	(94) 11.1319
(77) 12.4166	(95) 8.11199
(78) 9.71265	(96) 9.77227
(79) 10.4183	(97) 10.7874
(80) 10.9674	(98) 9.41655
(81) 8.64545	(99) 10.357
(82) 9.1853	(100) 11.6222
(83) 9.82722	
(84) 9.38411	
(85) 10.9321	
(86) 9.90419	
(87) 9.44898	
(88) 10.3746	
(89) 8.58701	
(90) 11.4266	

(*a*)

FIGURE 4.13
Data sets for three laser cutting machines.

4.5 FINDING RELATIONS BETWEEN DEPENDENT AND INDEPENDENT VARIABLES (MODEL BUILDING)

We now begin discussion of the class of experiments labeled 5 in Fig. 4.1. Here, rather than the estimation of "point values" of parameters that we have considered so far, we are interested in establishing *functional*

Variable: AAERNFIL.LSRDIMNB (Length = 100)

(1) 12.0385	(19) 8.09182	(37) 11.2467	(55) 11.6345
(2) 11.8215	(20) 11.0184	(38) 10.1286	(56) 10.837
(3) 10.3952	(21) 11.2673	(39) 11.9089	(57) 9.98616
(4) 10.5878	(22) 9.7048	(40) 8.47428	(58) 10.0743
(5) 12.3428	(23) 10.377	(41) 10.8922	(59) 11.8642
(6) 10.4097	(24) 9.81664	(42) 10.9906	(60) 10.287
(7) 9.05883	(25) 10.4164	(43) 10.9902	(61) 9.64558
(8) 10.6181	(26) 10.4704	(44) 10.0444	(62) 10.2181
(9) 10.3423	(27) 12.2337	(45) 10.5072	(63) 11.2048
(10) 9.20256	(28) 9.59394	(46) 9.50081	(64) 10.7631
(11) 9.86257	(29) 10.0771	(47) 11.1402	(65) 8.77416
(12) 10.8151	(30) 11.0896	(48) 10.7373	(66) 10.6129
(13) 11.1342	(31) 11.147	(49) 12.3174	(67) 11.087
(14) 10.8535	(32) 11.1278	(50) 11.0215	(68) 10.5529
(15) 10.2257	(33) 11.0949	(51) 9.74006	(69) 12.3588
(16) 12.111	(34) 12.6188	(52) 10.0383	(70) 8.35685
(17) 10.9202	(35) 10.4658	(53) 9.87976	(71) 9.55559
(18) 9.86548	(36) 9.75003	(54) 10.6648	(72) 10.8127

(73) 12.5759	(91) 11.5318
(74) 9.762	(92) 9.44169
(75) 10.1592	(93) 10.0481
(76) 10.6915	(94) 10.2478
(77) 13.1858	(95) 10.6513
(78) 11.3612	(96) 10.4894
(79) 9.57137	(97) 11.2768
(80) 11.4987	(98) 11.4503
(81) 10.8677	(99) 10.7371
(82) 11.6036	(100) 10.7147
(83) 11.3069	
(84) 9.17783	
(85) 9.8168	
(86) 10.4948	
(87) 10.4668	
(88) 9.84579	
(89) 8.65453	
(90) 10.238	

(b)

FIGURE 4.13
(Continued).

relationships between independent and dependent variables. For example, in the laser beam cutting machines we studied in Sec. 4.4, suppose we found that one of the two competitive designs was clearly superior with respect to variability. But now we want to see which design factors in this machine contribute to the variability, so that we can perhaps make design adjustments which will further reduce the scatter. You might think that a *theoretical* analysis of the machine would be adequate to identify

Variable: AAERNFIL.LSRDIMNC

(1) 14.3307	(19) 10.8781	(37) 6.77499	(55) 12.9937
(2) 9.64343	(20) 12.3336	(38) 6.99512	(56) 9.01024
(3) 13.0776	(21) 13.0312	(39) 11.5566	(57) 11.9922
(4) 7.55223	(22) 6.88846	(40) 13.4944	(58) 12.9887
(5) 9.36933	(23) 10.4613	(41) 9.2141	(59) 10.7438
(6) 11.4805	(24) 12.9426	(42) 12.1585	(60) 8.34109
(7) 12.087	(25) 9.60007	(43) 6.45246	(61) 11.4128
(8) 12.1129	(26) 9.55555	(44) 9.42498	(62) 11.1329
(9) 12.2989	(27) 7.24971	(45) 9.70552	(63) 10.4092
(10) 13.9337	(28) 9.47159	(46) 9.39475	(64) 8.09174
(11) 11.4393	(29) 10.4274	(47) 6.73852	(65) 9.2935
(12) 11.7944	(30) 11.5686	(48) 11.5053	(66) 11.6042
(13) 8.27672	(31) 11.3312	(49) 10.4213	(67) 13.6064
(14) 7.51226	(32) 10.0365	(50) 12.5851	(68) 11.5397
(15) 9.23886	(33) 8.55189	(51) 11.8274	(69) 10.6168
(16) 10.6589	(34) 10.7625	(52) 9.34851	(70) 8.52915
(17) 11.4846	(35) 12.8194	(53) 8.8205	(71) 11.8716
(18) 12.7413	(36) 6.82924	(54) 10.7403	(72) 11.0462

(73) 10.5053	(91) 12.4803
(74) 14.6866	(92) 10.8469
(75) 9.24455	(93) 11.3628
(76) 14.4671	(94) 10.0386
(77) 9.10517	(95) 7.71882
(78) 8.92766	(96) 13.8159
(79) 12.346	(97) 8.65681
(80) 9.64923	(98) 9.83358
(81) 11.7835	(99) 9.87297
(82) 11.1221	(100) 10.0912
(83) 13.5154	
(84) 13.1892	
(85) 10.3038	
(86) 10.1197	
(87) 10.0955	
(88) 12.5872	
(89) 11.0277	
(90) 6.36069	

(c)

FIGURE 4.13
(Continued).

and model the necessary relations, and such an analysis should, of course, always be attempted. Unfortunately, many manufacturing processes, even though they run satisfactorily and make a profit, are so complex that theoretical study alone is inadequate to identify and/or explain the needed functional relations. When this is the case, properly designed and executed experiments have been shown to provide a powerful tool for understanding and improving industrial products and processes.

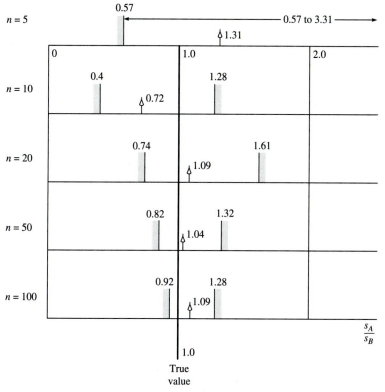

FIGURE 4.14
Effect of sample size on comparison of standard deviations.

Statistically based methods for planning, executing, and analyzing such experiments were first developed in the 1920s by R. A. Fisher,[1] working at the agricultural research station Rothamsted, near London, England. Since that time, many books on this subject ("Statistical Design of Experiments") have appeared and the methods have been successfully employed in a wide variety of applications, although most of the earlier work was concentrated in agriculture, medicine, and biology. Until recently, engineering applications (outside of chemical engineering) were relatively scarce and/or unpublicized. Only large companies which could afford to employ their own professional statisticians made regular use of

[1] R. A. Fisher, *The Design of Experiments,* Oliver & Boyd, London, 1935.

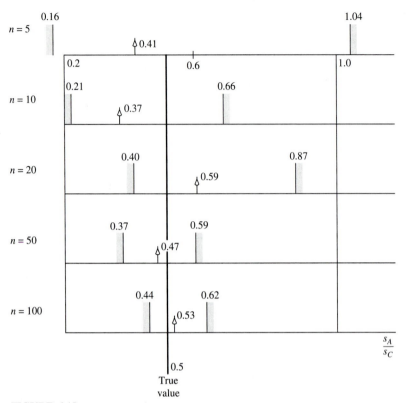

FIGURE 4.15
Effect of sample size when standard deviations are unequal.

the techniques, since the methods were not routinely taught in under-graduate engineering curriculums.

Simplified and abbreviated versions of the methods, with statistical jargon held to a minimum, so that the typical bachelor of science engineering graduate could learn and use them with a reasonable expenditure of time and effort, were not widely available. Part of the explanation of this situation lies in the concern that a superficial understanding of *any* technique may lead to its misuse; thus only the intelligentsia should be allowed to practice it. While such concerns are legitimate, they are also regularly overdone. Practical experimentation always includes a large component of common sense, and it is not unrealistic to expect this universal feature of good engineering to prevent the gross misapplication of statistical methods in most cases. We also need to say that even statistical "experts" make mistakes, so we need not deny ourselves useful tools because we don't have time to become expert in some specialty. Mistakes, in fact, have a long history as valuable *learning* experiences.[1] So as long as the mistakes are not catastrophic and/or embarrassingly frequent, we shouldn't pay excessive prices in the (vain) hopes of avoiding them.

A "quantum jump" in the importance and industrial application of these methods in the United States occurred in the 1980s. Some years earlier, Japanese industry recognized the unmet potential of these methods for improving the quality, cost, and speed of development of products and processes. These movements were led by design and manufacturing engineers rather than statisticians, and emphasized practical results rather than theoretical rigor. They realized that design and manufacturing considerations are best considered simultaneously (in parallel) rather than sequentially (in series), and that well-planned experimentation during the product/process development cycle was a potent tool for improvement. Large numbers of engineers were given intensive short courses presenting simplified routine methods for practical design of industrial experiments. Although certainly some of these practitioners made mistakes, they also had sufficient success that Japanese products came to dominate many markets and other countries soon found that they had to adopt similar measures in order to compete.

Another development that is making statistical experiment design more accessible to nonspecialists is the increasing availability of personal-computer-based software which is easily learned and used. The STATGRAPHICS package which I have been using in this text has a section, which I will demonstrate later, that implements some of the more common design tools. Packages[2] that are *totally* devoted to experiment design are also available and offer a wide range of techniques from the simplest to the most sophisticated. Some of these packages[3] use an "expert system" approach to build in warnings, guidance, and safeguards to minimize the chance that the methods will be misused.

Before getting into the details of some of these methods in the next few sections, we give a brief overview of their scope. In general, we are trying to find relationships between the independent variables (factors we are able to adjust) in a process and one or more dependent variables which reflect the quality of the product or its rate of production. If we run an experiment which "exercises" the process by changing each

---

[1] H. Petroski, *To Engineer Is Human,* St. Martin's Press, New York, 1985. R. R. Whyte (ed.), *Engineering Progress through Trouble,* Institute of Mechanical Engineers, London, 1975.

[2] Experiment Design Made Easy, Advanced Experiment Design: Stat-Ease Inc., 2021 E. Hennepin Ave., #191, Minneapolis, MN 55413, phone (612) 378-9449.
Experimental Design, Screen, Box-B, Factorial Design: Statistical Programs, 9941 Rowlett, Suite 6, Houston, TX 77075, phone (713) 947-1551.

[3] ECHIP: Echip Inc., 7460 Lancaster Pike, Hockessin, DE 19707, phone (302) 239-6227.

independent variable over some range, we may be able to process this data to synthesize a mathematical model relating the dependent variable to all the independent variables. Such a model, if reliable, has several uses. When we have little theoretical understanding of a process, we often start with *screening experiments.* Here we treat all variables even suspected of being significant, as independent variables, but we let each take on only two values in the experiment, a "high" value and a "low" one. This is necessary to keep down both cost and time when we are initially dealing with many variables, only a few of which will turn out to be really important. A properly designed screening experiment often quickly identifies the handful of variables that really matter. [Note, however, that if we don't use all available "process knowledge" when initially making up our list of variables, we might overlook some truly vital ones. This underlines again the importance of *engineers* (people who have the process knowledge) learning how to design experiments. If statistics experts *are* involved in the process, they need to closely consult with the process people to achieve a successful team effort, and, of course, the best statistical consultants have always operated in this fashion.]

Once screening experiments have narrowed the field of independent variables, more detailed studies of those remaining variables may be necessary. Some processes may exhibit a "peak" value of some important quality factor for a certain critical combination of independent-variable values. Detailed experiments may be able to discover this optimum so that we can try to operate the process there. Sometimes optimum (or high-quality) behavior can be obtained in a region where this high quality is only slightly affected by variation in one or more independent variables, giving a so-called robust design. If we can locate such combinations of values, we can relax the variability requirements on some of the independent variables and achieve a high-quality end product, using "cheap" raw materials or components.

We have just emphasized the applications that involve the design and manufacture of products, because of their great economic importance. Empirical (experimentally based) models are useful in other contexts also. When both the dependent variable and all the independent variables can be measured on numerical scales, we can actually obtain mathematical functions relating them. These functions can be very useful in compactly summarizing large amounts of experimental data needed in various engineering or commercial operations. For computerized operations a formula for computing some result is usually much quicker and less wasteful of memory than a table look up of the raw data would be. Empirical formulas obtained by curve-fitting experimental data can also sometimes provide clues as to the nature of the *theoretical* phenomena underlying the physical behavior.

4.5.1 Models Involving a Single Independent Variable (Simple Regression)

First we consider models relating a dependent variable to a single independent variable, both being numerically measurable. This simple situation is, of course, quite common, and one can give many examples: How does the tensile strength of an alloy steel vary with temperature? How does the discharge coefficient of an orifice vary with the fluid flow rate? What is the variation of the offset voltage of an operational amplifier with temperature? Even though we may have already decided that we are interested in the effect of only *one* independent variable, we need to remember that variables of no direct interest to us may have small, but not necessarily negligible, effects on our chosen dependent variable. This consideration leads us to a short discussion of the experimental technique called *randomization.*

In our present context, randomization has to do with the *sequencing* of the test runs we plan to make at various levels of the independent variable. Often there seems to be an obvious or natural sequence, and we want to show that this obvious sequence is not always the best one to implement. When we suspect that one or more "extraneous" variables as just described above are present, we should try to sequence the runs so that their effect on the dependent variable is of a random nature, so as not to distort the true effect of the independent variable we are studying. For example, if we are studying the effect of the quenching temperature on the strength of heat-treated steel, we should consider that the properties of the oil used in the quenching bath will gradually change as the oil is used. Thus the *last* specimens quenched will be feeling not only the effect of the quenching temperature but also some unknown effect of oil degradation. We can help to "cancel" this effect by using a test sequence other than the obvious one of a monotonically increasing set of quenching temperatures. Thus if we want to explore the range from 400 to 800°C with a total of 9 observations, rather than using the obvious sequence of 400, 450, 500, ..., 700, 750, 800, it might be wiser to use something like 400, 800, 450, 750, 500, 700, 550, 650, 600, attempting to combine new oil/old oil with high temperature/low temperature in such a way as to cancel the oil degradation effect.

The new sequence just shown is probably an improvement, but a little thought shows that now the *midrange* of temperatures is tested only with old oil and so our results will be biased by this effect. Rather than trying to use deterministic thinking (which often introduces a *new* bias into the results), we often schedule the runs in a *random* way, and this technique is called *run-sequence randomization.* Here, once we have decided on the range to be covered and the actual values to be tested, we select the sequence in some random fashion. This can be accomplished in various ways. We could just write the numbers 1 to 9 on nine slips of paper and, blindfolded, draw them from a box. A more scientific way employs a table of random numbers.[1] The STATGRAPHICS software allows construction of such a table by using the random number generator for the *discrete uniform distribution*. Figure 4.16 shows the

---

[1] E. B. Wilson, Jr., *An Introduction to Scientific Research*, McGraw-Hill, New York, 1952, p. 286.

(1) 1	(19) 16	(37) 9	(55) 2	(73) 24	(91) 30	(109) 9
(2) 14	(20) 15	(38) 19	(56) 22	(74) 16	(92) 18	(110) 11
(3) 21	(21) 1	(39) 10	(57) 2	(75) 15	(93) 30	(111) 10
(4) 29	(22) 16	(40) 18	(58) 13	(76) 21	(94) 8	(112) 10
(5) 11	(23) 30	(41) 14	(59) 27	(77) 13	(95) 12	(113) 29
(6) 4	(24) 26	(42) 6	(60) 10	(78) 14	(96) 23	(114) 6
(7) 13	(25) 2	(43) 7	(61) 29	(79) 3	(97) 9	(115) 3
(8) 8	(26) 20	(44) 3	(62) 27	(80) 23	(98) 16	(116) 6
(9) 8	(27) 6	(45) 29	(63) 5	(81) 10	(99) 19	(117) 15
(10) 16	(28) 4	(46) 14	(64) 1	(82) 20	(100) 4	(118) 22
(11) 2	(29) 23	(47) 17	(65) 4	(83) 18	(101) 28	(119) 22
(12) 2	(30) 29	(48) 25	(66) 29	(84) 4	(102) 4	(120) 17
(13) 13	(31) 21	(49) 21	(67) 28	(85) 14	(103) 11	(121) 30
(14) 6	(32) 5	(50) 18	(68) 9	(86) 26	(104) 16	(122) 11
(15) 4	(33) 2	(51) 21	(69) 12	(87) 30	(105) 13	(123) 4
(16) 8	(34) 26	(52) 3	(70) 23	(88) 1	(106) 13	(124) 6
(17) 12	(35) 6	(53) 24	(71) 25	(89) 11	(107) 22	(125) 10
(18) 6	(36) 1	(54) 3	(72) 26	(90) 2	(108) 30	(126) 30

(127) 18	(145) 29	(163) 5	(181) 4	(199) 24
(128) 23	(146) 3	(164) 1	(182) 5	(200) 24
(129) 15	(147) 17	(165) 4	(183) 9	
(130) 10	(148) 13	(166) 4	(184) 14	
(131) 25	(149) 12	(167) 2	(185) 6	
(132) 18	(150) 27	(168) 25	(186) 7	
(133) 18	(151) 11	(169) 12	(187) 8	
(134) 27	(152) 23	(170) 1	(188) 23	
(135) 26	(153) 1	(171) 29	(189) 2	
(136) 1	(154) 25	(172) 13	(190) 15	
(137) 30	(155) 20	(173) 22	(191) 29	
(138) 9	(156) 21	(174) 2	(192) 1	
(139) 9	(157) 20	(175) 7	(193) 21	
(140) 21	(158) 20	(176) 18	(194) 2	
(141) 19	(159) 5	(177) 1	(195) 1	
(142) 19	(160) 21	(178) 9	(196) 19	
(143) 28	(161) 28	(179) 20	(197) 19	
(144) 4	(162) 17	(180) 28	(198) 6	

FIGURE 4.16
Random number table for run-sequence randomization.

table produced by requesting 200 items from a discrete uniform distribution whose values are to be restricted to the range of 1 to 30. The table allows run-sequence randomization for experiments with up to 30 runs. To use it for our nine-run experiment, enter the table at a point chosen "randomly" (shut your eyes and jab your pencil). Then proceed through the table (in the numbered 1 to 200 sequence), watching for the numbers 1 to 9. As each number is encountered (ignore repeats), that determines its position in the experimental sequence. When I "jabbed my pencil," I hit the 29th entry (23) and started there, finally getting the sequence 5, 2, 6, 1, 9, 7, 3, 4, 8.

It might seem that we could use this sequence for *all* nine-run experiments, and it would not be disastrous if we actually did this. But it is better ("more random") to always generate a new sequence with a new "pencil jab." (If you reach the 200th entry before getting a complete sequence, just go back to entry 1 and continue on in the table.) You may be wondering about the significance of the 200-item sample used to generate the table. Since the goal was a table useful for experiments with up to 30 runs, one might initially think a sample of 30 would be adequate. A little reflection (or an actual trial!), of course, reminds us that a 30-item sample may *not* contain all the numbers from 1 to 30. Thus in constructing Fig. 4.16 I simply tried larger and larger samples until I got all the numbers from 1 to 30 at least once in the table. I could have gone to larger tables (which would give better randomness when the table is used over and over), but I was constrained by simple considerations of book space. Note also that if *you* used STATGRAPHICS to duplicate my table, you would probably get a useful table (check to make sure numbers 1 to 30 are all there), but it *wouldn't* be the same as mine because the software uses a *different* seed in the random number generator every time it is used, unless you intentionally enter the same seed. This also means that we can generate "new" random number tables any time we wish.

It may be helpful at this point to give a few more examples of situations where run-sequence randomization should be used. When experiments involve a human operator, several effects related to human behavior may cause subtle biases in results, and we should try to eliminate or minimize these by choice of a random sequence. If a procedure is physically and/or mentally fatiguing, runs made later in a sequence may show a degradation of technique, biasing the data. On the other hand, if an operator is inexperienced, performance may *improve* in later runs, due to learning. In testing apparatus at different power levels, when we change from one level to another, we generally wait for *equilibrium* to be restored before recording the new data point. Since the approach to steady state after a transient is an *asymptotic* process for most physical systems, the achievement of true equilibrium is time-consuming and difficult to precisely detect. Thus we will usually take the new data point somewhat *before* the new steady state is achieved. If our experimental sequence is the obvious one of a monotonic increase from the lowest to the highest power level, then our data will be biased in one direction. This effect can be minimized by using a random sequence, where the high and low levels are "scrambled" and the approach to equilibrium occurs from both sides, rather than only one side. Finally, we need to admit that if run-sequence randomization results in intolerable expense or inconvenience, we may opt to forgo it unless its benefits are overwhelmingly obvious.

Next we need to consider the types of models (relating dependent and independent variables) that we might want to use in practical problems and the mathematical methods used to get the numerical values of model parameters. The simplest model is the straight-line relation $y = mx + b$. For a given set of x, y data, how do we find the "best" values for parameters m and b? Most readers will have encountered elsewhere the classical answer to this question—the *least-squares* method (in fact, we have already used it earlier in this book; see Fig. 3.4). Here we set up an expression for the difference (called the *residual*) between the measured dependent variable and the model-predicted value at each data point. We *define* the best model (best values of m and b) as that which makes the sum of the squares of all the residuals as small as possible. While of course this definition of *best* is not the only possible one, it *is* a reasonable definition and exhibits some outstanding mathematical advantages. The main such advantage is that for a large and useful class of models (we will discuss this shortly when we get to multiple regression), the mathematics needed to find the optimum model parameters is linear algebra, which is easily computerized.

When our model is the simple straight line, the equations to find m and b are well known[1] and most readers will have a hand calculator which implements this operation. Statistical packages such as STATGRAPHICS of course also provide such facilities and include additional useful calculations and graphics. To get some practice with this general class of experiments, we will make up a computer-simulated data set and then subject it to the standard analysis methods to see how to use them and how to interpret the results. As usual, computer simulation provides the great advantage (compared with any *real* experiment) that

---

[1] E. O. Doebelin, *System Modeling and Response: Theoretical and Experimental Approaches,* Wiley, New York, 1980, pp. 300–320.

Obs. #	TEMP	s_{temp}	s_{mono}	s_{rand}	s_{err}
1 (4)	400	50.0	50.0	49.4	−0.608
2 (2)	450	52.0	51.8	51.8	−1.409
3 (7)	500	54.0	53.6	52.8	−1.826
4 (8)	550	56.0	55.4	54.6	0.487
5 (1)	600	58.0	57.2	58.0	−0.286
6 (3)	650	60.0	59.0	59.6	0.033
7 (6)	700	62.0	60.8	61.0	1.529
8 (9)	750	64.0	62.6	62.4	−1.080
9 (5)	800	66.0	64.4	65.2	1.020

FIGURE 4.17
Data set for simulated heat treatment experiment.

we know in advance what the correct answers are, so that we can easily tell whether the analysis methods are working well or badly.

Our data relate to the heat-treating situation mentioned above. Let's assume that there is a perfect linear relation between strength s (in kilopounds per square inch) and quenching temperature T (in degrees Celsius) and that the effect of run sequence on oil degradation is also directly proportional to the run number N, according to the formula

$$s = 34.0 + 0.04T - (N - 1)(0.2) \tag{4.10}$$

Using this formula, we can manufacture the set of "experimental" data shown in Fig. 4.17. The column labeled s_{temp} gives the strengths that would be obtained if there were no oil degradation effect, s_{mono} is for a monotonic testing sequence with the oil effect present, and s_{rand} is the (preferred) plan where we use the randomized sequence selected earlier from the table of random numbers. One nice thing about computer simulation is that not only can we make up "perfect" data but also we can then "contaminate" them with any kind of imperfections we wish, in an attempt to mimic the peculiarities of the real world. The column labeled s_{err} is gaussian random "noise" with mean value zero and a standard deviation of 1.0 kpsi. We can add this to any of our data to simulate random measurement errors and/or random fluctuations in the true value of specimen strength. Such effects could be either large or small relative to the basic data values in the other columns; we can just multiply s_{err} by any number (less than or greater than 1.0) to easily adjust the importance of these random phenomena.

If we want to fit a straight line to the TEMP versus s_{temp} data, or the TEMP versus s_{mono} data, complicated least-squares methods are hardly necessary since all the points fall exactly on a straight line. It is, however, a useful check of the computer software you are using (I am again using STATGRAPHICS) to exercise it with such perfect data. As expected, the software returns a slope of 0.0400 and an intercept of 34.00 for the TEMP versus s_{temp} data while giving 0.0360 and 35.60 for the TEMP versus s_{mono} data (see Fig. 4.18a and b). In a real-world experiment, using the unrandomized s_{mono} data leads (because of the unmodeled oil degradation effect) to *errors* in the computed model relating strength to temperature. That is, we are led to believe that strength is related to temperature along a straight line having a slope of 0.0360 and an intercept of 35.60 when the correct values are 0.0400 and 34.00. To show numerically the advantage of run-sequence randomization, we ask for a fit of the TEMP versus s_{rand} data and get 0.0388 and 33.92—not perfect but a clear improvement.

Fitting this last data produces, instead of the simple straight-line plots, the more complex graph of Fig. 4.19a and a printout of not only the intercept and slope, but also some statistical indicators of how good the fit of the model to the data actually is. First let's discuss the two sets of dashed lines on either side of the (solid) best-fit line. The innermost pair of lines corresponds to a 95 percent (the software allows use of any percentage) *confidence interval* while the outermost pair represents a 95 percent *prediction interval.* If we use the model to predict strength from chosen values of temperature and if the temperature, say 700, occurs many times, we are 95 percent sure that the true *average value* of the strength is within the confidence interval, about 60.5 to 61.5 on the graph. On the other hand, if we consider a *single* occurrence of the 700 temperature, we are 95 percent sure that the true value of the strength of that *individual specimen* lies within the prediction interval, about 59.5 to 62.5 on the graph. Formulas for computing these confidence and prediction intervals are available in statistics texts.[1] Most engineers these days would be using, perhaps on their own personal computer, a software package which produces results like these automatically. So our emphasis, in this example and elsewhere in this text, is on the results and their application rather than derivations and details of calculations.

A graph of residuals (differences between model-predicted values and measured values) versus the independent variable is sometimes

---

[1] R. L. Scheaffer and J. T. McClure, *Statistics for Engineers,* Duxbury Press, Boston, 1982, pp. 266, 267.

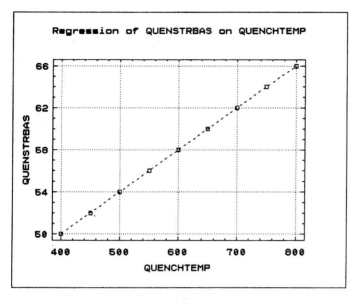

(*a*)

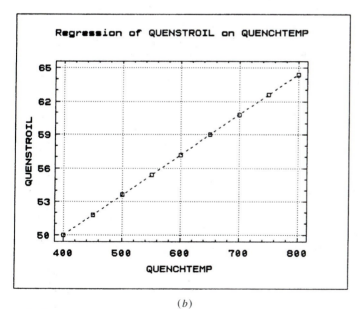

(*b*)

FIGURE 4.18
Straight-line fits of heat treatment data.

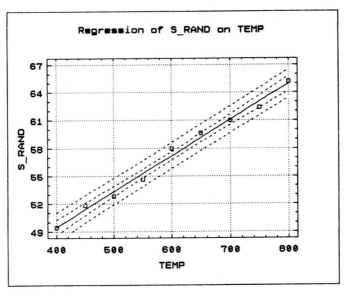

(*a*)

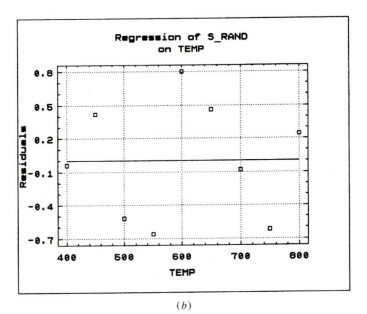

(*b*)

FIGURE 4.19
Fit of run-sequence randomized data.

useful for diagnosing deficiencies in the *form* (straight line, parabola, cubic, etc.) of the assumed model and is often provided by the software, as in Fig. 4.19*b*. The seemingly random scatter of the residuals is a desirable result, implying that the assumed model form is reasonable. When this graph shows obvious trends or patterns, it usually means we have chosen an unsuitable form of model, and study of the nature of the patterns often is a guide to choice of a better model. Note that more complex models almost *always* will allow us to reduce the sum of the squared residuals, but that does *not* necessarily mean that the model is "better." In our present example, if we took an eighth-degree polynomial as the model, we could get (without using least squares) a curve that went *exactly* through all nine data points. This would *not* be a better model since it implies an undulating relation between temperature and strength, whereas we *know* that the relation is really a straight line and the undulations are due to the randomized effect of the oil degradation.

Regression software usually computes the *correlation coefficient R* and the *coefficient of determination* R^2, which are defined as follows. We first compute the average of all the dependent-variable values; this is 57.2 for s_{rand}. Each observation (measured value of dependent variable) deviates from this average by some amount. We next compute the sum of the squares of all these deviations; in this case it is 228.0. We now take the sum of the squares of the differences between each model-predicted value and the average (57.2) of all the measured values; we get 225.8. Note that if the model *perfectly* predicted all the measured values, then these two numbers (228.0 and 225.8) would be the same. The coefficient of determination R^2 is defined as the ratio $225.8/228.0 = 0.99$, which is very close to the perfect-fit value of 1.0, implying that the model does a good job of explaining the variation in strength related to the quenching temperature. In statistical jargon we say that the model explains 99 percent of the variation in the dependent variable. Once we have R^2, the correlation coefficient R is, of course, a simple calculation; it is 0.995. (Whereas R^2 is always positive, when the line slope is negative, we assign the correlation coefficient the *negative* square root of R^2. Thus $0 < R^2 < 1.0$ and $-1.0 < R < +1.0$.)

Finally, statistical significance tests using the t and F distributions are also usually calculated. "Large" values of t and F indicate a high probability that the estimated parameters or the overall model are significant. Many engineers rely mainly on the *graphical* evidence of the quality of the model's predictions. In our present example, Fig. 4.19*a* alone really allows quite a good evaluation of the model's quality. When there is more scatter in the data than we have in Fig. 4.19, then we may be readier to augment visual judgment with numerical criteria such as R^2, particularly if we are trying to *compare* two alternative models and choose the better. However, when the scatter gets really large, we need to be careful to not accept a model which has the "best" (but still quite small compared to 1.0) R^2 value, when it is graphically clear that the model is not really predicting a valid relationship.

The studies above can be made a little more realistic by adding some random "noise" to the measured values of strength, by using the gaussian data called s_{err} in Fig. 4.17. We first try a "small" amount of noise, $0.1s_{err}$, add this to both s_{mono} and s_{rand}, and rerun the regression program, getting the graphs of Fig. 4.20. The benefits of run-sequence randomization are still significant since the line parameters (33.98, 0.0387) in Fig. 4.20a (the randomized sequence) are closer to the correct values than those (35.30, 0.0364) of the monotonic sequence of Fig. 4.20b. Note that the "wrong" relation of Fig. 4.20b, however, is visually a better fit than the more correct one of Fig. 4.20a, and has a "better" R^2 value (0.9997 compared to 0.9890). Thus the effects of run-sequence randomization are subtle. If we judged the relative quality of these two models simply by the goodness of fit, we would choose the wrong model.

Next we add a "large" amount (s_{err} rather than $0.1s_{err}$) of random noise to s_{temp}, s_{mono}, and s_{rand}, and we rerun the calculations, getting the graphs of Fig. 4.21a to c and the attached table. The data for s_{temp} are, of course, unrealistic since it includes no effect of oil degradation; but it is useful for our purposes here in illustrating the difficulties of model making. From the graphs and table we see that while all three cases produce line slopes and intercepts that are fair estimates of the correct values, it is now quite difficult to tell whether any of the three models is better than the others, including even the artificial data of s_{temp}. You are, of course, not to take this as a renunciation of our advice to use run-sequence randomization; it is still a recommended practice. We just wanted to show (as is so difficult to do with real experiments but so easy with computer simulations) that when the data exhibit considerable scatter, development of good predictive models may not be a routine procedure with only one possible answer.

We saw earlier in this chapter that using larger sample sizes can improve estimates of parameters such as the mean value and standard deviation, so we expect a similar improvement in the model-fitting studies we are now explaining. This can be demonstrated most clearly by using the $s_{temp} + s_{err}$ data shown in the table of Fig. 4.21, where we do not have the confusing effect of oil degradation. I simply made up a data set of 101 points, using Eq. (4.10) (without the oil degradation term), added the same amount of gaussian noise (standard deviation = 1.0) as in s_{err}, and requested the usual straight-line fit; the result was the graph of Fig. 4.22. Comparing this with Fig. 4.21a, we see that the prediction interval for a single observation (outermost tolerance bands) is about as bad (large) as before, but the confidence interval for the mean values is much better (narrower). The estimate of model parameters is also much improved,

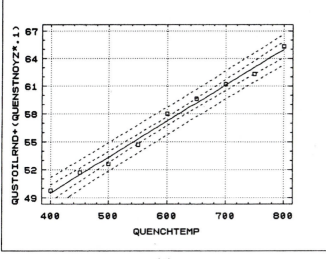

(a)

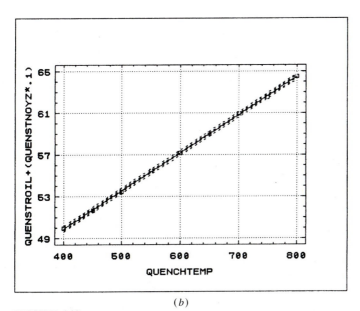

(b)

FIGURE 4.20
Heat treatment data with small added random noise.

Graph	Data set	Slope	Intercept	R^2
4.21 *a*	$s_{temp} + s_{err}$	0.0446	31.01	0.9774
4.21 *b*	$s_{mono} + s_{err}$	0.0406	32.61	0.9729
4.21 *c*	$s_{rand} + s_{err}$	0.0428	31.30	0.9621
4.18 *a*	"PERFECT"	0.0400	34.00	1.0000

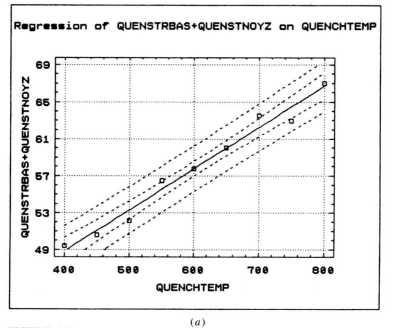

(*a*)

FIGURE 4.21
Heat treatment data with large added random noise.

34.33 and 0.0394. When we look at R^2, however, it is smaller ("worse")—0.9581. This is because the model (a smooth curve) *cannot* reduce the size of the residuals when they are caused (as they are in this example) by random noise. [We *could* make the residuals all zero (and R^2 equal to the perfect value 1.0) by fitting a 100th-degree polynomial to the 101 points, but this would be both numerically difficult and foolish in principle.] We see again that just because one model has a larger R^2 than another, it is not necessarily better. Finally, we should mention that when we add more points to improve model reliability, the points must be "spread out" over the full range of the independent variable. Adding, say, 92 points in the

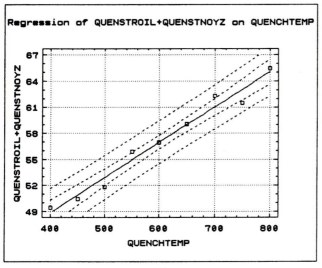

(*b*)

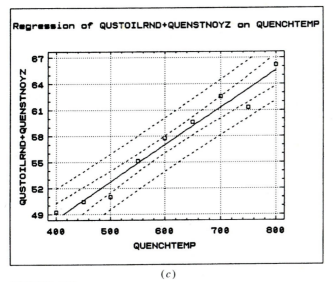

(*c*)

FIGURE 4.21
(Continued).

near neighborhood of 600°C to the nine-point model of Fig. 4.21 will *not* give the desired effect.

Up to this point, the model we have been trying to fit was *exactly* the same form as that underlying the measured data; all that least-squares analysis did was to find the best numerical values for the

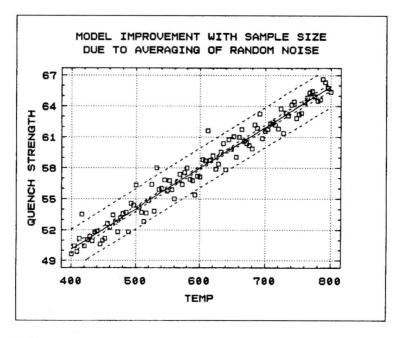

FIGURE 4.22
Improvement in modeling with larger sample size.

model parameters (line slope and intercept). In a real-world experiment we usually *don't* know the form of the "correct" model, especially when studying manufacturing processes, where the measures of product quality depend in a nebulous way on a large number of process variables. To get some feel for how least-squares methods work under these more realistic (and difficult) conditions, let's set up a simulation where we try to fit a *wrong* model to a set of measured data. Rather than make up an entirely new situation, let's take the s_{rand} data of Fig. 4.17 and add to each observation a term proportional to the square of temperature, creating a new dependent variable which we will call s_{para}:

$$s_{para} = s_{rand} + (1.0E-5)TEMP^2 \tag{4.11}$$

Figure 4.23 shows a table of these new data and a graph comparing our earlier straight-line data s_{rand} with the new parabolic data. Note that, to the eye, one of these curves does *not* appear clearly straighter than the other; this is masked by the randomized effect of oil degradation. Such masking of deterministic trends by random influences of one kind or another is, of course, common in real-world data.

We now pretend that we have just measured the s_{para} data and have been asked to find a good model relating s_{para} to TEMP. Since the scatter

TEMP	s_{rand}	s_{para}
400	49.4	51.4
450	51.8	53.8
500	52.8	55.3
550	54.6	57.6
600	58.0	61.6
650	59.6	63.8
700	61.0	65.9
750	62.4	68.0
800	65.2	71.6

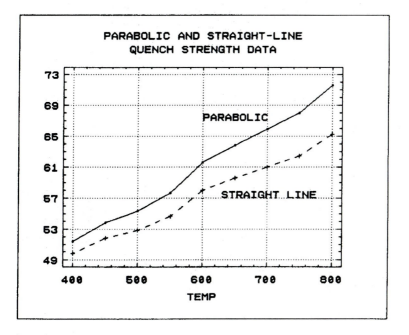

FIGURE 4.23
Heat treatment model with squared term present.

in this data is sufficient to hide any obvious curvature, most experimenters at this point would try to fit a straight-line model. That is, the *true* relation between strength and temperature is parabolic, but our raw-data graph does not make this apparent, so we try to fit an incorrect straight-line model. The resulting line (Fig. 4.24) has slope of 0.0503 and an intercept of 30.85, and an R^2 of 0.9933 and is visually a good fit to the data. Based on such results, most analysts would judge the curve fit successful and accept the straight-line model as useful for predicting the relation between strength and temperature.

We, of course, *know* that the data have a fairly large (see Fig. 4.23) component related to TEMP^2 since this is simulated data which we made up. Since the least-squares procedure not only failed to alert us to the presence of a second-order term but also gave wrong values for the slope and intercept of the first-order term, we might rightly be developing some concern about its utility. We need, however, to remember that all we *really* want, in practical terms, is a reliable way of predicting strength values from temperatures, and the line of Fig. 4.24 clearly does this. We are simply observing the well-known fact that *many* different functions (as long as we allow adjustable parameters) can be made to nicely fit any smooth data over some restricted range. Now, if we tried to *extrapolate* our relation below temperature 400 and/or above 800, the predictive quality of our model would deteriorate; but within the range of data on which the model was based, it really does give a useful relationship. Thus, when there is no underlying physical theory which says the relation should have a certain form, we are free to use whatever form we prefer (usually the simplest) so long as we get good fits.

Even though a straight-line fit of this data seems adequate, with the quick and easy computerized data analysis available today on personal computers, it is common to at least explore more complex models to see whether significant improvement can be made without excessive model complexity. For smooth data, as in our current example, added complexity in the model usually means adding higher-degree polynomial terms in TEMP. This simple strategy even has some theoretical foundation in that smooth functions can be fitted with Taylor-series approximations which exhibit increasing accuracy as we add the higher-degree polynomial terms. To carry out the necessary least-squares calculations, we need to use *multiple,* rather than *simple,* regression, even though we are still dealing only with a single independent variable. For the time being, however, let's just "blindly" use the available software to get some results and put off the background discussion until the next section, which will be devoted completely to multiple regression.

Once one has entered the raw data [n observations each of the independent (x) and dependent (y) variables], one can simply request models which include any number of terms in the polynomial $b_0 + b_1 x + b_2 x^2 + b_3 x^3 + \cdots$, and the desired least-squares fit will be calculated.

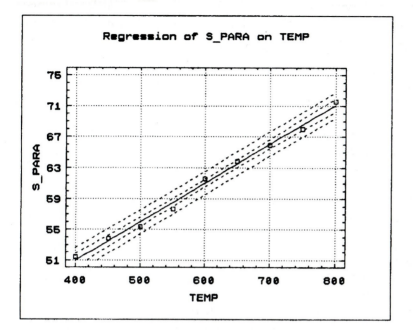

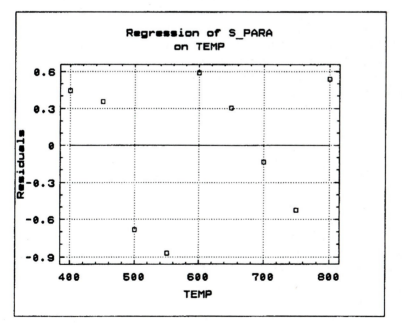

FIGURE 4.24
Straight-line fit to parabolic data.

Doing this for the same data as in Fig. 4.24 but requesting a second-degree polynomial, we get $R^2 = 0.9922$, the graph of Fig. 4.25, and the coefficients in the equations below.

True relation	Strength = 34.00 + 0.0400 TEMP + 0.00001 TEMP^2
First-degree model	Strength = 30.85 + 0.0503 TEMP
Second-degree model	Strength = 35.41 + 0.0343 TEMP + 0.000013 TEMP^2

From the graphs of predicted values and residuals, and from the numerical values of R^2, we can see that the quality of the fit is about the same for both the first-degree and second-degree models, but the second-degree model gives good estimates of the true coefficient values. If, as in the real-world situation, we didn't know that the relation was really second-degree, we probably would choose the simpler first-degree model *and there would really be nothing wrong with this choice*.

When random influences, like the randomized oil degradation effect present in the data we just analyzed, are quite small, then the least-squares fitting methods give much clearer guidance as to which terms in a model are really correct. Let's make up a data set in which the dependent variable is given exactly by the true relation above (no oil degradation, no random noise added). If we try to fit this second-degree data with a first-degree model, we get quite a good fit ($R^2 = 0.9995$), but now the residuals (see Fig. 4.26) show an obvious nonrandom pattern, which strongly suggests that the true relation is curved rather than straight. (We should always look for patterns of any kind in the residuals even though the patterns are sometimes obscured or completely hidden by large random effects.) Having spotted this pattern, we try the next more complex polynomial (second-degree) and get the perfect results of $R^2 = 1.0$ and coefficient values of 34.00, 0.0400, and 0.000010, demonstrating the accuracy of our computer software and the least-squares method for perfect data. Carrying this study one step further, we try to fit a third-degree polynomial to these same data. We again get perfect results for the first three coefficients and nearly zero ($-3.E-18$) for the coefficient of the cubic term, telling us clearly that this cubic term is *not* needed. Remember from our earlier example, however, that this indication of which terms are unnecessary is often partially or totally defeated by the presence of random effects in the data; thus we cannot always depend on it for help in choosing the proper model form.

We earlier gave a definition of the coefficient of determination, R^2. When we use available software to fit models which have more than 1 parameter that needs to be found (multiple regression), we sometimes encounter a quality-of-fit index called *adjusted* R^2. In the STATGRAPHICS software used for regression in this book, this measure is *preferred* to the "ordinary" R^2 which we defined earlier. Thus

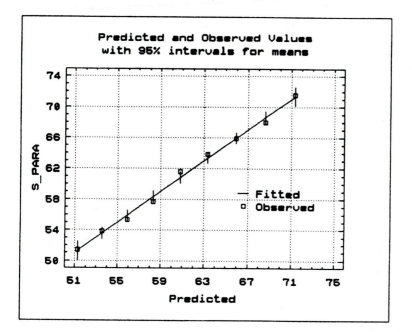

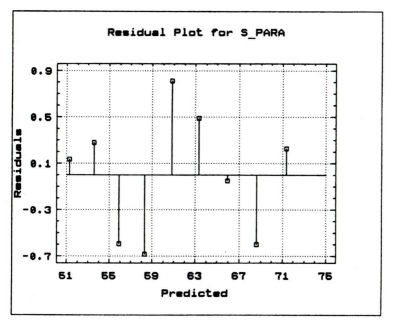

FIGURE 4.25
Parabolic fit to parabolic data.

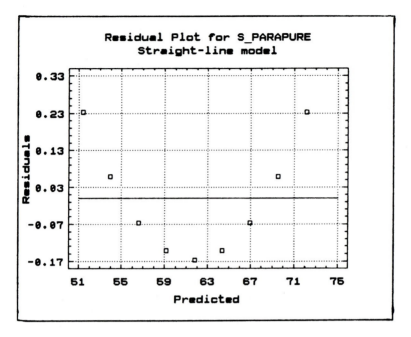

FIGURE 4.26
Residual plot shows model needs nonlinear terms.

any R^2 values quoted in this book for multiple regressions will be the adjusted values. We can get the "ordinary" values from STATGRAPHICS easily if we wish, but when the table which summarizes all the important regression results appears, it has the adjusted value in it, not the ordinary value. The authors of STATGRAPHICS prefer the adjusted values for the following reasons. Whenever we complicate our chosen model by adding more terms, each with its own unknown coefficient, the value of the ordinary R^2 will increase with each added term, implying that the model is "better." Actually, we should choose a more complex model *only* if the improvement in fit is sufficiently great to justify the added model complexity. The adjusted R^2 definition[1]

$$R^2_{\text{adj}} \overset{\Delta}{=} 1 - \frac{n-1}{n-p}(1-R^2)$$

where n is the number of observations (runs) in our experiment, p is the

---

[1] Neter, Wasserman, Kutner, *Applied Linear Statistical Models,* 2d ed., Irwin, Homewood, IL, 1985, p. 241.

number of model parameters to be found, and R^2 is the "ordinary" value, attempts to quantify this judgement with regard to the tradeoff between accuracy and complexity. It does this by reducing the numerical value of R^2_{adj} as p gets larger. Unless this reduction is overcome by a sufficiently large increase in the ordinary R^2 value, an increase in p will result in a *decrease* in R^2_{adj}, indicating that the more complex model is *not* better than the less complex one. To give an indication of the actual adjustment present in a typical example, let's take $R^2 = 0.90$ and $p = 6$. For $n = 10$, 20, and 30 we get, respectively, adjusted R^2 values of 0.775, 0.864, and 0.879. For a more complicated model ($p = 7$) these numbers become 0.700, 0.854, and 0.874.

Although the polynomial functions we have been using as examples are certainly reasonable models for much measured data and are widely used, the least-squares methods are easily applied to functions of many other forms, so long as the function is linear in the unknown coefficients. At least one[1] software product has automated the rapid comparison of alternative function forms for quality of fit. This program automatically fits and ranks (according to R^2 value) 221 different curves to the same set of x, y data. Since, as we have shown earlier, we should not rely completely on R^2, complete statistical data and graphs are also provided to aid in making a final judgment. A good discussion of the general problem of choosing the best functional forms is available in the literature.[2]

4.5.2 Models Involving Several Independent Variables (Multiple Regression)

When the dependent variable of interest is influenced by more than one independent variable and we want to study this relation, the statistical tool of multiple regression is often useful. The most common and mathematically tractable form of this method requires that the model assumed be *linear in the coefficients* that will be adjusted to minimize the sum of the squares of all the residuals. Mathematically, the relation must be expressible in the form

$$y = b_0 + b_1 f_1(x_1) + b_2 f_2(x_2) + b_3 f_3(x_3) + \cdots \qquad (4.12)$$

The functions $f_i(x_i)$ can be of any form; they need not be linear in x_i. The

[1] TableCurve, Jandel Scientific, 65 Koch Rd., Corte Madera, CA 94925, phone (800) 874-1888.

[2] C. Daniel and F. S. Wood, *Fitting Equations to Data*, Wiley, New York, 1980.

"best" values of the coefficients b will again be found by using a least-squares criterion. We first decide, as best we can, on the form of model [the $f_i(x_i)$ functions] and then gather data consisting of n observations of y and each x_i. Then we define the residual, at each observation, equal to the difference between the measured y and the model-predicted y. We add the squares of all the residuals, forming a measure, call it E, of the *total error*. To minimize E, we must choose the b's so as to make all the partial derivatives of E with respect to each b equal to zero, as taught in basic calculus courses. If there are r coefficients to be found, we need at least r observations to be able to find the b's. If there are exactly as many observations as there are b's, then least-squares methods are not needed and ordinary linear algebra methods for solving r equations in r unknowns can be used. Usually we have more observations than coefficients, and then least-squares methods[1] of linear algebra are needed. When we are using statistical software, such as STATGRAPHICS, all these details are "behind the scenes" and we need only to choose model forms, properly gather the experimental data, enter them into the program, and interpret the results. Many of the problems and methods we encountered while using simple regression in the previous section will be applicable to multiple regression also.

For those who are new to this subject, it is important at this point to discuss a little more fully the nature of the models allowed in the use of linear multiple regression. (When models linear in the coefficients are *not* used, various *nonlinear* regression methods are available. Their use is not as straightforward and, fortunately, not so often needed, but we will discuss them later, at least briefly.) Let's first point out that in Eq. (4.12) the x_i's could all be the *same* independent variable, say x_1, giving for example the case discussed in the previous section which used polynomials for the functions. In this section of course we are interested in relations involving *several* independent variables. For example,

$$y = b_1 x_a x_b + b_2 x_c^3 + b_3 \sin x_d \qquad (4.13)$$

Here the actual independent variables are x_a, x_b, x_c, and x_d. To make this fit the pattern of Eq. (4.12), we simply define $x_1 \stackrel{\Delta}{=} x_a x_b$, $x_2 \stackrel{\Delta}{=} x_c^3$, and $x_3 \stackrel{\Delta}{=} \sin x_d$. We can see from just this one example that we are allowed to use a very wide (but *not* unlimited) variety of functions in our models. Consider next the function

$$z = c_0 \frac{x_a^{c_1}}{x_b^{c_2}} \qquad (4.14)$$

---

[1] T. C. Hsia, *System Identification: Least-Squares Methods*, Lexington Books, D. C. Heath, Lexington, MA, 1977, p. 19.

Here the coefficients (the c's) do *not* enter the equation in a linear fashion, but a standard transformation (take the logarithms of both sides) gives us a usable form:

$$\log z = \log c_0 + c_1 \log x_a - c_2 \log x_b \qquad (4.15)$$

If we now define $y \triangleq \log z$, $b_0 \triangleq \log c_0$, $b_1 \triangleq c_1$, $b_2 \triangleq -c_2$, $x_1 \triangleq \log x_a$, and $x_2 \triangleq \log x_b$, we will be back to the form of Eq. (4.12). It is important to point out, however, that the least-squares method will now minimize the sum of the squared deviations for log z, *not* z itself. The set of b values found for this minimization will *not* be the same as that which would minimize the z deviations; but this does not mean that the procedure is foolish, since we will *check* the quality of the fit for z. If the fit is satisfactory, we don't really care that it was obtained unconventionally. If the fit is not acceptable, we have seen earlier that there can be many reasons for this, and we should try various fixes before blaming the trouble on the nonrigorous least-squares method. If necessary, we can finally go to one of the nonlinear regression techniques, which deal directly with the original model rather than transforming it, but have some undesirable features, such as requiring you to supply initial guesses for all the coefficient values. They also are iterative methods, which means that they may or may not converge to a valid (or *any*) answer and are computationally expensive, whereas the linear methods are one-step and quick and always give the same single answer.

As a simple first example, we return to the data of Fig. 4.17 where we studied the effect of quench temperature on the strength of a metal. In that earlier study we considered the oil degradation effect as a minor, but disturbing, influence on the relation of prime interest—that between temperature and strength. Recall that we employed the technique of run-sequence randomization to reduce the effect of oil degradation on the accuracy of modeling the strength-temperature relation. Suppose we now change our focus and decide that we want to study the nature of the oil degradation effect rather than merely trying to cancel its disturbing effects. That is, we want to find a model which quantitatively relates strength to *both* temperature and run number. We again will take advantage of the simulation approach and use Eq. (4.10)—the true relation we are trying to discover—as an aid to learning about multiple regression.

Because our purpose now is to study (not cancel) the oil degradation effect, it would seem unnecessary to use run-sequence randomization in gathering the experimental data. Since Fig. 4.17 has both kinds of data (s_{mono} and s_{rand}), let's try both to find out the effect. When I asked STATGRAPHICS to fit the s_{mono} data with a model of the same form as Eq. (4.10), it refused and gave this error message: "Variable $N - 1$ is a linear combination of other independent variables." When I did the same with the s_{rand} data, the routine ran normally and returned the perfect results $b_0 = 34.00$, $b_1 = 0.0400$, $b_2 = -0.2000$, and $R^2 = 1.00$; so it appears that run-sequence randomization is not only desirable but also necessary here. The difficulty with the s_{mono} data is revealed in the STATGRAPHICS error message (if you have some background in linear algebra); let me give a graphical interpretation for those who lack this background. Equation (4.10) is, of course, the equation of a plane, and this is graphed (using the b values just given above) in Fig. 4.27a. When the dependent variable is a function of only two independent variables, we get the simplest example of a useful general tool called the *response surface*. (This can be extended to examples with any number of independent variables.[1]) When we graph the s_{mono} and s_{rand} data on the same axes, we get Fig. 4.27b and c, respectively. The least-squares routine has the task of fitting a plane to these data. When the data all fall along a single line in space, as in Fig. 4.27b, there are an *infinite number* of planes which all contain this line; thus the routine is unable to find a unique answer and returns the error message. The observations in Fig. 4.27c "exercise" the model in a more general way and thus define a *single* plane which minimizes the sum of the squared residuals. This sort of problem is not limited to our specific example, so we should be generally on the lookout for it. Note that the two independent variables are not *caused* by each other; they are just *associated*. That is, for the unrandomized sequence, changes in TEMP and $N - 1$ both follow a straight-line pattern; thus they are proportional (linearly dependent) although *not* physically (causally) related.

Perfect data such as the s_{rand} data just analyzed is unrealistic, so we next contaminate it with the gaussian noise s_{err} of Fig. 4.17, giving $R^2 = 0.97$ and the following detailed results:

Independent variable		Coefficient and its 95% confidence interval		t-value	Significance level
Constant	b_0	31.04	(26.93 to 35.14)	18.5	0.000
TEMP	b_1	0.04493	(0.038 to 0.052)	15.7	0.000
$N-1$	b_2	-0.25857	(-0.61 to 0.092)	-1.8	0.121

As usual, random noise in the data prevents us from discovering the correct model parameters perfectly. We have not previously shown or

---

[1] G. E. P. Box and N. R. Draper, *Empirical Model Building and Response Surfaces*, Wiley, New York, 1987.

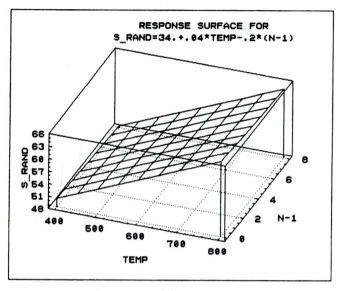

(*a*)

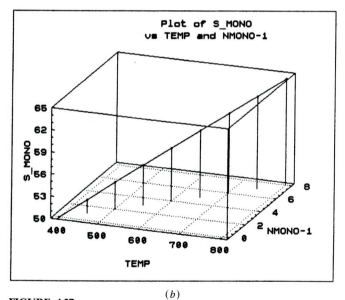

(*b*)

FIGURE 4.27
Graphical interpretation of linear dependence.

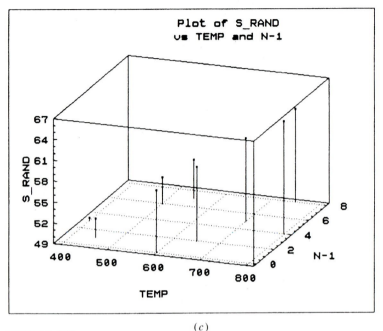

(c)

FIGURE 4.27
(Continued).

discussed the t values and significance levels, so let's do that now. Each coefficient has its own standard deviation (not shown in the preceding table); the corresponding t value is found by dividing the coefficient by its standard deviation. The significance level is obtained from a table of the t distribution. (This is all "behind the scenes" in the computer software.) An independent variable is considered statistically significant if its significance level is sufficiently *small*. The interpretation goes as follows. If the coefficient were truly zero (not really present in the true relation), what would be the probability of getting a computed coefficient value as large (in *absolute* value) as that actually found? This probability is the number found in the significance-level column of our table. Thus if b_2 were really zero, the probability of getting a value as large as 0.2587 would be 0.121 (12.1 percent). This is a roundabout way of saying that b_2 is probably not zero. The zero significance levels for the other two coefficients are really small *nonzero* numbers (<0.001) and indicate that these two coefficients are more significant than b_2.

What are these significance levels good for? We need to keep recalling that our real goal is to decide whether to keep an independent variable in our model or discard it as not useful in improving the fit of the model to the data. The significance levels just discussed do not alone allow us to make such decisions. We need to consider *all* the available evidence, which includes R^2, residuals (individually and summed), and a variety of graphical displays. Figure 4.28a shows observed and model-predicted values together with 95 percent prediction intervals for individual observations. [Remember that 95 percent intervals for *mean* values (which are also available on request from the software) would be narrower.] Three-dimensional displays as in Fig. 4.28b are also possible but become unavailable as soon as we have more than two independent variables. In comparing residual plots (like Fig. 4.28c) for competitive models, we want a random appearance, and the better model has the smaller residuals. Sometimes an accurate fit is more important (for some practical reason) in one region of the independent variables than in another. Then we might prefer a model which had small residuals there, even if the fit was worse elsewhere. Note that relying solely on R^2 in such a case is not wise since it is a global rather than local criterion.

Figure 4.29 shows another display of the residuals that is sometimes useful, a normal (gaussian) probability plot. In a good model the residuals should be random effects rather than show any deterministic trends or biases. While randomness does *not* imply any particular distribution function, gaussian tendencies *are* quite common in real-world data, so the software is set up to check the residuals for fit to the gaussian function. In comparing two competitive models that were otherwise of similar predictive quality, the one conforming more closely to gaussian residuals would often be preferred. Our data was intentionally contaminated with gaussian noise, so the good fit shown in Fig. 4.29 should not be surprising.

The *component-effects* plots of Fig. 4.30 are our final examples of graphical displays useful in evaluating and comparing models. These graphs can be used for models with any number of independent variables. The graphs are constructed by first computing the average value of the independent variable under consideration; let's use $N - 1$ for our explanation (its average value is 4.0). We subtract this average value from each individual value, multiply this difference by the b value for that independent variable (-0.259 for $N - 1$), and plot (against the independent variable) the straight line given by this result. Having plotted this straight line, we then plot the residual at each value of the independent variable, *relative* to this line. The purpose is to show the contribution of the individual independent variables to the total value of the dependent variable, as in Eq. (4.12), except that we subtract the *average* contribution (giving the positive and negative values of Fig. 4.30) and we superimpose the residuals to show whether the contribution is really meaningful compared with the random variation. Thus Fig. 4.30a shows

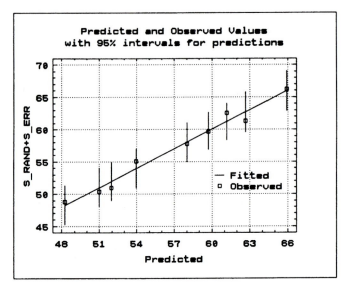

(a)

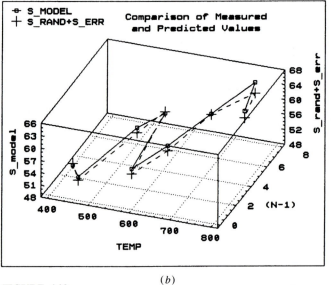

(b)

FIGURE 4.28
Various graphs helpful in evaluating the quality of the model.

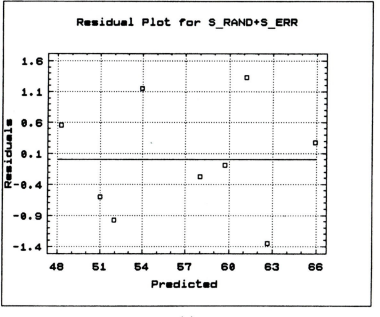

(c)

FIGURE 4.28
(Continued).

clearly that TEMP has a major effect on strength while the oil degradation term $N - 1$ is almost lost in the random noise in Fig. 4.30b. If we didn't know that there really *is* an oil degradation effect and if we had to rely entirely on our analysis of the measured data, we might choose to ignore it when the indications are this marginal.

As we might expect (and as we have seen in earlier examples), our fitted models get closer to the true relation when random noise is less. When we contaminate s_{rand} with $0.1s_{err}$ (rather than $1.0s_{err}$), we get $R^2 = 0.9996$ and the detailed results below:

Independent variable		Coefficient and its 95% confidence interval		t-value	Significance level
Constant	b_0	33.70	(33.29 to 34.11)	201.1	0.000
TEMP	b_1	0.04049	(0.0398 to 0.0412)	141.5	0.000
$N - 1$	b_2	−0.206	(−0.171 to −0.241)	−14.4	0.000

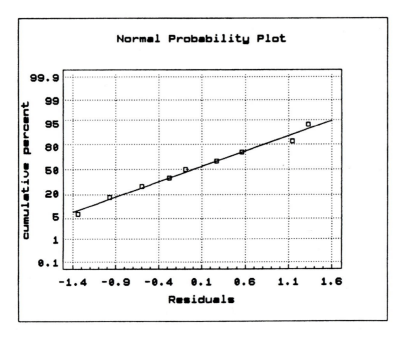

FIGURE 4.29
Graph to check whether residuals are close to gaussian.

We see that the model parameters are now closer to the true values, and the $N - 1$ term (oil degradation effect) is clearly useful in the model. To explore the effect of including "wrong" terms in the model, I included a term in the square of temperature, getting $R^2 = 0.9995$ and the table below:

Inde-pendent variable		Coefficient and its 95% confidence interval		t-value	Signif-icance level
Constant	b_0	33.62	(31.24 to 36.00)	36.4	0.0000
TEMP	b_1	0.04078	(0.0326 to 0.0490)	12.8	0.0001
$N - 1$	b_2	−0.206	(−0.165 to −0.246)	−13.1	0.0000
$TEMP^2$	b_3	−2.4E−7	(−1.E−5 to 1.E−5)	−0.09	0.9303

The term in $TEMP^2$ is clearly to be rejected, based on both its contribution to the dependent variable and its poor significance level.

To again illustrate the difficulty caused by excessive random variation in data, let's analyze s_{rand} contaminated with $2s_{err}$. This gives $R^2 = 0.887$ and the table on page 224.

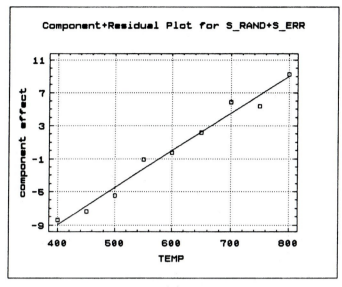

(*a*)

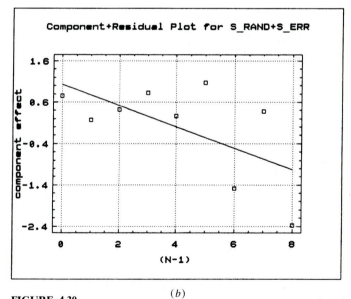

(*b*)

FIGURE 4.30
Component-effects plots.

Independent variable		Coefficient and its 95% confidence interval		t-value	Significance level
Constant	b_0	26.4	(−21.1 to 73.8)	1.43	0.213
TEMP	b_1	0.056	(−0.108 to 0.22)	0.88	0.421
$N-1$	b_2	−0.316	(−1.1 to 0.49)	−1.00	0.360
TEMP^2	b_3	−4.8E−6	(−1.4E−4 to 1.3E−4)	−0.09	0.930

While the coefficients are now both inaccurate (we know this *only* because this is simulated data) and uncertain, the fitted model might still be of some practical use, as we can see from Fig. 4.31*a*. Deleting the TEMP^2 term (see Fig. 4.31*b*) should (and does) have little effect. When there is this much random variation, deleting also the $N-1$ term (Fig. 4.31*c*) is barely noticeable graphically since its true contribution is small compared with that of TEMP. In a real-world experiment (where of course we *don't* know the true relation), if we relied entirely on graphs like Fig. 4.31, we might in this case have chosen to accept the simplest model, which includes only a constant and the effect of TEMP. Considering in addition the tabulated significance levels, we see that there is a good chance that oil degradation *is* of some importance and, thus we would include it in our model.

4.5.3 Factorial Experiment Plans and Orthogonal Arrays

Whether we use multiple regression, analysis of variance (ANOVA), or Taguchi methods when studying relations among variables, we need to plan our experimentation to make it most effective and efficient. In Sec. 4.5.2 we introduced multiple regression concepts and did enough numerical examples to give you some concrete experience. In terms of planning, we saw how run-sequence randomization could be used. At this point we have only mentioned ANOVA and Taguchi methods; they will be discussed in more detail later. To get ready for these topics and to also develop multiple regression in more detail, we need first to familiarize you with the concepts of *factorial experiments* and *orthogonal arrays*.

In statistics, the word *factor* is used synonymously with the term *independent variable*; thus a four-factor experiment has four independent variables. To exercise our experimental apparatus, we need to set the independent variables at different values; these are called *levels*. Levels can be numerical (quantitative) values (set the temperature at 400, 500, and 600°C), or they can be categorical (qualitative) values, such as machining the test parts by using three different cutting fluids, brand A, brand B, and brand C. When we choose the number of factors and the

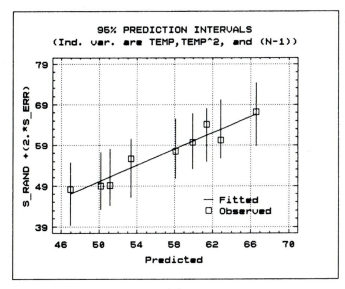

(*a*)

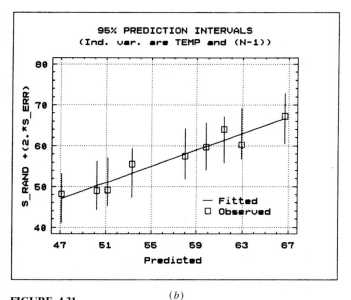

(*b*)

FIGURE 4.31
Data with large random noise.

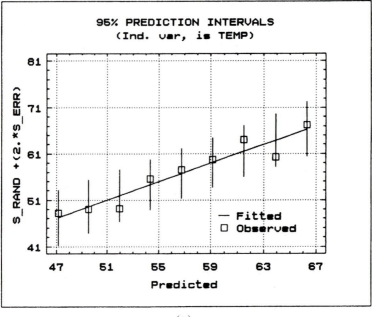

FIGURE 4.31
(Continued).

levels for each, we are starting to define the scope of our experiment; but the choice of the *combinations* of levels to be studied is also a critical decision. *Full-factorial* experiments study *all* possible combinations of the chosen levels and quickly become prohibitively expensive as the number of factors and/or levels increase. *Fractional-factorial* experiments attempt to deal with this problem by carefully selecting a subset of all the possible combinations so as to get the most useful information with the least expenditure of time and effort. Note that a four-factor experiment with three, four, two, and three levels for the factors gives $(3)(4)(2)(3) = 72$ runs for a full-factorial study. Since the effect of a factor can only be felt if it *does* change, the minimum number of levels is two. For preliminary (screening) experiments, where we are unsure which independent variables really matter, there are often a large number (10 or 20) of potential factors and we choose to exercise each at only two or three levels. In terms of our initial experience with multiple regression, two-level experiments would let us estimate straight-line relations but *not* those involving curvature (two points determine a straight line). To detect the presence of curvature, at least three levels are needed.

When a large number of factors are involved, even the minimum of two levels leads to huge full-factorial experiments. In a study of excessive shrinkage in extruded thermoplastic casings for automobile speedometer cables,[1] 15 factors in the manufacturing process were initially identified for study, giving a full-factorial experiment with 32,768 runs! A fractional-factorial design (based on a Taguchi orthogonal array) required only 16 combinations to be run. The manufacturing process was such that a minimum of 300 ft of casing had to be produced at each of the 16 conditions, so a total of 48,000 ft was committed to the experiment. Each 3000-ft length was randomly sampled at four locations and the shrinkage measured. Analysis of the data showed that 70 percent of the shrinkage effect was caused by 2 of the 15 variables. Furthermore, the model allowed prediction of an *optimum* combination of factors which, when implemented, reduced the shrinkage in the actual manufacturing process by a 5-to-1 ratio. Titles of some other studies from the referenced source will give you some idea of the types of problems being solved with such methods:

Power Steering Hose Quick Connect Study
Theoretical Engine Design
Fork Lift Truck Vibration
Foam System Optimization
Printed Circuit Assemblies Study
Heat Exchanger Product Design
Gold Plating of Pin and Socket Contacts

In each case, a model relating some aspect of product quality to various process parameters was found and then used to adjust the operating point of the process for optimum results.

As you might guess, highly simplified fractional-factorial experiments are only capable of producing simple models; however, we have documentation of literally thousands of practical applications where these simple models were adequate to enable engineers to significantly improve quality and/or cost. When we refer to these as simple models, we usually mean that they include only the *main effects* and not the *interactions*. If we think in terms of multiple regression and a simple three-factor model of the form

$$y = b_0 + b_1 x_1 + b_2 x_2 + b_3 x_3$$
$$+ b_4(x_1 x_2) + b_5(x_2 x_3) + b_6(x_1 x_3) + b_7(x_1 x_2 x_3) \qquad (4.16)$$

the *main effect* of a factor x_1 would be the model term $b_1 x_1$, while a term

---

[1] J. Quinlan, *Product Improvement by Application of Taguchi Methods,* American Supplier Institute, 6 Parklane Blvd., Suite 411, Dearborn, MI, 1985, (313) 336-8877.

$b_4x_1x_2$ would be called an *interaction* between x_1 and x_2. That is, interaction refers to a situation where the effect on the dependent variable of changes in one factor depends on the value of some other factor. Mechanical engineer M. S. Phadke,[1] one of the expert industrial practitioners of the Taguchi approach, says "Fortunately, in most practical situations the additive model (no interactions) provides an excellent approximation."[2]

If we now concentrate on models which include only the main effects, then the concept of *orthogonal arrays* provides a powerful tool for frugally exploring relations that involve many factors. When we first begin to plan an experiment and have decided, at least tentatively, on a list of factors (independent variables) to explore, we use our knowledge of the physical process to also decide on the *range* of values of interest to us for each factor. Figure 4.32 shows one way of displaying this sort of data, for a five-factor experiment where all the factors are quantitative. Such a bar-graph display is useful for summarizing the *scope* of the experiment in compact form. A critical next step is to decide on the list of factor combinations, called *runs* or *trials,* that will actually be exercised. Figure 4.33 shows a possible choice for the data of Fig. 4.32. How do we choose these combinations to realize the most efficient experimental plan? When interactions are small relative to main effects and we are content to estimate *only* main effects, the use of orthogonal arrays gives experimental plans that produce maximum results with minimum effort. (When, initially or after some preliminary experimentation, we suspect the presence of significant interaction, more complex plans are available.[3,4])

To explain orthogonal arrays, let's use the three-factor model of Eq. (4.16), with only two levels of each factor, as a simple example which allows some convenient graphical interpretations. Since there are eight unknown coefficients, an experiment with eight runs would give us eight linear algebraic equations in eight unknowns, allowing solution (without the use of least-squares methods) for the model coefficients. Note that a three-factor, full-factorial experiment at two levels for each factor would give $(2)(2)(2) = 8$ runs, and the combinations to be used would also be explicitly fixed. If the physical relation does *not* involve any interactions,

---

[1] M. S. Phadke, *Quality Engineering Using Robust Design,* Prentice-Hall, Englewood Cliffs, NJ, 1989.

[2] Ibid., p. 50.

[3] Ibid., p. 174.

[4] K. R. Bhote, *World Class Quality: Using Design of Experiments to Make It Happen,* Amacon, New York, 1991.

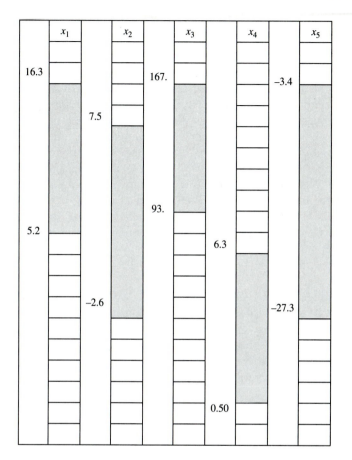

FIGURE 4.32
Display showing ranges of independent variables.

Run	x_1	x_2	x_3	x_4	x_5
A	15.4	6.4	129.	0.54	−12.3
B	11.6	−2.2	165.	1.45	−3.2
C	9.8	1.5	98.	6.30	−26.5
D	7.7	4.6	114.	3.45	−7.7
E	5.5	0.5	148.	4.66	−19.5

FIGURE 4.33
One possible choice of factor combinations.

then Eq. (4.16) doesn't need the terms involving b_4, b_5, b_6, and b_7, the simplified model needs only four terms, and the experiment to find them needs only four runs. Since eight combinations are still possible, we now need to choose four of these to actually run our experiment. At this point it is profitable to start thinking graphically. To facilitate this, we note that in a two-level experiment, we can call the levels $+1$ and -1 without really restricting the generality of our results. We want to think of our factors as having an average value, about which their variation takes place, and we want to use the number 0 for the average, the number $+1$ to denote an increase in the factor, and the number -1 to denote a decrease in the factor. This sort of "coding" is regularly used in statistical analysis, and we can demonstrate its applicability by considering x_3 of Fig. 4.32 and taking the two actual levels as 167 and 93, with whatever units are proper for that physical quantity. By simply *defining* for x_3 a new scale which has its zero at 130, and units 37 times the original units, our "new" x_3 now has levels -1 and $+1$. Such a conversion can clearly be applied to any of our factors.

When an experiment has only three factors, we can graphically display the "sample space" of all three with a set of rectangular coordinates, as in Fig. 4.34. Any combination of factor values plots as a single point in this sample space. (Note that we *cannot* display the dependent variable y when there are more than two factors.) For a full-factorial experiment, the eight runs mentioned earlier are represented by the eight corners of the cube. If we assume no interactions, then we need only four runs, but we have not yet chosen *which* four of

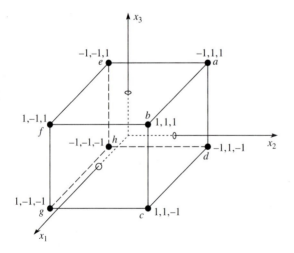

FIGURE 4.34
Sample space for a three-factor experiment.

the eight available. Since this sort of question was answered long ago and can be found in many books, one approach would be to simply look it up and get on with the experiment. Phadke's book[1] catalogs 33 pages of orthogonal arrays which allow one to design two-, three-, and four-level experiments for as many as 63 factors. For our present problem, his L_4 array suggests we use points a, c, f, and h in Fig. 4.34. We want to now give some insight into *why* these combinations are the "best", and this will also bring out the significance of the word *orthogonal*.

Let's again simulate an experiment where we know the correct answer in advance. Suppose the true relation among the variables is given by

$$y_{int} = 1.0 + 2.0x_1 - 3.0x_2 + 0.5x_3$$
$$- 0.2x_1x_2 - 1.0x_2x_3 + 2.0x_1x_3 + 0.1x_1x_2x_3 \qquad (4.17)$$

where both main effects and fairly strong interactions are present. Figure 4.35 shows values computed from this formula for all eight runs of a full-factorial experiment. To see how strong the interactions are, I also show y_{noint}, which is computed from Eq. (4.17) with all four interaction terms omitted. We now want to explain why points a, c, f, and h should be used rather than some other combination, and we also show a simple calculation method for finding the coefficients without using either linear

Cube corner	x_1	x_2	x_3	y_{int}	y_{noint}
a	−1	1	1	−6.4	−3.5
b	1	1	1	1.4	0.5
c	1	1	−1	−1.8	−0.5
d	−1	1	−1	−1.2	−4.5
e	−1	−1	1	1.4	2.5
f	1	−1	1	9.6	6.5
g	1	−1	−1	2.8	5.5
h	−1	−1	−1	2.2	1.5

FIGURE 4.35
Data set for factorial experiment.

---

[1] Phadke, pp. 286–319.

equation solvers or least-squares routines. We first give the calculation method and then show how it works. Since this method *assumes* there are no interactions, we will get correct results when we use the y_{noint} data in Fig. 4.35 but wrong results when y_{int} is used.

$$b_0 = \frac{y_a + y_c + y_f + y_h}{4} \tag{4.18}$$

$$b_1 = \frac{(y_f + y_c) - (y_a + y_h)}{4} \tag{4.19}$$

$$b_2 = \frac{(y_a + y_c) - (y_f + y_h)}{4} \tag{4.20}$$

$$b_3 = \frac{(y_a + y_f) - (y_c + y_h)}{4} \tag{4.21}$$

At this point you should do these calculations to see that they produce exactly the correct coefficients in the model without interactions. If we ran our multiple regression software on this set of data, it would also compute the correct coefficients (four equations in four unknowns). To see why our simple calculation method works, we note that for each factor there is a run which contributes a certain increasing effect on y and another run which contributes exactly the *opposite* (negative) effect. Thus if we add the y's from two such runs, the contributions to y from that factor will exactly *cancel*. The constant term b_0 of course does *not* have such an effect; thus in Eq. (4.18) a simple averaging of the four readings gives us the b_0 coefficient. In Eq. (4.19) the x_2 and x_3 effects in $y_f + y_c$ cancel since each contributes *both* a positive and an equal negative term. This same thing occurs in $y_a + y_h$. The effect of x_1, however, is doubled (it appears twice) in each of the expressions in parentheses, so we need to divide by 4 to get b_1. Similar reasoning can be applied to Eqs. (4.20) and (4.21). The term *orthogonal* is applied to experiment plans such as this in which the combinations are chosen so that when we try to calculate the individual effect of one factor, we find that all the *other* factor effects are balanced (canceled). Once you see the idea behind this scheme, it is fairly easy to invent such orthogonal experiment plans as long as there are not too many factors and/or levels. For complicated experiments most engineering analysts would simply look up an applicable array in a reference text, such as Phadke, mentioned above.

When strong interactions *are* present and we assume a model which has none, we are still able to compute our model's coefficients, but the model will *not* be able to accurately predict the dependent variable for factor combinations other than the ones used to define the model. In our present example, we can compute a main-effects-only model from the y_{int} data (runs *a, c, f,* and *h*) in two ways. We could use multiple regression software (reverts to four equations in four unknowns and least-squares analysis not needed), or we could use the scheme of Eqs. (4.18) to (4.21). Using multiple regression, we get the model

$$y = 0.9 + 3.0x_1 - 5.0x_2 + 0.7x_3 \qquad (4.22)$$

You should evaluate this model for all eight runs of Fig. 4.35 to see that it gives perfect results for a, c, f, and h but very inaccurate predictions for all the other combinations and is thus *not* a valid model. Using Eqs. (4.18) to (4.21), we get exactly the same results (try it). Real-world relations among dependent and independent variables do not, of course, fall into neat "black-or-white" categories of "no interactions" and "strong interactions." In Eq. (4.17), if we choose the coefficients of the interaction terms to be small compared to those on the main effects, then our simplified modeling scheme *does* produce a model useful over the entire range of factor levels and combinations. We have said before, and we now repeat, that thousands of practical applications have shown that many industrial processes *can* be usefully modeled without considering interactions. So this simple approach should not be rejected until it is proved inadequate in a specific case.

When the number of factors is small, then, of course, the added expense of testing for interactions may not be prohibitive and would allow us to perhaps *prove* that interactions were negligible rather than assuming and hoping that they were. In our present example we would need eight runs, rather than four, to do this. Using multiple regression on the y_{int} data of Fig. 4.35, we get (as you might guess) a perfect calculation of the eight coefficients. We could also, as we have in earlier examples, contaminate our data with random noise, only to find the usual result that small amounts of noise cause little trouble but too much noise can prevent us from finding a useful model. There is also a computing scheme analogous to Eqs. (4.18) to (4.21) available.

$$b_0 = \frac{y_a + y_b + y_c + y_d + y_e + y_f + y_g + y_h}{8} \qquad (4.23)$$

$$b_1 = \frac{(y_b + y_c + y_f + y_g) - (y_a + y_d + y_e + y_h)}{8} \qquad (4.24)$$

$$b_4 = \frac{[(y_b + y_c) - (y_a + y_d)] - [(y_f + y_g) - (y_e + y_h)]}{8} \qquad (4.25)$$

$$b_7 = \frac{[(y_b + y_e) - (y_a + y_f)] - [(y_c + y_h) - (y_d + y_g)]}{8} \qquad (4.26)$$

I have not given formulas for b_2 and b_3. You should derive these from Fig. 4.34, our earlier explanation, and the pattern apparent in Eq. (4.24). Similarly derive formulas for b_5 and b_6, by using b_4 as a "pattern." Note that in these interactions between two factors, say, x_1 and x_2, we could derive *two* formulas for b_4—one which considered the interaction as the effect of x_1 on x_2 and another which considered it as the effect of x_2 on x_1. These two formulas should be (and *are*) equivalent and give the same number for b_4. The "triple" interaction term b_7 can be found in three ways, all equivalent. In Eq. (4.26) I expressed it as the effect on the product x_1x_2 of changes in x_3. The expression in brackets at the left sums (the minus sign is needed since the product x_1x_2 is negative at a and f) the contributions to this product for x_3 equal to $+1$, while that at the right gives this sum for x_3 equal to -1. To check the correctness of your derivations and/or the given results, compute numerical values for all the model coefficients. They should agree perfectly with those in Eq. (4.17).

Having gone through some simple examples, we now want to give some more general information on orthogonal arrays.[1] In Fig. 4.34 we used runs a, c, f, and h as our orthogonal array. Actually, some other combinations would be equally correct and useful. Note that a and c lie on one face of the cube at the ends of a diagonal while f and h are found on a parallel face and an "opposite" diagonal. Since the x's all enter the expression for y in a symmetric manner, we can define our four runs from opposite diagonals on *any* two parallel cube faces. That is, b, e and g, d are just as good as the runs we used earlier, as are a, h and f, c, etc. Simple computing schemes can be derived for any of these cases by using the same approach that we used on the $(a, c)(f, h)$ combination. If we think algebraically rather than graphically, *any* four points on or within the cube would give four equations in four unknowns, but here we would rely on routines of linear algebra, rather than our intuitive graphical thinking, to get a solution.

When we need to consider experiments with more factors and/or levels, we require a more specific definition of what makes an array of experiment runs an orthogonal array. In general, to be considered orthogonal, an array must be such that when we look at any pair of columns, every combination of levels appears the same number of times for each pair of columns. When this is true, the averaged effect of any one factor can be determined while the levels of all other factors are varying. Figure 4.36 shows an orthogonal array which allows study of seven factors at two levels each with only eight experimental runs. Check it to see that it satisfies the definition just given. It would be good practice here, as in any sequence of experimental runs, to randomize the order in which the runs are actually made, rather than following the A, B, C, ...

[1] N. Logothetis and H. P. Wynn, *Quality Through Design: Experimental Design, Off-Line Quality Control, and Taguchi's Contributions,* Clarendon Press, Oxford, 1989, p. 93.

Run	x_1	x_2	x_3	x_4	x_5	x_6	x_7
A	-1	-1	-1	-1	-1	-1	-1
B	-1	-1	-1	+1	+1	+1	+1
C	-1	+1	+1	-1	-1	+1	+1
D	-1	+1	+1	+1	+1	-1	-1
E	+1	-1	+1	-1	+1	-1	+1
F	+1	-1	+1	+1	-1	+1	-1
G	+1	+1	-1	-1	+1	+1	-1
H	+1	+1	-1	+1	-1	-1	+1

FIGURE 4.36
Orthogonal array for seven 2-level factors.

order shown in the table. This can be done as we showed earlier (Fig. 4.16). Orthogonal arrays, such as this one, can be used to plan experiments with *fewer* independent variables than shown in the table, but we must still use as many rows, to preserve the orthogonality. That is, following Fig. 4.36, we could plan experiments with eight runs, but *any* number of factors from two to seven. Experimental plans for less than seven factors, however, would not be the most frugal possible, since we can find other arrays which require only one more run than we have factors (as we saw in the four-run, three-factor example of Fig. 4.34). When runs exceed factors by more than 1, we can compute some interaction effects in addition to the main effects. Both main and interaction effects can again be computed in two ways: by using ready-made multiple regression software or "intuitive" schemes like Eqs. (4.18) to (4.21) and (4.23) to (4.26). When there are more than two or three factors, the intuitive schemes (especially for the interactions) are less obvious; so from here on we will rely on the computerized least-squares methods.

Next we want to do an example using three-level factors. Such experiments require more runs to study a given number of factors than two-level schemes would, but such experiments have the advantage of being able to detect curvature in the relation between dependent and independent variables. That is, we can now compute three numerical values for the effect of a factor, and these three points need *not* fall on a straight line when they are plotted against the independent variable. We have, then, a crude but useful way of detecting peaks and valleys in the relation. A peak may represent a process operating point where some important quality parameter achieves a maximum value, so we want to identify it if it is present and then try to operate the process in this region. A valley might show an operating region of minimum cost, which again is of great practical interest. Two-level experiments are incapable of detecting such features. Note that if our three selected levels are too close together, the effect of curvature will be slight and perhaps unnoticeable. If we spread the levels too far apart, we may miss localized peaks entirely. Since at the beginning of an experiment we usually don't even know whether much curvature is present, choosing the spread between levels may involve some trial and error.

Figure 4.37 shows an 18-run orthogonal array useful for up to eight factors, seven of which are at three levels and one of which is at two

Run	y_1	y_n	x_1	x_2	x_3	x_4	x_5	x_6	x_7	x_8
1	3.5	-0.65	-1	-1	-1	-1	-1	-1	-1	-1
2	-2.0	-0.07	-1	-1	0	0	0	0	0	0
3	-3.5	-0.42	-1	-1	+1	+1	+1	+1	+1	+1
4	-0.5	-0.99	-1	0	-1	-1	0	0	+1	+1
5	8.0	0.97	-1	0	0	0	+1	+1	-1	-1
6	-0.5	1.13	-1	0	+1	+1	-1	-1	0	0
7	-1.5	-0.54	-1	+1	-1	0	-1	+1	0	+1
8	0.0	-0.48	-1	+1	0	+1	0	-1	+1	-1
9	-0.5	-0.16	-1	+1	+1	-1	+1	0	-1	0
10	2.5	-0.52	+1	-1	-1	+1	+1	0	0	-1
11	10.0	0.02	+1	-1	0	-1	-1	+1	+1	0
12	-2.5	0.28	+1	-1	+1	0	0	-1	-1	+1
13	1.5	-0.66	+1	0	-1	0	+1	-1	+1	0
14	4.0	1.29	+1	0	0	+1	-1	0	-1	+1
15	13.5	-0.35	+1	0	+1	-1	0	+1	0	-1
16	5.5	0.44	+1	+1	-1	+1	0	+1	-1	0
17	-6.0	-0.17	+1	+1	0	-1	+1	-1	0	+1
18	10.5	0.39	+1	+1	+1	0	-1	0	+1	-1

FIGURE 4.37
Orthogonal array for seven 3-level factors and one 2-level factor.

levels. A variety of such mixed-level arrays are available in the literature.[1] We again simulate a real-world experiment by assuming we know the relation among the variables to be given by

$$y = 3.0 + 2.0x_1 - 3.0x_2^2 + 0.5x_3 - 1.0x_4 - 2.0x_5 + 3.0x_6 + 2.0x_7^2 - 4.0x_8$$

$$(4.27)$$

Instead of using the actual numerical values for the various factor levels, we code them as −1 (the low level), 0 (the average level), and +1 (the high level). For the two-level factor x_1 of course we have only the −1 and +1 levels. Using the 18 runs shown in Fig. 4.37, we can easily calculate the corresponding y values, which we then treat as measured values to be fed into the multiple regression routine to see if it can extract the coefficients. I also include a column (y_n) of gaussian random noise (mean value 0, standard deviation 1.0) which we can use to contaminate y to make the simulation more realistic. Using STATGRAPHICS multiple regression, I first analyzed the perfect y data, using a model of exactly the correct form and all 18 experimental runs. As usual, the software produced perfect results, extracting exactly the correct coefficient values of Eq. (4.27). Knowledge of the correct form of model is, of course, unrealistic, so we need to make our simulation more true to life. When we don't know exactly what form of model to use, we often start with only terms that are linear in the independent variables. This is not unreasonable since any smooth nonlinear function *can* be approximated, for small changes near a selected operating point, by a linear function given by a Taylor-series expansion. Submitting this model to multiple regression produced $R^2 = 0.80$ and the following results.

Independent variable	Coefficient	Standard deviation	Significance level
Constant	2.33	0.57	0.003
x_1	2.00	0.57	0.006
x_2	0.00	0.69	1.000
x_3	0.50	0.69	0.490
x_4	−1.00	0.69	0.183
x_5	−2.00	0.69	0.018
x_6	3.00	0.69	0.002
x_7	1.30E−16	0.69	1.000
x_8	−4.00	0.69	0.000

[1] Phadke, op. cit., p. 77.

The near-zero coefficients and the significance levels near 1.0 (recall that a large significance level means a *small* significance) for the x_2 and x_7 terms tell us that linear terms in these factors have little effect on y. Note that a conclusion that these two factors have no effect and thus should not have been included in the initial choice of possible factors would lead us astray in this case. They *do* have an effect, but it is not a linear one. Since in real-world experiments we wouldn't have known ahead of time that we should try squared (rather than linear) terms, this example shows us how careful we need to be in applying multiple regression to situations where we have no "physics" to guide our selection of independent variables.

Since the calculations just made used the perfect y data, we need to see now the effect of including some noise. Repeating the multiple regression using all linear terms, but now with $y + y_n$ as the "measured" dependent variable, gave $R^2 = 0.76$ and this table:

Inde-pendent variable	Coefficient	Standard deviation	Signif-icance level
Constant	2.30	0.63	0.006
x_1	2.11	0.63	0.009
x_2	0.07	0.78	0.929
x_3	0.82	0.78	0.321
x_4	−0.69	0.78	0.399
x_5	−2.22	0.78	0.019
x_6	3.06	0.78	0.004
x_7	−0.36	0.78	0.655
x_8	−3.99	0.78	0.001

With this more realistic data, we still get fairly strong signals that linear terms in x_2 and x_7 are inappropriate, but the signals are not as clear-cut as before. Assuming that we were patient enough to try some other versions of these two factors (rather than discarding them entirely), and that at some point we decided to try squared terms for each, we would get $R^2 = 0.99$ and this table:

Inde-pendent variable	Coefficient	Standard deviation	Signif-icance level
Constant	3.17	0.33	0.000
x_1	2.11	0.15	0.000
$x_2 \cdot x_2$	−3.39	0.31	0.000
x_3	0.82	0.18	0.001
x_4	−0.69	0.18	0.004
x_5	−2.22	0.18	0.000
x_6	3.06	0.18	0.000
$x_7 \cdot x_7$	2.09	0.31	0.000
x_8	−3.99	0.18	0.000

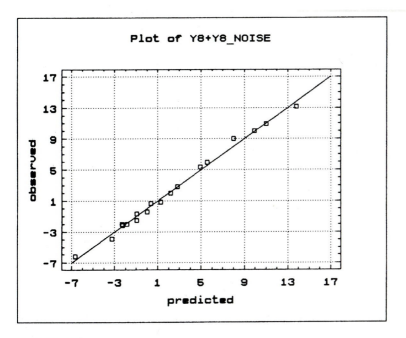

FIGURE 4.38
Good model obtained from noisy data when model form is correct.

The good fit implied by the R^2 value is confirmed by the graph of Fig. 4.38. If we increased the magnitude of the noise on y, these results (as we have seen in earlier such studies) would deteriorate. As we have said before, to preserve the orthogonality which allows the simple intuitive calculation of main effects, we must use *all* the 18 runs shown in Fig. 4.37. This restriction does *not* apply when we use least-squares routines of multiple regression, as we just did above. To get a solution, we need only provide data for as many runs as there are unknown coefficients, in our case nine. Thus any 9 (or more) of the 18 runs could be submitted for multiple regression analysis, and solutions would be possible. These solutions would, in general, give *different* values for the coefficients, and it is not obvious which combinations, if any, would be best to use if we decided to have fewer than 18 runs. This is particularly true in the real-world situation, where the correct form of the model is not known.

4.5.4 Experiments with Qualitative (Nonnumerical) Factors: Analysis of Means, Multiple Regression, and Analysis of Variance

We mentioned earlier—and now we want to discuss in more detail—that class of experiments where one or more of the independent variables cannot be assigned numerical values. Such variables are often called *qualitative, categorical,* or *classification* variables. (Remember that the *dependent* variable must *always* be measurable in numerical terms.) In studies aimed at improving metal-cutting efficiency in machining operations, we might want to include in our experiments the effect of the cutting fluid used. Such fluids are generally purchased ready-made, rather than mixing ingredients according to some recipe, thus our only control over this factor would be to use different grades of fluids from a particular vendor and/or to try fluids from different vendors. We thus have no numbers to attach to the various levels of this factor; we can only describe them as brand A, brand B, brand C, etc.

The classical way of dealing with qualitative factors involves the technique called *analysis of variance,* commonly abbreviated as ANOVA (pronounced uh-no′-vuh). Before explaining this technique, I want to show a simpler approach which often is adequate by itself to identify the individual importance of several qualitative factors affecting some dependent variable. This method usually goes by the name *analysis of means,* although it is sometimes not given a specific name and thus is often not easily found in the index of a statistics text. Whereas analysis of variance is uniformly given the acronym ANOVA, the less widely used analysis of means does not seem to have a standardized abbreviation. For our own purposes in this text, we will coin and use the term *ANOME* (pronounced uh-no′-me). We will use an example from the field of metal machining as a vehicle for explaining the general concept of analysis of means. We begin by noting that the word *analysis* in both ANOVA and ANOME is to be interpreted in a specific way, not as in the common usage of the term as just some general kind of investigation or study. The specific meaning which is appropriate here is the one for which the word *synthesis* would be an antonym. That is, we want *analysis* to imply the breaking down of a whole into its component parts, whereas *synthesis* implies a combining of parts into an integrated whole. The quantity which we wish to break down into its component parts is either the total mean value of our dependent variable (ANOME) or the variance of that dependent variable from its total mean value (ANOVA). The "component parts" are the individual contributions of the various factors (independent variables). Thus, both ANOME and ANOVA share the common goal of finding which factors are the most important, but they go about it in different ways.

In our machining example, the dependent variable will be the surface finish of the machined part, expressed as a single number R_q, the root-mean-square (rms) roughness in microinches.[1] Our experiment

---

[1] E. O. Doebelin, *Measurement Systems,* 4th ed., McGraw-Hill, New York, 1990, p. 364.

involves four factors which we believe have some effect on the measured roughness, and each of these factors will be explored at three levels. We want to try three different cutting fluids, three different cutting-tool materials, workpiece material from three different suppliers, and three different brands of machine tools (lathes) which we have in our plant. Just as we discovered earlier in using multiple regression (with quantitative factors), useful results can be obtained for qualitative factors *without* applying the concepts of orthogonal arrays and fractional-factorial designs. In most cases, of course, these concepts significantly enhance the efficiency of our experimentation, so we use them whenever we can. In our present example, a standard orthogonal array which allows study of four three-level factors with only nine runs is available from the literature (e.g., Phadke), so we will use this approach.

Following our usual practice, we manufacture some data so that we will know in advance the correct results and can thus more easily judge how our analysis methods are working. That is, we assume that we already know the contribution of each level of each factor to the total surface roughness. We also assume that we are interested only in the main effects and that interactions are negligible. (The orthogonal array selected above *requires* such assumptions. If we are not willing to work under these restrictions, more comprehensive arrays with more complex analysis methods are available and should be used.) Let's now list the (assumed known) individual effects of each factor:

Cutting fluid		Tool mat'l		Part mat'l		Lathe	
Level	Effect	Level	Effect	Level	Effect	Level	Effect
1	5.0	1	2.0	1	3.0	1	1.0
2	3.0	2	1.0	2	4.0	2	1.0
3	4.0	3	1.0	3	5.0	3	1.0

Using the above individual effects and the factor combinations in our orthogonal array, we can calculate the total surface roughness for each experimental run by simple addition, producing Fig. 4.39, which we then pretend is the result of laboratory measurements. The column R_n is gaussian random noise (mean $= 0.0$, standard deviation $= 0.5$) which we can add to R_q in later analysis if we wish to use more realistic data.

To extract the effect of each factor on the total surface roughness, we can use the same general approach used earlier with multiple regression in Eqs. (4.18) to (4.21), except that we don't have numerical values to use for the factor levels. We first compute the grand average of all nine R_q values in Fig. 4.39; this is 10.33. To now find the individual contributions of cutting-fluid effects to this total, we use the balancing (orthogonality) properties as follows. If we add the R_q values for runs 1, 2, and 3, then the effect of fluid 1 is included 3 times but the effects of the other three factors average out (cancel), because all three levels of each appear once in this sum. The contribution of fluid 1 alone (which we call f_1) is then obtained by subtracting the grand average from the average R_q for the three runs that used fluid 1. In equation form,

$$f_1 = \frac{11. + 11. + 12.}{3} - 10.33 = 1.00 \qquad f_2 = \frac{10. + 10. + 8.}{3} - 10.33 = -1.00 \qquad f_3 = \frac{12. + 9. + 10.}{3} - 10.33 = 0.00 \qquad (4.28)$$

$$t_1 = \frac{11. + 10. + 12}{3} - 10.33 = 0.67 \qquad t_2 = \frac{11. + 10. + 9}{3} - 10.33 = -0.33 \qquad t_3 = \frac{12 + 8. + 10.}{3} - 10.33 = -0.33 \qquad (4.29)$$

$$w_1 = \frac{11. + 8. + 9.}{3} - 10.33 = -1.0 \qquad w_2 = \frac{11. + 10. + 10.}{3} - 10.33 = 0.00 \qquad w_3 = \frac{12. + 10. + 12.}{3} - 10.33 = 1.00 \qquad (4.30)$$

$$m_1 = \frac{11. + 10. + 10.}{3} - 10.33 = 0.00 \qquad m_2 = \frac{11. + 8. + 12.}{3} - 10.33 = 0.00 \qquad m_3 = \frac{12. + 10. + 9.}{3} - 10.33 = 0.00 \qquad (4.31)$$

The contributions of fluids 2 and 3 (f_2 and f_3), of tool materials 1, 2, and 3 (t_1, t_2, and t_3), of workpiece materials 1, 2, and 3 (w_1, w_2, and w_3), and of machines 1, 2, and 3 (m_1, m_2, and m_3) are similarly calculated, as you can check in the above equations.

If we now compare the results given in the above equation set with the known correct effects of each factor level which we assumed at the beginning of this exercise, there appears to be some discrepancy. That is, the assumed fluid effects were, respectively, 5, 3, and 4 whereas we have computed 1, −1, and 0. A little reflection shows that we *have* computed useful results—they just need a proper interpretation. The numbers 5, 3, 4 are the *absolute* contributions to the total roughness while the numbers 1, −1, 0 are the contributions relative to the average. That is, fluid 1 increases the roughness by 1 unit above the average, fluid 2 decreases it by 1 unit below the average, and fluid 3 has no effect relative to the average. Similarly, the absolute contributions of the three machines are each 1 unit, but since these are all the same, they have *no* (0.0) effect on the average. That is, so far as surface roughness is concerned, we can use any of the three machines, and there will be no advantage or penalty. We see now that our computed effects *are* useful both for comparing the relative importance of the four factors and for "calibrating" the various factor levels. That is,

1. Tool material 1 increases roughness about 67 percent as much as fluid 1 does. Tool materials 2 and 3 decrease roughness about 33 percent as much as fluid 2 does.

Run	Fluid	Tool mat'l	Part mat'l	Lathe	R_q, μin	R_n
1	1	1	1	1	11.0	−0.15
2	1	2	2	2	11.0	−0.46
3	1	3	3	3	12.0	0.21
4	2	1	2	3	10.0	−0.08
5	2	2	3	1	10.0	−0.13
6	2	3	1	2	8.0	0.80
7	3	1	3	2	12.0	−0.73
8	3	2	1	3	9.0	−0.08
9	3	3	2	1	10.0	−0.15

FIGURE 4.39
Data table for surface roughness experiment.

2. Workpiece material 3 increases roughness as much as material 1 decreases it, and both these effects are of a magnitude equal to those of fluids 1 and 2.

3. Tool materials 2 and 3 reduce roughness about 50 percent as much as material 1 increases it.

If our results are a good approximation to the real behavior, one of their most useful applications is the prediction of *optimum* operating conditions for the machining process. That is, we tried only 9 of the 81 combinations possible in a full-factorial experiment, and there is no assurance at all that the 9 values of surface roughness found include the optimum (minimum) value possible with these fluids, tools, part materials, and machines. According to our results, we should get the best surface finish (minimum roughness) by using fluid 2, tool material 2 or 3, part material 1, and any one of the three machines. For tool material 3, this happens, by coincidence, to be run 6 of our experiment. Usually this coincidence would not occur, and we would need to run a *validation experiment* at the predicted optimum conditions to verify that the predicted low value of roughness is actually obtained. Note that if we had *not* already run the optimum conditions, our results also would predict *how much* improvement we should expect in the surface finish. In our case the predicted improvement is $-1.00 - 0.33 - 1.00 = -2.33$, reducing the roughness from 10.33 to 8.00. This numerical prediction is quite useful since it tells us whether the optimum condition is a *significant* improvement and therefore worth striving for.

More extensive validation experiments can sometimes be justified if the economics warrant. Here we would run not one point, but a whole new set of factor combinations. If our original results are truly valid, we should be able to predict from them the results of our new set of combinations. Sometimes we may even *plan* for such validation experiments before we measure any data at all. That is, in allocating our total experimental budget, we expend only a fraction of it on the initial experiment and reserve the rest for later validation runs. When exploring unfamiliar phenomena, we often learn much from initial exploratory work that makes subsequent planning much more effective. Some authors[1] recommend that only about 25 percent of the total budget be invested in the initial experiment design.

The example of ANOME (analysis of means) just completed can be used as a pattern for all ANOME studies since it includes all the major steps involved. We used the "perfect" data and thus got "perfect" results. In a real-world experiment, of course, we should not expect such perfection. If, in Fig. 4.39, we added noise R_n (or some fraction or multiple of it) to R_q, then our computations would not produce exact predictions of the various factor effects. However, as we have seen in earlier data contamination studies, as long as the random-noise effects are small relative to the true values, we are still able to calculate useful results, even though they will not be perfect.

We want to next show that multiple regression methods can be used for problems (like that just treated above) where the independent variables are qualitative. Our purpose is to make you aware that a *single* tool (multiple regression) can be applied to several different classes of practical problems, thus reducing the number of statistical concepts that you need to learn and stay proficient in. Remember, this is a book for engineers (occasional users of statistics), not statisticians (specialists). If you use statistics occasionally, it is a considerable advantage to need to stay familiar with only a small number of concepts and methods. The device which allows us to use the regression techniques illustrated earlier with quantitative independent variables on qualitative ones is the *dummy variable.*[2]

Using multiple regression, we will now carry through the same example used above to explain the ANOME approach. First we must

---

[1] G. E. P. Box, W. G. Hunter, and J. S. Hunter, *Statistics for Experimenters,* Wiley, New York, 1978, p. 304.

[2] Scheaffer and McClure, *Statistics for Engineers,* pp. 284, 295.

define dummy variables to represent the levels of all the factors. In any such problem we *always* define, for each factor, one less dummy variable than the number of levels present in that factor. Since our four factors each have three levels, there will be two dummy variables for each factor, giving a total of eight dummy variables to be defined. The scheme for defining dummy variables is always the same and goes as follows (we use our example to show the pattern which you can use for any new problem):

$$F_{d1} = \begin{cases} 1 & \text{fluid 1} \\ 0 & \text{fluid 2 or 3} \end{cases} \qquad F_{d2} = \begin{cases} 1 & \text{fluid 2} \\ 0 & \text{fluid 1 or 3} \end{cases}$$

$$T_{d1} = \begin{cases} 1 & \text{tool 1} \\ 0 & \text{tool 2 or 3} \end{cases} \qquad T_{d2} = \begin{cases} 1 & \text{tool 2} \\ 0 & \text{tool 1 or 3} \end{cases}$$

$$W_{d1} = \begin{cases} 1 & \text{work 1} \\ 0 & \text{work 2 or 3} \end{cases} \qquad W_{d2} = \begin{cases} 1 & \text{work 2} \\ 0 & \text{work 1 or 3} \end{cases}$$

$$M_{d1} = \begin{cases} 1 & \text{lathe 1} \\ 0 & \text{lathe 2 or 3} \end{cases} \qquad M_{d2} = \begin{cases} 1 & \text{lathe 2} \\ 0 & \text{lathe 1 or 3} \end{cases}$$

These eight dummy variables appear in the equation used to model the effect of the four factors on surface roughness R_q,

$$R_q = b_0 + b_1 F_{d1} + b_2 F_{d2} + b_3 T_{d1} + b_4 T_{d2}$$
$$+ b_5 W_{d1} + b_6 W_{d2} + b_7 M_{d1} + b_8 M_{d2} \tag{4.32}$$

Note that we have defined for each factor one less dummy variable than the number of levels, and that values of 1 have been assigned to levels 1 and 2. This assignment is arbitrary; we could equally well have assigned value 1 to levels 1 and 3, or 2 and 3. At first you might also think that this dummy-variable scheme is inadequate since the factors have *three* levels each and there are only *two* dummy variables defined for each factor. Actually the scheme is a proper one since, in Eq. (4.32), each of the three levels *does* have an effect on dependent variable R_q. That is, if level 1 is present in a run, then the b_1 term is present and the b_2 term is absent. If level 2 is present, this is reversed. If level 3 is present, *both* terms are missing. Thus Eq. (4.32) is sensitive to all three levels, but the effects are "coded" in a certain way. (If you were to define *three* dummy variables in such a situation, your multiple regression software would probably return an error message saying that one of these variables is a linear combination of the other two.)

Once the dummy variables have been defined, we can apply our usual multiple regression methods to find numerical values for all the b's in Eq. (4.32) and from them deduce the relative importance of the various factors in our experiment. Although our ANOME version of this experiment was based on use of a particular orthogonal array which defined explicitly the nine runs needed, we have more flexibility when using multiple regression. Since Eq. (4.32) has nine unknown coefficients, in theory *any* nine or more combinations of factor levels could be used to find them. To make our comparison of these two alternative approaches most direct, I will first use the *same* set of nine runs as in Fig. 4.39. However, I also want to explore the use of more runs, so in Fig. 4.40 I have added four more runs, so we can do calculations with as many as 13 runs if we wish. The four additional combinations chosen have no particular significance; I chose more or less at random from the total of 81 available. The random noise R_n in Fig. 4.40 is also a new sample, but it still has mean = 0 and standard deviation = 0.5.

First, using the first nine runs in Fig. 4.40 with no noise added, I computed a regression. Since this is nine equations in nine unknowns, least-squares methods are not needed and the solution obtained should fit exactly at the nine points used for input. As expected, we get $R^2 = 1.00$, and using the coefficients produced by this calculation, we see that Eq. (4.32) becomes

$$R_q = 11.0 + 1.0F_{d1} - 1.0F_{d2} + 1.0T_{d1} + (4E-15)T_{d2} - 2.0W_{d1}$$

$$- 1.0W_{d2} + (1E-15)M_{d1} + (3E-15)M_{d2} \qquad (4.33)$$

If we treat the $E-15$ coefficients as being zero, substitution of the F, T, W, and M values corresponding to the nine runs gives an exact reproduction of the R_q values, so the model is a perfect fit. The relative importance of the various factors and levels is easily deduced by noting the effect on R_q of their presence or absence. For example, if we change from fluid 1 to fluid 2, R_q decreases by 2.0 units, while changing from fluid 2 to 3 causes an increase of 1.0 units. Whether we use lathe 1, 2, or 3 clearly makes no difference. Part material 1 contributes 1.0 unit less than does part material 2, which contributes 1.0 less than does part material 3. Tool materials 2 and 3 have the same effect, while material 1 adds 1.0 unit to R_q. We see that all the factor effects are accurately identified by our model.

I next contaminated the R_q data by adding the noise R_n to them, and then I reran the regression. Since we still have nine equations in nine unknowns, we get a model which perfectly fits the given points, but these points are "wrong" because of the added noise. The software thus returns $R^2 = 1.00$, but the coefficients are incorrect, giving us another example of

Run	Fluid	Tool mat'l	Part mat'l	Lathe	R_q, µin	R_n
1	1	1	1	1	11.0	0.13
2	1	2	2	2	11.0	0.31
3	1	3	3	3	12.0	0.12
4	2	1	2	3	10.0	0.23
5	2	2	3	1	10.0	0.80
6	2	3	1	2	8.0	0.01
7	3	1	3	2	12.0	−0.39
8	3	2	1	3	9.0	0.11
9	3	3	2	1	10.0	0.06
10	2	2	2	2	9.0	−0.25
11	3	3	3	3	11.0	0.01
12	1	1	2	2	12.0	0.28
13	2	2	1	1	8.0	0.79

(*a*)

Run	F_{d1}	F_{d2}	T_{d1}	T_{d2}	W_{d1}	W_{d2}	M_{d1}	M_{d2}	R_q, µin	R_n
1	1	0	1	0	1	0	1	0	11.0	0.13
2	1	0	0	1	0	1	0	1	11.0	0.31
3	1	0	0	0	0	0	0	0	12.0	0.12
4	0	1	1	0	0	1	0	0	10.0	0.23
5	0	1	0	1	0	0	1	0	10.0	0.80
6	0	1	0	0	1	0	0	1	8.0	0.01
7	0	0	1	0	0	0	0	1	12.0	−0.39
8	0	0	0	1	1	0	0	0	9.0	0.11
9	0	0	0	0	0	1	1	0	10.0	0.06
10	0	1	0	1	0	1	0	1	9.0	−0.25
11	0	0	0	0	0	0	0	0	11.0	0.01
12	1	0	1	0	0	1	0	1	12.0	0.28
13	0	1	0	1	1	0	1	0	8.0	0.79

(*b*)

FIGURE 4.40
Surface roughness experiment with noisy data and added runs.

the danger involved in blindly accepting models based only on a large value of R^2. Our model equation with these wrong coefficients is

$$R_q = 11.02 + 1.19F_{d1} - 0.48F_{d2} + 0.39T_{d1} - 0.51T_{d2} - 1.59W_{d1}$$
$$- 1.01W_{d2} - 0.16M_{d1} - 0.15M_{d2} \qquad (4.34)$$

Adding even only one run to our experiment will force a least-squares (rather than exact) fit and will give improved values for the coefficients. Since our random noise has zero mean value, the more runs we add, the better the coefficient estimates become, because the noise averages out more and more. Next I ran a regression using all 13 "noisy" runs in Fig. 4.40, and I got $R^2 = 0.94$ and the model equation

$$R_q = 10.93 + 1.34F_{d1} - 0.69F_{d2} + 1.001T_{d1} + 0.23T_{d2} - 2.06W_{d1}$$

$$- 1.04W_{d2} + 0.21M_{d1} - 0.25M_{d2} \qquad (4.35)$$

These coefficients are, of course, not exactly right, but they are useful estimates which would improve if we used more than 13 runs, and of course they would be better, even with 13 runs, if the noise level were smaller. Once we have more runs than coefficients, we get least-squares fits *and* statistical tests of coefficient significance, as we earlier saw at Fig. 4.17. The significance level of the various coefficients can be useful in evaluating and possibly improving our model, so let's list them.

Coefficient	Significance level	Coefficient	Significance level
b_0	0.000	b_5	0.001
b_1	0.062	b_6	0.013
b_2	0.050	b_7	0.450
b_3	0.017	b_8	0.377
b_4	0.408		

Recalling that a large significance level indicates a small significance, we might consider dropping from our assumed model the dummy variables associated with coefficients 4, 7, and 8. When I did this, I got $R^2 = 0.93$ and the model equation

$$R_q = 11.01 + 1.33F_{d1} - 0.63F_{d2} + 0.85T_{d1} - 1.97W_{d1} - 1.09W_{d2} \quad (4.36)$$

All the significance levels were now 0.05 or less. Figure 4.41*a* shows the graphical quality of fit for Eq. (4.35) while Fig. 4.41*b* is for Eq. (4.36).

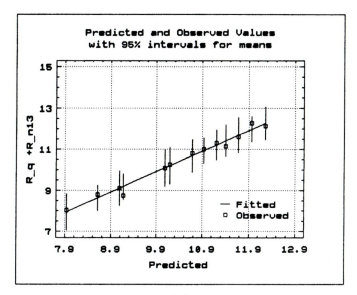

(*a*)

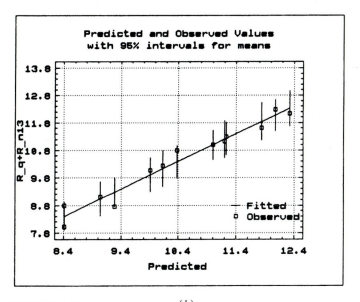

(*b*)

FIGURE 4.41
Comparison of model quality for (*a*) complex and (*b*) simpler models.

While the difference in R^2 is small, the graphs show a fairly clear advantage for the more complex model, so we would probably choose it. The 95 percent confidence intervals shown would be narrower if the random noise were less and/or the number of runs were greater.

While we have intentionally emphasized multiple regression as our statistical tool of choice for model-building experiments with quantitative, qualitative, or mixed factors, the ANOVA approach is in wide practical use and is described in many textbooks.[1] We now want to give you at least an introductory familiarity with this technique by applying it to the data of Fig. 4.39, using the noise-contaminated values for the surface roughness. The major goal of ANOVA is to consider a set of factors affecting some dependent variable and decide which of these are sufficiently important to warrant inclusion in a model. As with other statistical methods, we must be careful to distinguish between *statistical* significance and *practical* importance. Statistical significance requires only that a factor's effect be large *relative to the random fluctuations* in our data; it may or may not be practically important. That is, two factors may each have equally small significance levels (such as the *t*-test significance levels we used earlier), making them both statistically significant, but their individual contributions to the total value of the dependent variable may be very different, making one of them important and the other unimportant in a practical sense.

The details of ANOVA calculations and the interpretation of results depend somewhat on the type of experiment plan being used. Often, factorial or fractional-factorial plans will be selected for the reasons we have discussed earlier. In general, the procedure first computes the sum of the squares of all the dependent variable values. This is called the *grand total sum of squares* and in our example is 954.76. Part of this grand total is due to the average value of the dependent variable, and part of it is caused by the effects of the various factors. We want to "analyze" (decompose into component parts) that portion of the grand total sum of squares which is associated with the factor effects. To separate out the cumulative effect of all the factors together, we next compute the contribution of the average value of the dependent variable and subtract it from the grand total. The average of the nine $(R_q + R_n)$ values is 10.248, and 9 times its square (called the *sum of squares due to the mean*) is 945.15. The difference between the grand total sum of squares and the sum of squares due to the mean is the combined effect of all the

---

[1] Phadke, *Quality Engineering Using Robust Design*, pp. 51–59. T. B. Barker, *Quality by Experimental Design*, Marcel Dekker, New York, 1985, chap. 13. And Box, Hunter, and Hunter, *Statistics for Experimenters*, chap. 6.

factors and is called the *total sum of squares* (in our example it is 9.61). We next need to decompose this value into the parts due to each factor, so that we can see their relative importance. We do this by computing the difference between the total average value (10.248) and the average for the runs devoted to each level of a particular factor, squaring these differences and weighting them according to how many times the factor level appeared, and finally summing all these terms. The result is called the *sum of squares due to the factor.* Note that we are getting the "total" effect of the factor and are not extracting the effect of each level of that factor. To clarify the somewhat confusing word definition just given, let's show the actual calculation for the cutting-fluid factor *f* in our example:

$$\text{Sum of squares due to factor } f = 3\left(\frac{10.85 + 10.54 + 12.21}{3} - 10.248\right)^2$$

$$+ 3\left(\frac{9.92 + 9.87 + 7.20}{3} - 10.248\right)^2$$

$$+ 3\left(\frac{11.27 + 8.92 + 9.85}{3} - 10.248\right)^2$$

$$= 4.43 \tag{4.37}$$

(Note that this simple calculation of the effect of factor *f requires* that our array be *orthogonal,* so that the effects of all the other factors *cancel* in our calculation for factor *f*.) Similar calculations for factors *t* (tool material), *w* (workpiece material), and *m* (machine used) produce, respectively, sums of squares of 1.23, 3.90, and 0.05. The sum of these four factor sums of squares is 9.61, the same as the difference between the grand total sum of squares and the sum of squares due to the mean. (This equality is not a coincidence peculiar to this example; it is a consequence of the method used, and it always occurs.)

The factor effects 4.43, 1.23, 3.90, and 0.05 are the main results of such an analysis since they indicate the relative contribution of the four factors to the total surface roughness. These numbers, of course, are not identical to those we obtained earlier for the same example by using ANOME or multiple regression, since ANOVA here provides only the *total* effect of a factor (rather than each level) and also there is a *squaring* effect (since the definition of variance involves squares of quantities) that influences the numerical results. However, there should be a qualitative correspondence between the various methods as to the relative importance assigned to each factor. That is, ANOVA rates the four factors in decreasing importance as *f, w, t,* and *m* (4.43, 3.90, 1.23, 0.05). If we do the above ANOVA calculations using the perfect (no-noise-added) data, so that we can fairly compare with our earlier multiple regression results (which used the perfect data), we get the same *f, w, t, m* sequence but the numbers are 6.0, 6.0, 2.0, 0.0. Multiple regression gives the same sequence, but the numerical values were 2.0, 2.0, 1.0, 0.0.

There is some additional terminology associated with ANOVA, having to do with degrees of freedom, that may be useful to you. The number of *degrees of freedom* for a factor is defined to be 1 less than the number of levels; thus each of our four factors has 2 degrees of freedom. The number of degrees of freedom for an experiment, and for the grand total sum of squares, is defined equal to the number of runs, in our case nine. For the total sum of squares, the number of degrees of freedom is equal to 1 less than the number of runs, in our case eight. For orthogonal experiments, like our present example, it can be shown that

Degrees of freedom for total sum of squares

= sum of degrees of freedom for all factors

+ degrees of freedom for error (4.38)

The term *error* here refers to the difference between the actual measured values and values predicted by our model. In addition to the main goal of estimating the relative importance of the various factors, we want to estimate the statistical significance of each factor, i.e., how confident we are of its calculated value. In multiple regression, assuming we have more runs than coefficients to be calculated, the t test and associated significance level give us this information. In ANOVA, similar tests are available, and we again find that we require a certain minimum number of runs if such calculations are to be possible. Here the requirement is that there be at least 1 degree of freedom for error. In our present example, Eq. (4.38) shows that we have 0 degrees of freedom for error, so it is not possible to calculate significance levels. In a sense, using the relations in Eq. (4.38), we have "used up" all the available degrees of freedom (those given by the total sum of squares) in estimating the four parameters (2 degrees of freedom each), thus none are left to estimate error effects. (We earlier encountered the analogous situation in multiple regression when we estimated nine coefficients from nine runs, found that least-squares solution was not needed, got an exact fit of the nine points, but did *not* obtain any significance levels for our coefficients.) When we want to study *interactions* among factors, these introduce additional degrees of freedom and thus require additional runs to allow their calculation. The number of added degrees of freedom is given by the product of the degrees of freedom for each factor. Thus the interaction between a two-level factor and a three-level factor requires $(2-1)(3-1) = 2$ additional runs for its estimation. Recall that many practical studies require *no* interactions to be considered and that when interactions *are* needed, two-factor ($A \cdot B$) interactions are generally sufficient, higher-order interactions ($A \cdot B \cdot C$, $A \cdot B \cdot C \cdot D$, etc.) being difficult to validate and interpret in a useful fashion.

Commerical statistical software usually provides some kind of routine for performing ANOVA. The ANOVA routine in STATGRAPHICS is quite versatile in that it will accept many types of experimental plans, not just orthogonal ones. It also does not require "balanced" designs, i.e., plans for which each different run occurs the same number of times. This is of interest when we decide to *repeat* (replicate) one or more runs in an attempt to isolate the effect of random errors in the measurement of the dependent variable. If we decide to repeat only certain runs, rather than all those in the basic plan, the experiment is called *unbalanced* and the ANOVA calculation method must be modified.

Let's now use STATGRAPHICS to perform ANOVA on the surface roughness data we just analyzed "manually" above. Entry of the data from Fig. 4.39 into the software is quite straightforward. The dependent variable R_q is entered as a list of nine values (11.0, 11.0, 12.0, etc.). Each factor also is a list of nine values giving the level of that factor for each of the nine runs. STATGRAPHICS accepts either numbers (such as the 1, 2, 3 used in our example) or letters, such as A, B, C. The data entry for the cutting-fluid factor, e.g., is the list 1, 1, 1, 2, 2, 2, 3, 3, 3 (see Fig. 4.39), and the other three factors are similarly entered. The only other data entry is a selection of the confidence level (usually 95 percent) to be used in computing confidence intervals for the mean-value estimates. (When interactions are of interest, they are easily requested in STATGRAPHICS by simply entering an asterisk into the proper row or column of a factor table provided as a menu. Of course, you must have enough runs in your experiment to *allow* calculation of the interactions you request.)

When I attempted the above calculation, STATGRAPHICS refused to run this data and returned the error message "no degrees of freedom for error." While the ANOVA computation certainly *can* be done with this data (we did it manually earlier, getting the factor effects for *f, t, w,* and *m,* respectively, as 6, 2, 6, 0), one *cannot* compute any significance levels when there are no degrees of freedom available for the error, as there are here. Since this situation is unusual and undesirable in practical applications, the software programmer evidently wanted to discourage such use and warn us with the error message. However, I did want to see whether the software agreed with our hand calculations, so I "tricked" the program by submitting a new set of data which was simply the set used above expanded to 18 runs by *repeating* the 9 original runs. We now have 9 degrees of freedom for error, and the calculation will proceed; however, we need to divide the factor effects computed in this way by 2, since the repetition causes a doubling. This scheme was successful, producing factor effects of 12, 4, 12, 0, correspondingly perfectly with our hand calculations. Since this is perfect data, the significance levels (computed from an F test) were also perfect, respectively, 0, 0, 0, 1 for f, t, w, and m. The value 1 for m is proper since the m effect is *exactly* zero; it does *not* mean (as it usually would) that this factor effect is statistically not significant. The software also produces tables and graphs of the effects on the dependent variable of the various levels of all the factors. This is essentially an ANOME calculation, so the STATGRAPHICS ANOVA does *both* these studies for us. These means tables and graphs also include confidence intervals for the estimates. Figure 4.42 shows these ANOME results (the confidence intervals are zero because we use perfect data).

If we still use our perfect data but want to get results with a minimum of runs, we can take the first 10 runs from Fig. 4.40*a*, giving 1 degree of freedom for error. This calculation produces factor effects of 6.56, 2.06, 6.00, and 0.00. If we now contaminate these data with the noise R_n of Fig. 4.40*a*, these numbers change to 6.36, 1.42, 6.60, and 0.39 with significance levels of 0.20, 0.39, 0.19, and 0.63, respectively. Using all 13 runs of noisy data, we get 8.19, 2.00, 7.52, and 0.41 with significance levels of 0.003, 0.034, 0.003, and 0.275. While these results all differ, in a real-world experiment where we do *not* know the correct answers, they each would give us a good idea of the relative importance of the four factors and would thus be practically useful. To see what might be gained by more extensive experiments, I made up a data set of 18 runs by repeating the first 9 runs of Fig. 4.40*a* [using a *new* set of 18 R_n values, but with the same mean (0) and standard deviation (0.5)]. Running ANOVA for this data gave 15.5, 7.4, 11.3, and 0.3 with significance 0.0001, 0.00015, 0.0003, and 0.58 and the graph of Fig. 4.43*a*. I then reran this with the noise reduced to 10 percent of the original values and got 12.3, 4.27, 11.9, and 0.003 with significance 0.0001, 0.00015, 0.0003, and 0.58 and the graph of Fig. 4.43*b*. (Graphs for the other three factors were available but are not shown here.)

4.5.5   Stepwise Regression

Since we are recommending multiple regression for a wide variety of model-building tasks, we want to make you aware of a version of this method that is sometimes useful in speeding up consideration of various alternative models and choosing the most appropriate. This scheme, usually called *stepwise regression,* is most useful at the early stages of model building when we may have a large number of independent variables and we are not very clear as to which are really important. Let's also assume that we are at this point still looking only at main effects, not any interactions. (Stepwise regression *can* include as many interactions

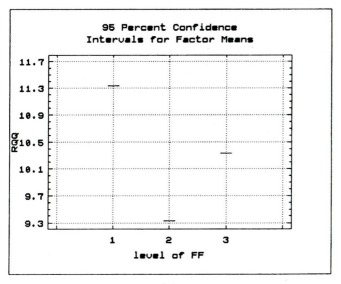

(a)

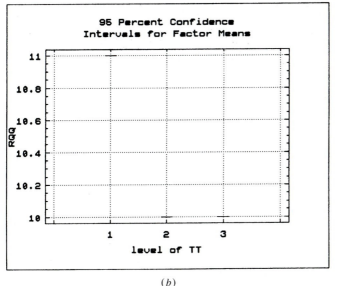

(b)

FIGURE 4.42
ANOME results for the four factors, noise-free data.

as you wish, but *you* have to decide which interactions you want to consider. Then stepwise regression can help you decide which ones to keep in your model.) The brute-force approach to such a problem would be to do a regression on every possible combination of the factors. If we

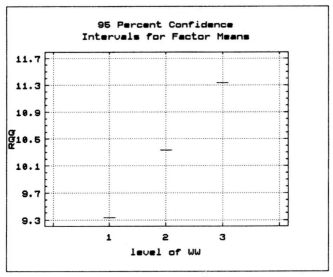

(c)

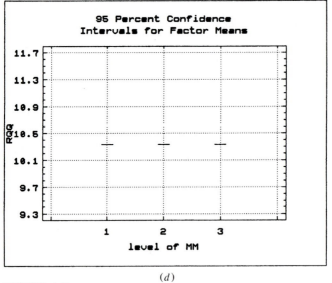

(d)

FIGURE 4.42
(Continued).

had, say, 4 factors, we would need to run 4 one-variable models, 6 two-variable models, 4 three-variable models, and 1 four-variable model. If we include the model which has only a constant (none of the four factors) in it, then we have a total of 16 models to study. In general, if we have k factors, there are 2^k models. If we include not only the factors

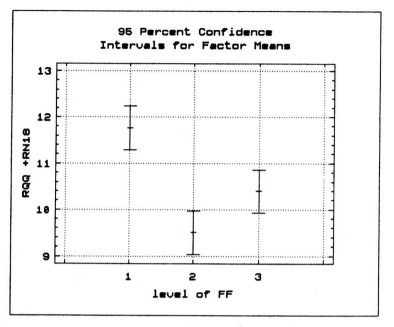

(*a*)

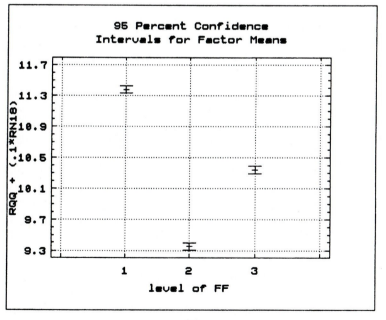

FIGURE 4.43   (*b*)
ANOME results for fluid factor with (*a*) very noisy data and (*b*) slightly noisy data.

themselves but also their possible interaction terms, the total number of models increases greatly, making the investigation of all possible models often impractical.

Our goal, as usual, is to find the simplest model which adequately explains the data. The purpose of a stepwise regression procedure is to somehow avoid the computational cost of doing all these regressions but still find the best one-variable, two-variable, three-variable, etc., models. Various methods for doing this have been developed and are available in commercial software. None are foolproof; in their attempts to economize they sometimes will miss the best model. They do, however, often find good models, and they can achieve great savings in computing time and cost, especially when the number of factors is large. When thoughtfully applied, stepwise regression is regarded as another useful tool in model building by many engineers, although some expert practitioners[1] prefer other approaches. We will now give a brief demonstration of the stepwise regression routine available in STATGRAPHICS. Most other statistical software provides capabilities that are similar in principle and application.

Our example data set has five basic factors V_1 through V_5, and the dependent variable is called Y_{v5}. We again simulate a real-world experiment so that we will know the correct model and can thus easily evaluate the success or failure of our model-building efforts. The correct equation relating the variables is

$$Y_{v5} = 310. + 2.(V_1) - 3.(V_2^2) + 4.(V_1 V_2) - 7.(V_4) + 9.(V_5) \qquad (4.39)$$

Notice that the correct model does *not* depend on V_3. We have intentionally included this feature to simulate the real-world situation where we mistakenly believe that a certain factor influences the dependent variable when it has no effect at all. Our experiment comprises the 20 runs given in Fig. 4.44, and again we provide both perfect data and a gaussian noise Y_{v5n} (mean = 0.0, standard deviation = 5.0) to be added when we want a more realistic calculation.

To check out our software, first we postulate a model of exactly the correct form, use the perfect (noise-free) data, and run an ordinary multiple regression, using all 20 runs. As expected, the routine returns exactly the correct coefficients of Eq. (4.39) above. Next we repeat this, using the noise-contaminated data (note that the noise is quite small compared with the basic values), and we get $R^2 = 0.9994$ and the coefficients 324.6, 1.75, −3.33, 4.07, −6.90, and 7.93. Fig. 4.45 shows the

---

[1] Daniel and Wood, *Fitting Equations to Data*, p. 84.

Run	V_1	V_2	V_3	V_4	V_5	Y_{v5}	Y_{v5n}
1	7.97	0.2	0.2	11.0	6.3	311.9	6.4
2	15.73	0.4	0.8	12.0	6.6	341.5	7.8
3	23.09	0.6	1.8	13.0	6.9	381.6	5.5
4	29.86	0.8	3.2	24.0	7.2	360.2	−3.5
5	35.87	1.0	5.0	15.0	7.5	484.7	2.4
6	40.96	1.2	7.2	16.0	7.8	542.4	3.7
7	45.01	1.4	9.8	27.0	8.1	530.1	−1.8
8	47.90	1.6	12.8	18.0	8.4	654.3	−1.3
9	49.57	1.8	16.2	19.0	5.7	674.7	2.5
10	49.98	2.0	20.0	30.0	6.0	641.8	1.3
11	49.11	2.2	24.2	21.0	6.8	740.0	11.0
12	46.98	2.4	28.8	22.0	8.0	755.7	−3.4
13	43.66	2.6	33.8	23.0	9.9	759.2	6.8
14	39.22	2.8	39.2	14.0	10.2	797.9	−7.1
15	33.77	3.0	45.0	25.0	10.5	675.3	5.0
16	27.47	3.2	51.2	26.0	9.8	592.0	3.2
17	20.46	3.4	57.8	17.0	9.1	557.4	−3.8
18	12.93	3.6	64.8	28.0	8.4	362.8	5.1
19	5.07	3.8	72.2	29.0	6.7	211.2	6.1
20	−2.92	4.0	80.0	20.0	12.0	177.5	−0.4

FIGURE 4.44
Five-factor model data for stepwise regression study.

overall quality of the fit and the contributions of the individual terms (the component-effect graphs) in the equation. Because the random noise is so small, the 95 percent prediction intervals are about the same size as the point symbols and do not show up clearly in Fig. 4.45b. The largest component effect is that of $V_1 \cdot V_2$, followed, in decreasing importance, by V_4, V_1, $V_2 \cdot V_2$, and V_5. The residuals (deviation of point from line) look larger on, say, the V_5 graph than on that for $V_1 \cdot V_2$, but they really are not; it is just the vertical scale of the graph that makes it appear so.

We now need to exercise the stepwise regression routine in a more realistic way. Let's suppose that the physics of our experimental

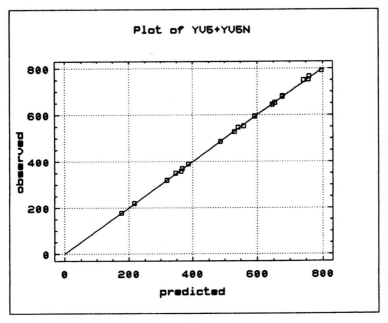

(*a*)

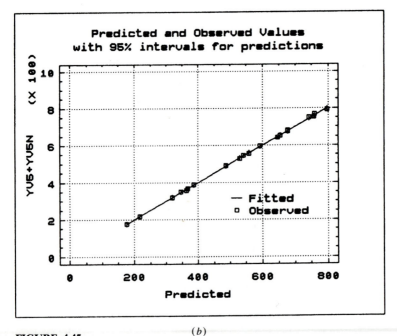

(*b*)

FIGURE 4.45
Results of ordinary multiple regression for low-noise data.

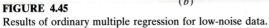

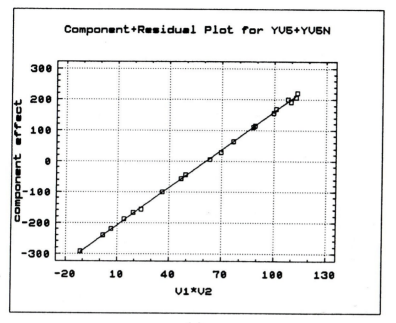

(*c*)

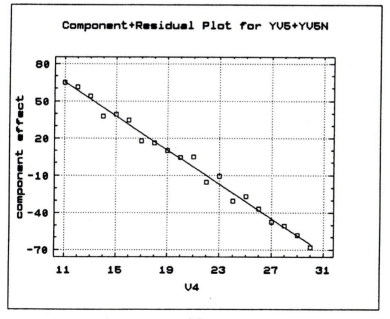

(*d*)

FIGURE 4.45
(Continued).

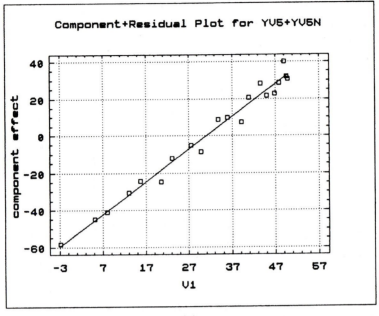

(*e*)

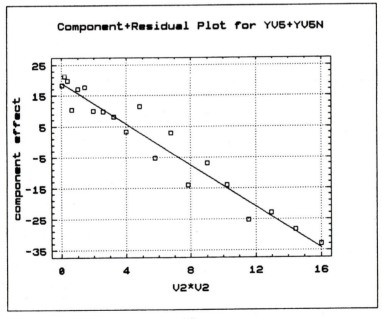

(*f*)

FIGURE 4.45
(Continued).

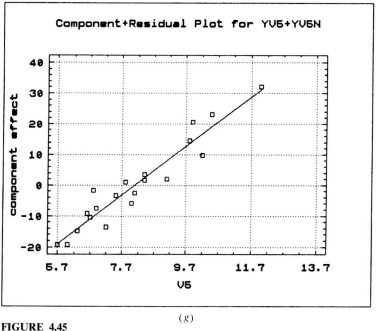

(g)

FIGURE 4.45
(Continued).

apparatus is poorly understood, so that we don't have much theoretical guidance regarding the relationship among our variables. In such circumstances one often starts with a model that includes all the independent variables but only to the first power and with no interactions at all, i.e.,

$$YV5 = B0 + B1(V1) + B2(V2) + B3(V3) + B4(V4) + B5(V5) \quad (4.40)$$

While we do not want to give a detailed explanation of the theory behind stepwise regression methods (most people using such software have no such background), we do need to at least outline the basic concept. Without actually running any regression calculations (which are the time-consuming computations we are trying to avoid), it is possible to compute quickly some quantities called *partial correlation coefficients* and *statistical F tests* for each of the independent variables in our model. The correlation coefficients are always in the range of -1.0 to $+1.0$. Values near 0 mean that this independent variable has little effect on the dependent variable, while absolute values near 1.0 mean that it has a strong effect. Thus we can use these numbers to select or discard independent variables. The statistical *F* tests provide essentially the same capability. A large F value means an important variable; a small value denotes an unimportant one. The STATGRAPHICS procedure requires you to select two specific F values which will then be used as criteria for entering or removing variables from the model. The value selected for removal must be equal to our less than the value selected for entering. The default value for both these is 4.0. This default value corresponds roughly to a significance level of 0.05. That is, if the F value is above 4, the chance of getting such a large value when the variable's effect is really 0 is less than 5 percent. We will use this default F value in all our upcoming calculations unless we state otherwise.

In STATGRAPHICS (and in most other stepwise packages) there is a choice among several different techniques for implementing the basic stepwise approach. We discuss briefly here two common ones, *forward selection* and *backward elimination.* In forward selection one starts with no variables in the model and then adds them, one at a time, as dictated by the computed F values relative to the entry-level value (such as the default 4.0). The computation of the F values is redone each time a new variable is added since the values change as the model changes. Forward selection thus starts with a one-variable model and adds the next most important variable until the entry-level F-value criterion is not met. Variables inserted in an earlier step may be *removed* in a later step if the changing F values dictate. If one chooses a small insertion value (say, 0.2) and an even smaller removal value (say, 0.02), the final model will probably include *all* the variables, and the procedure will produce a complete set (one-variable, two-variable, three-variable, . . . , n-variable) of models. If we use larger insertion or removal F values (such as the default 4), and if some of our variables are not very important, then the procedure will start its evaluation, continue through several models, halt, and announce that the best model has been found, and this model will not include all the variables. This is not a failing of the procedure; this is exactly what it is *supposed* to do. Sometimes we will want to "force" a complete set of models so that *we* can exercise some judgment in further selections. Other times we prefer to let the stepwise procedure do some of the selecting for us. By adjusting the enter or remove F values, we have control over this aspect of the model building.

As you might guess, backward elimination reverses the process used in forward selection. It starts with *all* the variables in the model and then deletes them, one by one, again based on the computed F values relative to the chosen insertion or removal values. We can, as before, exert some control over this process by using different numerical values for the insertion or removal F values, depending on our needs at different stages of model building. Our whole discussion of stepwise regression is obviously not rigorous or complete; more detailed explanations are available in the literature.[1] However, one can *use* the software without becoming an expert in its theory.

We will use forward selection now to do some model building based on our example. Of course we must enter the experimental data (the lists of values in Fig. 4.44) into our software. Our dependent variable will be the noise-contaminated $Y_{v5} + Y_{v5n}$. (This noise is quite small relative to the base values. If you want to use these data in your own software and prefer larger noise, you can just multiply all the noise values by some constant.) When I submitted the model of Eq. (4.40) for stepwise regression with the enter and remove F values both at 4.0, the best one-variable model used V_1 and the best two-variable model used V_1 and V_5. The procedure stopped at this point and announced that the V_1, V_5 model was best; i.e., no other variables had sufficiently large F values to make them worthy of entry into the model. The R^2 value was 0.81, the model equation was

$$Y_{v5} = -53.46 + 10.49V_1 + 31.76V_5 \qquad (4.41)$$

and the graphical display of the quality of fit gave Fig. 4.46a. The statistical significance levels of the constant, V_1, and V_5 coefficients were, respectively, 0.6271, 0.0000, and 0.0132, indicating that V_1 and V_5 are quite significant but the constant term is not. The graph itself is probably sufficient to indicate that this is *not* a satisfactory model, so we need to try some different models.

Just to make sure that use of the other variables did not greatly improve the model, I reran the regression, using an F enter of 0.2 and an F remove of 0.02 to force a complete set of models. The best one-, two-, three-, four- and five-variable models were, respectively, (V_1), (V_1, V_5), (V_1, V_2, V_5), (V_1, V_2, V_4, V_5) and $(V_1, V_2, V_3, V_4, V_5)$. Results were available for all these models, but the best (all five variables) still displayed such a poor graphical fit (Fig. 4.46b, $R^2 = 0.89$) that further modeling attempts are justified. As a matter of interest, the significance levels for the coefficients (constant through V_5) were 0.32, 0.27, 0.06, 0.07, 0.009, and 0.72. Note that V_3, which we *know* has no effect on the dependent variable, is indicated to be *more* significant than the constant, V_1, and V_5. This misleading indication emphasizes again how careful we need to be in interpreting statistical results.

Once the obvious first choice of model form [Eq. (4.40)] has been tried and found wanting, the *next* steps become much less scientific and more dependent on experience and judgment. However, if we have a speedy interactive computing facility available, we can quickly evaluate

---

[1] J. Sall, *SAS Regression Applications*, SAS Tech. Rep. A-102, SAS Institute, Box 8000, Cary, NC, 1981, pp. 4-1 to 4-27.

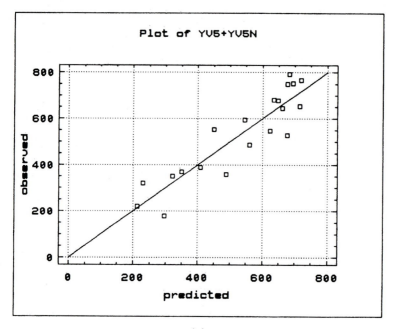

(a)

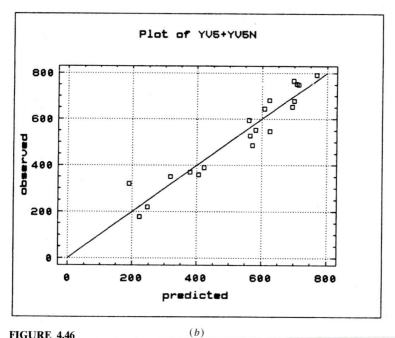

(b)

FIGURE 4.46
Stepwise regression: (a) best two-variable and (b) five-variable linear models.

many trial models (even crude guesses), learning from these results as we go, and hopefully we can arrive at a useful model in reasonable time. A reasonable next step would be to try some interaction terms and/or functions of the independent variables *other* than the first power. There are, of course, an *infinite* variety of functions to pick from; however, polynomials have a good record of practical success, and it seems reasonable to start with the second power (squares of the factors). As far as interactions are concerned, most experienced analysts feel that interactions of more than two variables are rarely fruitful, so we can limit our trials to the various combinations of two factors. Within these guidelines there is *still* a lot of room for choice, and it may be best to simply start trying a few simple models to "get our feet wet."

Since V_1 and V_5 were selected earlier as worthy of inclusion, I next tried to simply augment these two with a single interaction term. There are 10 possible two-factor interactions; we could try them one at a time. I first tried $V_1 \cdot V_2$ (lucky guess!), V_1, and V_5 in the stepwise regression (default F values), and I got the model

$$Y_{v5} = 227.4 + 3.33V_1 + 3.27V_1V_2 \qquad (4.42)$$

Note that V_5 was discarded by the stepwise procedure (at the $F = 4$ level of significance), probably because $V_1 \cdot V_2$ is such a strong contributor to the dependent variable. This model has $R^2 = 0.937$ and coefficient significance levels of 0.0000, 0.0064, and 0.0000 for the constant, V_1, and $V_1 \cdot V_2$ terms, respectively. The graph of Fig. 4.47 is also encouraging. I next tried the stepwise regression on the variable list V_1, V_2, V_3, V_4, V_5, $V_1 \cdot V_2$, $V_1 \cdot V_3$, $V_1 \cdot V_4$, $V_1 \cdot V_5$. It selected as the best model

$$Y_{v5} = 401.84 - 9.05V_4 + 3.73V_1V_2 + 0.05534V_1V_4 + 0.20489V_1V_5 \quad (4.43)$$

which has $R^2 = 0.9986$, all significance levels 0.0000, and the graph of Fig. 4.48. This model seems to be excellent in all respects, but it does *not* agree with our known correct equation! At this point we need to remember that if this were a real-world experiment, if we were faced with this kind of result, we would almost surely decide that we had found a very good model to represent the measured data. While the details of our model do not agree with the known correct form, the $V_1 \cdot V_2$ term, which is the largest effect, *does* have a coefficient close to the correct value. We simply need to recognize that *many* combinations of functions can be made to fit a set of data and that a failure to find the "correct" form (which, of course, we rarely know in a real-world problem) is not a fatal error. We *did* find a model which "explains" the data.

We earlier mentioned the desirability of running so-called validation experiments as a final step in model building. Recall that this involved

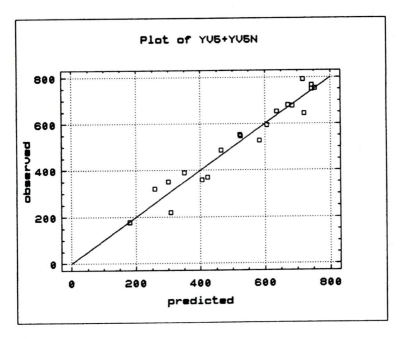

FIGURE 4.47
Two-variable model with one interaction.

testing the predictive accuracy of the model on combinations of factors which had not themselves been used in generating the model. This is often implemented by simply using only a fraction of the total runs for model building and reserving the rest to test the model so obtained. In our present example, I reran the variable list used to obtain Eq. (4.43) but used only the first 15 runs, reserving the last 5 runs for a validation test. This produced the model

$$Y_{v5} = 374.78 + 1.0701V_1 - 7.0411V_4 + 3.8207V_1V_2 + 0.1534V_1V_5 \quad (4.44)$$

which has $R^2 = 0.9990$, all coefficient significance levels less than 0.001, and the excellent graphical fit of Fig. 4.49a. Obviously this model's quality is quite comparable to that of Eq. (4.43). To see whether the model is any good at predicting the dependent variable for independent-variable combinations *other* than those used in the fitting, I substituted in the last five runs and made the graph of Fig. 4.49b. It is visually obvious here that good predictions *are* obtained. This sort of test is, of course, not exhaustive since we checked only a specific five combinations of independent variable from the infinite number possible within the multidimensional space defined by the ranges of the variables in Fig. 4.44.

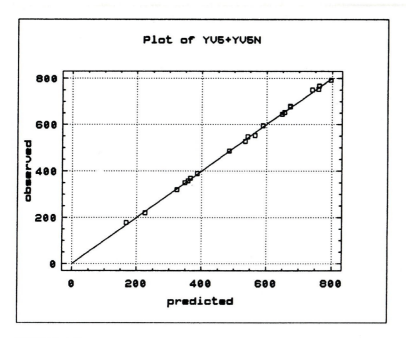

FIGURE 4.48
Four-variable model with three interactions.

In a practical application, we would try to identify any *critical* combinations of variables and try to verify our model at this subset of conditions.

4.5.6 Nonlinear Regression

All the model-building methods we have explained so far require that the equation relating the dependent variable to the various independent variables be linear in the unknown coefficients. Using this approach directly or applying some transformation which results in the proper form will handle a large percentage of the practical problems we encounter. When this linear approach does not work, various techniques of nonlinear regression have been developed and are available. We now want to give a brief introduction to the use of such methods.

We first need to emphasize that all the nonlinear methods differ in fundamental behavior from the linear techniques and are generally less satisfactory in several ways. Thus it is probably wise to try rather hard to find an acceptable linear form of model, even if a nonlinear one seems more appropriate at first. Assuming we have done this and are still not

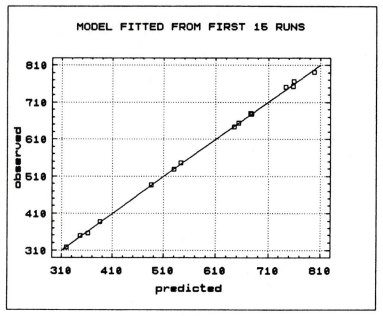

(a)

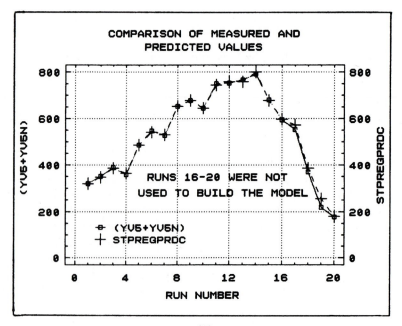

(b)

FIGURE 4.49
Model validation test.

satisfied with the results, it may be time to try some nonlinear method. The basic problem with all nonlinear regression methods is that they must use some sort of iterative search algorithm rather than the completely predictable one-step calculations of linear algebra. This means that there is *no* guarantee that the search for the "correct" coefficients will even converge, and if it does converge, that it will converge to the proper numerical values. Furthermore, we must provide as input to the nonlinear routine not only the *form* of our model but also a complete set of initial numerical estimates of every unknown coefficient. (Recall that in all our linear model-building examples we needed to provide only the form of the desired model, never any estimates of the coefficient values.) Quite often you will find that unless your initial guesses are fairly close to the correct values, the procedure fails to converge or converges on unacceptable values.

The above disclaimers should not be taken to mean that nonlinear methods should never be used, but that you should be prepared for some difficulties of a type not encountered with the linear methods. Most of the available nonlinear regression methods can handle both simple (one independent variable) and multiple (several independent variables) problems. For our brief treatment we will use a case with one independent variable. In the theoretical study of the dynamic response of physical systems described by linear ordinary differential equations with constant coefficients, we often predict time-response functions made up of sums of exponential terms, each term with its own multiplying constant and its own exponent. When we study such systems experimentally rather than theoretically, we obtain data in the form of time histories of the dependent variables, and we want to process these data to deduce the unknown coefficients and exponents of a function to fit the data. While the multiplying coefficients enter the equation in a linear fashion, the exponents do not, and there is no transformation that will correct this. This is a sufficiently important practical application that software *specifically* tailored to this task is available.[1] Rather than using this specialized software, we will instead use a *general-purpose* nonlinear regression that is part of STATGRAPHICS and many other standard packages—the *Marquardt algorithm.*[2]

Our physical example will relate to the step-function response of an overdamped second-order pressure transducer. Theory predicts that we

---

[1] RSTRIP, MicroMath Scientific Software, PO Box 21550, Salt Lake City, UT 84121, phone (800) 942-6284.

[2] D. W. Marquardt, "An Algorithm for Least Squares Estimation of Nonlinear Parameters," *Journal of the Society of Industrial and Applied Mathematics,* 2: 431–441, 1963.

should expect an equation with two exponential terms, each with its own time constant, and many experiments have verified this general form. Thus we are quite sure of the *form* of the desired model. The numerical values for any *specific* transducer are, however, unknown, and we wish to find them by using a nonlinear regression applied to the measured response curve. As usual, we will simulate a real experiment so that we know the correct answers. Our step-response curve is given by

$$P = 1.0 + Ce^{-t/\tau_1} - (C + 1.0)e^{-t/\tau_2} \qquad C = 0.5 \qquad \tau_1 = 2.0 \qquad \tau_2 = 5.0$$

$$(4.45)$$

Figure 4.50 gives a table of perfect data and also a gaussian noise with mean $= 0$ and standard deviation $= 0.05$ which we can add to simulate noisy data if we wish. These lists of dependent- and independent-variable values must be entered into the software, together with our best estimates of the unknown parameters C, τ_1, and τ_2. We must also provide numerical values for seven parameters which control the action of the algorithm. These are things like how many iterations to try before giving up, how many times you allow the function to be evaluated (each iteration involves several function calls), stopping conditions based on how small the residuals have become or how much the estimated coefficients are still changing, etc. Default values are provided for all

Time	Pressure	Noise	Time	Pressure	Noise
0.0	0.00	0.117	11.0	0.836	−0.050
1.0	0.075	0.053	12.0	0.865	0.066
2.0	0.178	−0.038	13.0	0.889	−0.079
3.0	0.288	0.034	14.0	0.909	0.071
4.0	0.394	0.042	15.0	0.926	0.036
5.0	0.489	−0.021	16.0	0.939	−0.044
6.0	0.573	−0.010	17.0	0.950	0.041
7.0	0.645	0.020	18.0	0.959	0.048
8.0	0.706	0.062	19.0	0.966	−0.022
9.0	0.758	0.013	20.0	0.972	−0.004
10.0	0.800	−0.037			

FIGURE 4.50
Pressure transducer step-response data.

seven parameters, and first-time users of the routine should simply use these until they get enough experience to try some adjustments.

As a first try, let's use the noise-free data and initial estimates for C, τ_1, and τ_2 of 0.2, 3.33, and 10.0. The routine worked perfectly, giving the correct parameter values 0.5, 2.0, and 5.0 and the graph of Fig. 4.51a. A rerun with the noise added gave 0.434, 1.956, 5.029, and Fig. 4.51b. To see the effect of poor initial estimates, I tried the noise-free data with the values 1, 1, 1 and got perfect results; however, when I added a *tiny* noise (10^{-6} times the basic noise), the routine would not run. Reducing the noise to 10^{-10} produced parameters 248.8, 2.88, 2.89 and the graph of Fig. 4.52. These results begin to show some of the peculiarities one can expect from nonlinear regression. As further exploration I tried the data with the "full-scale" noise, using initial parameter estimates that were about 50 percent too high and then 50 percent too low. Both trials ran, producing parameters 0.833, 2.58, 4.62 for the high values and -0.191, 6.45, 6.57 for the low values. While these two parameter sets differ widely and neither is very close to our correct values, the graphs of Fig. 4.53 both show reasonable fits, which most analysts would accept under the real-world conditions of *not knowing* the correct values. While some of these results might shake your confidence in the utility of these methods, keep in mind that no matter what kind of results the software might produce, we *always* will apply the simple but effective evaluation provided by a graphical comparison of the model with the measured data. This, together with the usual commonsense measures and possible theoretical interpretations associated with good engineering practice, in most cases should protect us from making any disastrous mistakes.

4.5.7 Response-Surface Methods

When we are doing experimental studies aimed at optimizing the performance of some process or product, the method of *response surfaces* may be helpful. Our goal in such studies is usually to find a model for the process behavior and, what is more important, to locate sets of operating conditions (product parameter values in the case of a product) that result in some sort of *optimum* condition. If the dependent variable being modeled is related to cost, the optimum might be a minimum value; if we are measuring some quality variable, we will be looking for a maximum. When we consider a simple case where the dependent variable depends mainly on only two independent variables, a useful geometric interpretation is available since we can graph the dependent variable vertically (z-axis) against the two independent variables (x and y axis) displayed on a horizontal plane. Such a graph forms some kind of *surface*, which is where the technique gets its name. Although it is most easily explained and applied to the case of two independent variables, the method is

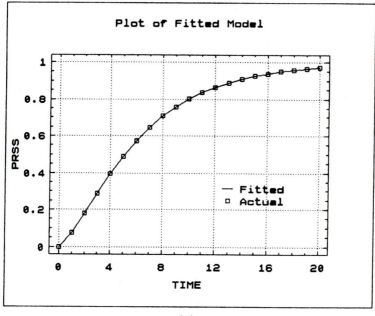

(a)

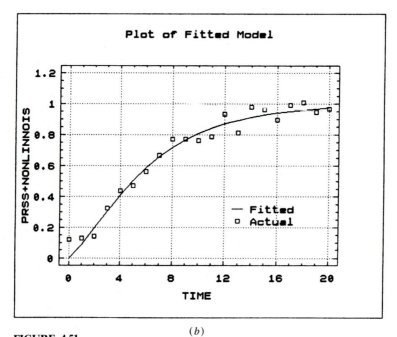

(b)

FIGURE 4.51
Nonlinear regression fit of (a) noise-free and (b) noisy data.

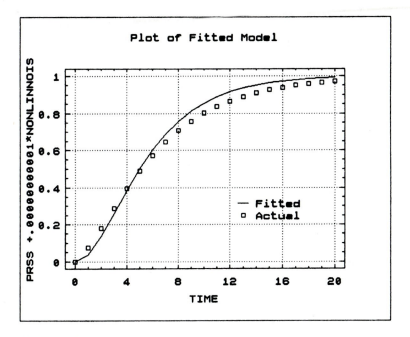

FIGURE 4.52
Effect of poor initial estimates on data with extremely small noise.

applicable to any number of variables. Here we usually hold all independent variables except two constant, and we plot three-dimensional surfaces (just as described above) to explore the effects of varying those two variables. Since there may be many such two-variable combinations that need to be studied, and complicating interactions may be present, such studies may be quite complex and require careful planning and interpretation. However, many successful applications have been carried out.[1]

We will use a simple example with two independent variables to explain the approach. Figure 4.54*a* shows a response surface for a dependent variable called *z*. Contour plots of surfaces are also useful, and Fig. 4.54*b* shows the one that goes with Fig. 4.54*a*. Both these graphs were created in STATGRAPHICS, by using the formula shown in the contour plot. This formula has no particular physical significance; I merely made up something which had a shape useful for my discussion of

---

[1] G. E. P. Box, *Empirical Model-Building and Response Surfaces,* Wiley, New York, 1987. Also Box, Hunter, and Hunter, *Statistics for Experimenters,* chap. 15.

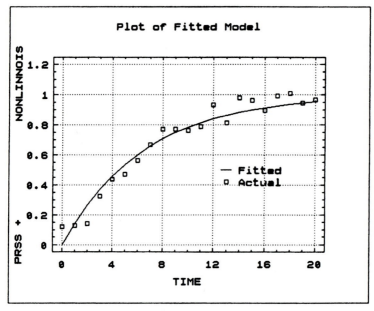

(a)

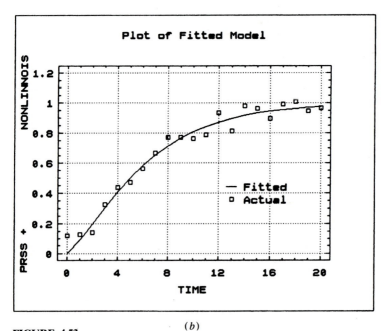

(b)

FIGURE 4.53
Noisy data with initial estimates (a) 50 percent high and (b) 50 percent low.

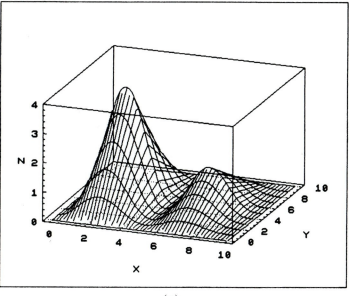

(*a*)

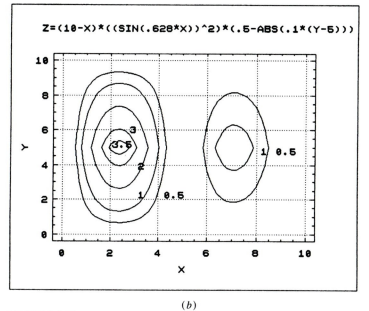

(*b*)

FIGURE 4.54
Plot of (*a*) response surface and (*b*) contour lines.

the general method. In a real-world situation of course we would *not* know the shape of this surface before running the experiment, unless we had a good theory predicting the relation among x, y, and z. Let's suppose that z represents some process quality that we would like to maximize. A common situation would be that we had been successfully "operating the process" at certain conditions (a pair of x, y values) but suspected that better conditions existed and wanted to find these. We could, of course, take the brute-force approach of blindly running a large number of x, y combinations to establish the "entire" surface and thus locate any existing peaks. Response-surface methodology provides a more efficient and systematic approach.

Before starting the two-variable example, I want to at least sketch how the more complicated multivariable cases might be dealt with. In the formula defining z in Fig. 4.54b, the numerical values (10., 0.628, 2., 0.5, 0.1, and 5.) might be not constants but rather some functions of *additional* independent variables, say u, v, w, etc. For example, the "number" 10. might really be the term $u \cdot v$ for the case where $u = 2$. and $v = 5$. If we fix all independent variables *other* than x and y, then we can always get plots like Fig. 4.54; however; there will be *many* such plots if we are interested in a wide range of operating conditions. Figure 4.55 shows how the shape changes when some of the numbers are changed. Also, we might need to look at plots in which x, y, w, etc., are fixed and we allow u and v to vary, giving perhaps totally different shapes to those response surfaces. Clearly, such a study can become quite complex and confusing, so we need all the help we can get in trying to understand the process behavior.

We return now to the two-variable example, and the sequence of steps in such studies is as follows. Using an x, y starting point corresponding to the current operating procedure for the process, we use multiple regression to fit a plane to the response surface in the neighborhood of the starting point. A plane, rather than a more complex surface, is used to reduce the necessary number of runs and to keep costs low. Also, at this stage, relatively crude approximations are adequate for the purpose of the fit. The main purpose of fitting the plane is to estimate the *direction of steepest ascent,* which will be used to plan the next stage in the experiment. To understand this steep-ascent idea, suppose in Fig. 4.54 that our starting point is $x = 9$, $y = 2.5$. We are searching for a peak and want to plan our next experimental runs so as to locate any peaks which might exist in our neighborhood. As we move away from the starting point, some directions of motion would be toward the peak and others would be away from it. On *any* smooth surface, at *any* given point, there is only one direction of motion for which z has a maximum rate of change as we change x and y. This is the path of steepest ascent, and it is always perpendicular to the contour line at that point. Since we don't

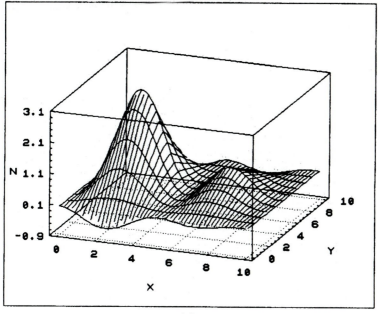

(a)

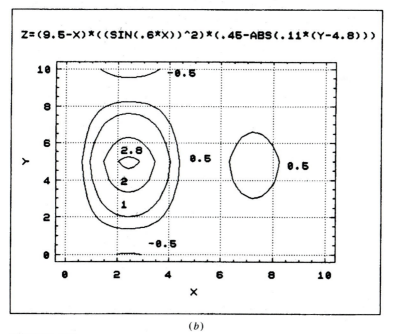

Z=(9.5-X)*((SIN(.6*X))^2)*(.45-ABS(.11*(Y-4.8)))

(b)

FIGURE 4.55
Changes in response surface due to parameter changes.

know the shape of the surface or the contour lines at this stage of the experiment, we have to somehow estimate this steep-ascent direction. This is done by fitting the best plane at this point and using the contour lines of this plane to estimate the steep-ascent direction.

Once we have fitted the plane and thus have its equation, we can set z equal to any chosen constant value to get the equation of the contour line that goes through that z value. The equation of the steep-ascent line is then easily obtained from analytic geometry since the slope (being perpendicular) is the reciprocal of the contour line's slope. We then plan our next set of runs to fall along the steep-ascent line, hoping that it will cut through any peaks and thus tentatively locate them. If we do locate a maximum along the steep-ascent line, the final step is to refine our process model in such a way as to locate the peak more reliably. This is done by planning the next stage of experimentation as a multiple regression fit of a second-order polynomial in the neighborhood of the estimated peak location. Recall that a plane *cannot* exhibit a peak and that the simplest polynomial which can is a second-order polynomial. If the true response surface is relatively smooth and our fitting polynomial is not used too far from the peak, the fit should be relatively good for predicting process behavior. Note that even if we didn't fit the polynomial, just the raw results (measured z values) allow us to see whether there is a peak and (at least roughly) where it is located. Note also that if there are *other* peaks (as in Fig. 4.54), we might have to repeat this whole procedure, starting from other initial points. It may *not* be obvious how to choose these points, but hopefully operating knowledge of a real process would give us some guidance.

Let's now actually carry through the numbers for our example. Figure 4.56 gives data for a nine-run experiment to fit a plane in the neighborhood of $x = 9$, $y = 2.5$, the chosen starting point. Since a plane requires only three parameters to specify, we are using more runs than absolutely necessary; however, we might want to use more runs to get a better fit. Multiple regression of these data gave the following equation for our plane:

$$z = 2.0314 - 0.22584x + 0.070501y \qquad (4.46)$$

This regression had a low R^2 of 0.67, and Fig. 4.57 shows also a relatively poor fit; but we will use it anyway, since fitting a plane to a curved surface is expected to be crude. (The obvious *trend* in the residuals on this graph shows the problem.) On this plane, the contour line at its "center" ($x = 9$, $y = 2.5$) has $z = 0.17405$. The equation of this contour line is given by

$$0.17405 = 2.0314 - 0.22584x + 0.070501y$$
$$y = -26.326 + 3.2028x \qquad (4.47)$$

Run	x	y	z	Noise
1	8	1.0	0.1812	–0.00851
2	8	2.5	0.4530	–0.00817
3	8	4.0	0.7248	–0.00418
4	9	1.0	0.0348	0.00391
5	9	2.5	0.0871	–0.00124
6	9	4.0	0.1393	–0.00947
7	10	1.0	0.0	–0.00235
8	10	2.5	0.0	–0.00769
9	10	4.0	0.0	–0.00687

FIGURE 4.56
Data set for fitting a plane.

The slope of this line in the xy plane is 3.2028 (72.7°), making the slope of the (perpendicular) steep-ascent line 162.7° (−0.3122). Since the steep-ascent line goes through (9, 2.5) and has a known slope, we can find its equation as

$$y = 5.3101 - 0.3122x \qquad (4.48)$$

We now want to space some experiment runs along this line, hoping to cut through a peak. I decided to try five runs using the x, y pairs (5, 3.75), (6, 3.44), (7, 3.12), (8, 2.81), and (9, 2.50). I simulated these runs, using the formula from Fig. 4.54 to get z values, and then added some gaussian noise ($\sigma = 0.005$) for realism. The z values, including the noise, were 4.75E−6, 0.4724, 0.8466, 0.5096, and 0.0871, giving the graph of Fig. 4.58. It is clear there that our search for peaks has been successful, although the actual peak is not necessarily at $x = 7$ because of the coarse spacing of the points.

Once we are close to a peak, we can locate it more accurately by running a final set of trials covering an area around the tentative peak location. Figure 4.59 shows the x, y values used as well as the resulting "measured" z values (noise included). The existence of a peak is now clear, and we have it pretty well localized although there is still some uncertainty due to the coarseness of the point spacing. This last experiment is fairly extensive (15 runs) since we are now pretty sure that we are "close to home," and we will use these same data to fit a second-degree polynomial, which has six coefficients and thus *requires* at

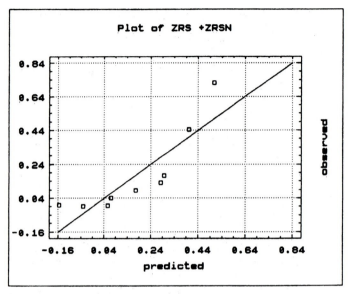

(a)

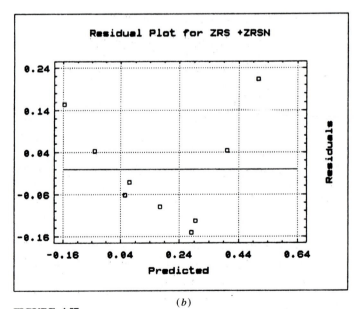

(b)

FIGURE 4.57
Quality of fit for the plane.

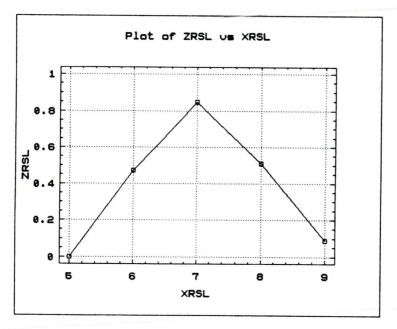

FIGURE 4.58
Evidence of a peak along the steep-ascent line.

least six runs. Multiple regression of these data gives $R^2 = 0.73$ and the surface equation

$$z = -8.573 + 2.394x + 0.4145y - 0.000207xy - 0.1685x^2 - 0.04134y^2$$

$$(4.49)$$

All the coefficient significance levels are 0.008 or less, except for the xy term which is 0.99, so we can neglect the xy term if we want a slightly simpler model. Figure 4.60 shows the polynomial surface of Eq. (4.49). If we want to improve the fit of this surface and are willing to restrict its extent in x and y, we could run more points near the peak and rerun the regression. Whether this added expense is worthwhile depends on the goals of the project and the resources assigned to it.

Response-surface methods may be useful in implementing an important design concept, called *robust design,* for products and processes. In robust design, we strive to operate at peak points and wish also to find or create "broad" peaks (see Fig. 4.61). When a peak is broad, changes in the pertinent independent variables do not cause large changes in the peak value. Thus the process (or product) quality is relatively *insensitive* to moderate changes and maintains its high value

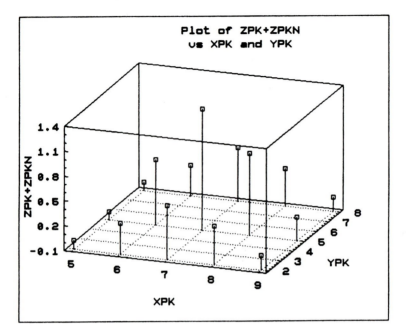

FIGURE 4.59
Fifteen-run experiment to document the peak more precisely.

when changes occur. This means that we can *relax* quality requirements on raw materials and components which go into a product (thus lowering cost) and still maintain overall quality. This scheme of producing low-priced, high-quality products from "low-quality" components may seem too good to be true, and of course it cannot *always* be achieved, but much of Japan's recent success in international markets has been built on this and allied concepts gathered under the general title of *Taguchi methods.*[1]

4.5.8 Concluding Remarks on Sec. 4.5

The various parts of Sec. 4.5 addressed the general area of experimenta-tion which can be described as model building, i.e., the study of relationships among dependent and independent variables in some

---

[1] Phadke, *Quality Engineering Using Robust Design.*

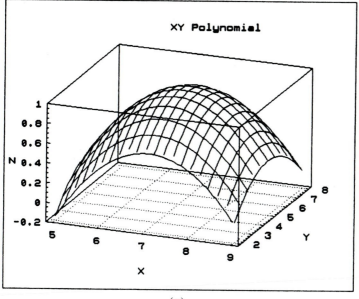

(*a*)

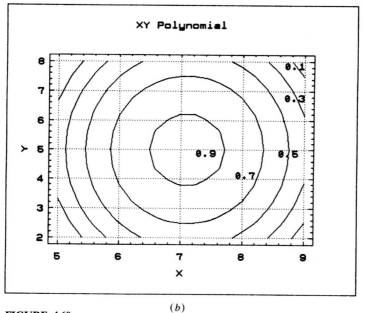

(*b*)

FIGURE 4.60
Polynomial fit to 15-run experiment data.

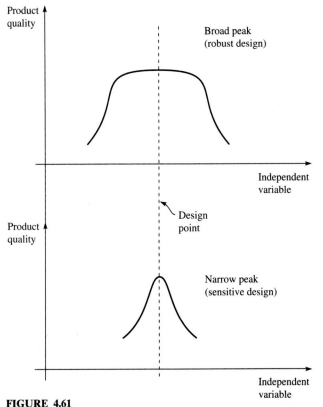

FIGURE 4.61
Broad peak gives robust product design.

process or product. We explained several different approaches to such questions, indicating the suitability of particular methods for different versions of the general problem. However, for most engineers, such experimentation is not a full-time job, and it is impractical for them to become expert in the application of a whole array of statistical techniques. For this reason, we tried to show that a *single* technique—multiple regression—can be adapted to the solution of most practical problems of this sort. We believe that this approach can be at least a partial solution to the widely recognized[1] deficiency of many engineers in

[1] A. Penzias, "Teaching Statistics to Engineers," *Science,* 244 (4908): 1025, June 2, 1989. R. G. Batson, "Statistical Training: A National Necessity," *Engineering Education,* September/October 1989, pp. 598–601; "Statistically Speaking," *Engineering Education,* September/October, 1989, p. 596.

this important area of experimentation. Most undergraduate curricula cannot justify an entire required course in statistics. Thus we need to develop a mini-course approach where *carefully selected* topics are embedded in already established engineering applications courses (such as the project labs to which this book is addressed) and taught from an applied viewpoint by engineering faculty.

The de-emphasis of ANOVA methods in this text is unconventional, but in my opinion, necessary and desirable in a condensed, but still practically useful presentation for engineers who are occasional users of statistics. Acknowledgement of certain advantages of regression over ANOVA can even be found in some texts[1] written by and for statistics specialists. I also believe that many ANOVA methods developed long ago were mainly aimed at producing *simple calculations* in an era when complex regression computations were considered impractical. Today, user-friendly statistics software and ubiquitous personal computers encourage the use of more complex theories and their associated number-crunching, making regression, in my opinion, the method of choice in most of those cases where both ANOVA and regression might be applied.

Turning now to some technical details of Sec. 4.5, we carefully explained in Sec. 4.5.1 the advantages of *run-sequence randomization* in actually carrying out any experimental plan. Later in Sec. 4.5, such as in those paragraphs devoted to factorial and fractional-factorial plans, we wanted to show most clearly the *patterns* inherent in such plans, and thus in displaying tables of runs, we did *not* show the random sequence that would actually be used in practice. Since this is an important detail, I am taking this opportunity to forcefully call it to your attention. The run-sequence randomization methods explained in Sec. 4.5.1 can and should be applied to *all* experiments unless there are compelling reasons not to.

At this point I also want to discuss and reference a few real-world applications to augment the text discussions, which often used made-up data for the important pedagogical advantages that this provides. Our first such study[2] is from the engine control area of automotive engineering. Most automotive engines these days are computer-controlled to optimize their performance in the face of conflicting requirements for fuel efficiency, pollution control, driveability, etc. To be able to design these

---

[1] J. Neter, W. Wasserman, and M. H. Kutner, *Applied Linear Statistical Models,* 2d ed., Irwin, Homewood, IL, 1985, p. 520.

[2] J. A. Tennant et al., *Development and Validation of Engine Models via Automated Dynamometer Tests,* Paper 790178, SAE, 1979.

control systems, engineers must have valid models for the relationships among many engine variables. To date, theory has been unable to predict these relations with the accuracy needed, so an experimental approach is necessary.

The experiment plan begins with the selection of 23 test points which cover the range of operating conditions felt to be pertinent for the specific engine. Each test point is defined by a specific engine speed and torque. There were six speed values from 600 to 3800 RPM. Torque values included the minimum values at 600, 900, 1300, and 1800 RPM and the maximum at 1800, 2500, and 3800 RPM. Other torque values range from 20 to 200 ft·lb. Next, 90 combinations of air/fuel ratio A/F, spark advance SA, and exhaust-gas recirculation EGR were selected. By combining the 23 speed and torque points with these 90 points, 2070 engine settings were defined in all. For each combination of A/F, SA, and EGR, the engine was stepped through the 23 speed and load points in a specific sequence, remaining at each point for 23 seconds before stepping to the next. The test time for each sequence was about 9 minutes, and the total experiment took about 12 hours.

Six different sequences (i.e., testing order) of speed and load points were used for studying the combinations of A/F, SA, and EGR so that there would be a variety of temperature histories prior to each set point and therefore the effect of temperature on emission levels and fuel consumption would be reflected in the test data. During each of the 2070 tests, 62 engine parameters were recorded, and about 40 of these were plotted to evaluate the quality of the data and detect any instrumentation problems. After some preliminary editing of the data, multiple regression was used to obtain models for predicting several variables. The fuel flow rate was found to be accurately ($R^2 = 1.0$) predicted by a model using RPM and the fuel injector pulse width (PW) as factors in a polynomial model with terms up to the second power. The airflow rate model ($R^2 = 0.99$) used RPM up to the third power and manifold absolute pressure (MAP) up to the third power. Hydrocarbon emissions required the most complex model. It used RPM up to the third power, torque up to the third power, air/fuel ratio up to third power, spark advance up to to the third power, exhaust gas recirculation up to the third power, and spark plug temperature up to the first power to achieve $R^2 = 0.93$.

As is good general model-building practice, the model obtained from the test program described above was then subjected to *validation testing* at operating conditions different from those points used to do the regression. The referenced paper describes this procedure and gives many other useful details which we choose not to pursue here. This example, although treated in abbreviated fashion, should give you some feel for the complexity encountered in some practical applications.

Our next example is from the field of instrument calibration, a common application of regression methods. The report[1] on which we base our brief discussion includes many useful details which we omit here. The National Bureau of Standards [now the National Institute for Standards and Technology (NIST)] studied the performance of four different types of positive-displacement flowmeters used to measure cryogenic liquids. The test plan used five different temperatures, five different pressures, and four flow rates from 50 to 100 percent of maximum flow. Combinations to be tested were selected by using fractional-factorial concepts combined with some practical restrictions dictated by the test apparatus, resulting in a total of 24 trials. The readings of the flowmeter were compared with the "true" flow rate given by the calibration apparatus, and the flowmeter error thus defined was separated into a systematic error (bias) and a random error (imprecision). The bias error, which can be corrected by calibration, is taken as the dependent variable in the model used to describe flowmeter performance.

Preliminary study of the data (graphs of bias error versus pressure) showed by visual examination that pressure had no apparent effect, so it was discarded as an independent variable in the model. The model used as a general form for all the flowmeters was

$$\text{Bias error} = B_0 + B_1 T + B_2 T^2 + B_3 G + B_4 G^2 + B_5 \theta \qquad (4.50)$$

Here, B_0 is the average value of the bias, T is the test temperature, G is the mass flow rate indicated by the meter, and θ is an elapsed-time parameter used to allow modeling of the gradual wear that occurs as a meter is used. One of the meters tested showed *no* dependence on any of the factors, so its bias model includes only the B_0 term. Other meters required various combinations of terms, but none required all the terms.

4.6 ACCELERATED TESTING

Most technical libraries will contain a large number of books devoted to the statistical methods of experiment design we have introduced in the earlier sections of this chapter. Text material on accelerated testing methods is much more sparse. I was able to find only one book devoted entirely to this subject and one other which included a chapter on these methods.[2] Since I feel that accelerated testing is an important part of the

---

[1] J. A. Brennan et al., *An Evaluation of Positive-Displacement Cryogenic Volumetric Flowmeters,* NBS Tech. Note 605, 1971.

[2] W. Nelson, *Accelerated Testing: Statistical Models, Test Plans, and Data Analyses,* Wiley, New York, 1990. Lipson and Sheth, *Statistical Design and Analysis of Engineering Experiments,* chap. 5.

design and manufacturing "revolution" that is occurring in the United States during the 1980–2000 period, I wanted to include some introductory material in this book so that you would at least be aware of some of the major features of the techniques.

Nelson[1] gives a list of 14 purposes for which accelerated testing might be used. Here we concentrate on two major items from this list that relate directly to the means of designing and producing low-cost, high-quality products with a short lead time for the design/manufacture cycle. That is, it is not sufficient in today's competitive international markets to produce quality products; they must also have reasonable prices, and they must be available before the market has turned to newer and better versions produced by competitors. Our first purpose for accelerated testing is the timely identification (and correction) of flaws in the design and/or manufacturing process proposed for a new product. No matter how careful and skillful we might be in designing a product and the processes used to manufacture it, there are nearly always flaws that can be discovered only by testing the actual product. These flaws need to be found and corrected as quickly as possible so that we can get a reliable stream of high-quality products to the market before our competitors establish dominance. Such accelerated testing is thus part of the design process and occurs before we freeze the product design and the process design. Testing may have to be done on prototype products since the final manufacturing processes may not be available at this stage, so we need to be careful in interpreting test results.

The second purpose of accelerated test methods that we choose to cover here is sometimes called *environmental stress screening*. This procedure is an ongoing part of the manufacturing process and has the goal of detecting (*before* they are shipped) products which would fail some time after the customer received them. That is, even after we have refined and "frozen" the design and manufacturing process using accelerated testing as just described above, there will *still* be some faulty products produced, and we don't want these to get into the hands of our customers, for many good reasons. For example, most products have warranties, and if a product fails within the warranty period, we must repair or replace it, an extremely expensive procedure which also generates ill will among customers. Environmental stress screening subjects each product produced to some kind of "overstress" which is carefully designed to force incipient failures to occur during the screening process at our plant rather than later in the customer's hands. Obviously, such overstress cannot be so great as to damage the *good* products, so this screening must be skillfully designed. While such screening is mainly intended to weed out bad products which occur normally even when our

---

[1] Nelson, op. cit., p. 23.

design and manufacture are proper, it may also detect process and/or design flaws that may develop over time and thus can serve to trigger investigation and corrective action.

Having given a brief overview of the practical importance of accelerated testing, we can now pursue a few of the details. When a product is put into actual service with customers, there will still be failures from time to time. Products designed and built with a philosophy of zero failures usually will be economically uncompetitive. Customers will tolerate a certain level of failure if it is "low enough" and if there is a price advantage. Just what level of failure will be acceptable depends on the product and requires a careful judgment call by the responsible engineers and managers. When failure results in human injury or expensive equipment damage, we obviously need to be appropriately conservative. It thus makes sense to talk about the *life* of a product, since we just saw that it will not usually be infinite. If we subject a sample of products to their expected service environment (either the *actual* environment with customers or a simulation of this environment in our plant), as time goes by, failures will occur. If we mentally consider a large sample (say, 10,000), so that a reasonably smooth bar graph will result, we can define a *failure rate* as the percentage of the total sample that failured during a specified time interval (say, 1 month), and we plot a graph of failure rate versus time as in Fig. 4.62*a*. For an infinitely large sample, this display becomes a smooth curve as in Fig. 4.62*b*.

Depending on the product and environment, the failure-rate curve can take on different shapes. The shape in Fig. 4.62*b* is often called the *bathtub curve* and is commonly quoted as being applicable to many practical situations. However, Nelson[1] warns that, in his experience, it applies in only about 10 to 15 percent of the cases and that it is more common for the failure rate to either increase or decrease monotonically. For those cases where the bathtub shape does apply, the practical interpretation is that there is an early high failure rate of short duration (called *infant mortality*), followed by a long period of low and nearly constant failure rate called *normal life,* ending in a high terminal failure rate called *wearout.* In all forms of accelerated testing, including the two types just discussed above, we want to force failures to occur *earlier* than they would under the service environment. When we are developing the product design and manufacturing process, this acceleration of failures allows us to speed up the development and get to market more quickly. When our product and/or process design has been finalized and we are producing for the market, failure acceleration through stress screening keeps poor products from reaching the customer.

When we want to quickly gather information on the life of a

---

[1] Ibid., p. 70.

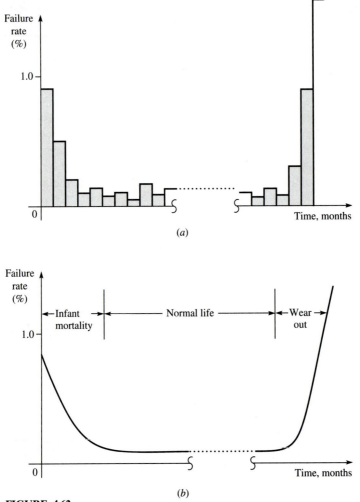

FIGURE 4.62
Variation of failure rate with product age (the "bathtub" curve).

product, there are three basic avenues open to us. If the failure is due to the accumulation of an excessive number of operating cycles of the product, we can test the product at a cyclic rate higher than that experienced in the service environment. If an automobile hood latch is operated about once a week in normal service, we can test it at 10 cycles/min to compress many years of actual service into a few hours of test time. Some cyclic situations need to be considered more carefully. For example, the compressor in a household refrigerator may cycle on and off perhaps once an hour. If we try to accelerate this rate to, say, once a minute, we may get misleading results because the failure modes of the compressor may involve temperature effects which would operate quite differently at the high cyclic rate than at the low.

A second basic approach is to use a large sample size. When the product is exposed to the normal service environment, suppose we test 10 items and get no failures in the first 100 hours. If we had used a 100-item sample, we might have gotten some failures in the same 100-hour test time since the larger sample has a higher probability of containing some bad items. Even though both tests use the normal (rather than a more severe) environment, the large-sample test is rightly called accelerated because the small-sample test would require a *longer* time to develop the same level of information about product failure.

The third basic approach is to increase the severity of one or more of the failure-causing factors above the level associated with the normal service environment. This will cause more failures earlier, and thus we obtain failure information more quickly. The problem, of course, is that we are really interested in product life under normal service conditions, *not* under overstress conditions. We thus need some kind of relation between life and stress to convert life data at high stress to life predictions at normal stress. The determination of these life-stress relations is a critical problem in the intelligent use of accelerated testing.

The means of accelerating failure for the two classes of applications we are discussing is usually an increase in some product stress beyond the level experienced in the service environment. We here use the word *stress* in a very general way to refer to any physical effect acting on the product to induce failure. This includes actual mechanical stress (pounds per square inch), temperature, corrosive atmospheres, humidity, thermal cycling, high voltage, solar radiation, etc. The simplest situation is where all (or most of) the observed failures can be attributed to a single cause. Things become more complicated when there are several possible causes of failure. Nelson explains methods for dealing with both the simple and complex cases, but here we will concentrate on products with a single failure mode. To intelligently perform and interpret accelerated tests of the kind we are considering, we must choose and then apply two different types of mathematical models. We need a model which relates average life to stress for the operative failure mode, and we need a statistical model showing how life is distributed for a constant stress value.

For readers with some background in machine or structure design, the stress-life relation associated with *fatigue failure* may be familiar, so we use this as a first example. Earlier (Sec. 2.7) we worked with the statistical distribution of life at a *constant* stress amplitude. We now need to examine the relation between *average* life and stress amplitude. This relation has been established by experimental testing since no theory so far proposed has been found adequate to predict this behavior. The testing consists of running large numbers of specimens (say, 100) at each of several stress amplitudes (say, 30,000, 50,000, 70,000, 90,000 and

110,000 psi) until they all fail. For our present purpose we need only some kind of average life at each stress level. This can be defined in various ways; we here use the median life [we could alternatively use the average (mean) life]. When we use the median, there is a 50 percent probability that failure will occur before this life and a 50 percent probability that failure will occur after. For the 30,000-psi stress of Fig. 2.18, the median life is about 2,620,000 cycles. We would need median-life numbers for the other stress levels, and then we could plot curves relating this life to stress. These curves are, of course, the standard *S-N* curves presented in materials and machine design texts. By plotting both axes logarithmically, it is found that the curves become nearly straight lines, so this is conventional (see Fig. 4.63*a*). Once one establishes this general type of behavior, it is possible to use accelerated

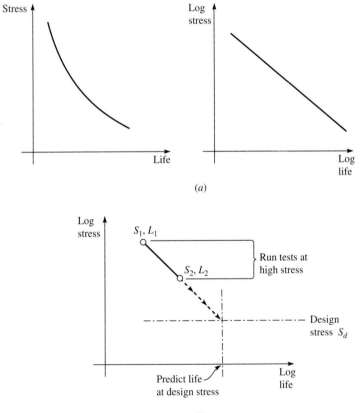

(a)

(b)

FIGURE 4.63
Stress versus life behavior for fatigue failure.

testing as shown in Fig. 4.63*b*. We run specimens at stresses S_1 and S_2, both higher than the product design stress, and determine median lives L_1 and L_2. We can then extrapolate this data to predict the median life at product design stress S_d. This is accelerated testing since testing time at the two high-stress levels can be considerably shorter than that needed to test directly at the design stress. If experience with this class of materials allows us to estimate (without further experiment) the *slope* of the extrapolating line, we could *further* accelerate the experimentation by running tests at only *one* stress level. Our mechanical stress example can be generalized as follows. If we know (analytically or graphically) the life/stress relation for a failure mode, test data at high stress/short life conditions can be used to predict life at design stress/design life conditions. This constitutes accelerated testing since we obtain the desired information in a shorter time.

The mechanical fatigue stress example just given is a specific application of a general life-stress relation called the *inverse power law*. This is not, of course, a *universal* relation but has been successfully employed for a variety of problems such as life versus voltage of electrical insulations and transformer oils, life versus load for rolling-contact bearings, life versus voltage for incandescent and flash lamps, and life versus temperature range for metal fatigue due to thermal cycling. In its general form, the inverse power law can be stated as

$$L = \frac{C_1}{S^{C_2}} \qquad (4.51)$$

where L is the life, C_1 and C_2 are constants needed to fit the function to any specific data, and S is the stress effect which causes failure. Taking logarithms of both sides, we get

$$\ln L = \ln C_1 - C_2 \ln S$$
$$y = b + mx \qquad (4.52)$$

which we see is a straight-line relation, as exemplified by Fig. 4.63. Clearly, once C_1 and C_2 are known, accelerated test data at one high-stress, short-life point can be used to predict design life at design stress.

Another life-stress relation that has found wide applicability is the *Arrhenius law*. While the inverse power law has no theoretical base, the Arrhenius law does have some connection to underlying physical effects. It is used for failure (or degradation) modes related to the effects of absolute temperature. Swedish chemist S. A. Arrhenius proposed in 1889 a law relating chemical reaction rates to the absolute temperature of the reacting materials. Although this was an empirical law, it agrees very closely with kinetic molecular theory, which relates reaction rates to the number of molecular collisions per unit of time.[1] These collision rates are theoretically known to increase with absolute temperature. In a physical process leading to failure of a component or material, we assume the failure is the result of some chemical reaction which has a degrading effect on performance. When this degradation reaches a certain critical level, we define failure to have occurred. Higher temperature causes the chemical reaction to proceed more quickly, so failure occurs sooner at high temperature. The Arrhenius relation found in chemistry texts contains several basic physical parameters such as activation energy and Boltzmann's constant. When it is used in accelerated testing, we merely adopt the *form* of the relation but treat the parameters as empirical constants to be found by experiments on the specific application under study. The Arrhenius law, in this form, has been successfully applied to the effect of temperature on the useful life of electrical insulations and dielectrics, semiconductor devices such as transistors and integrated circuits, battery cells, lubricants and greases, and plastic materials.[2]

The Arrhenius law used in accelerated testing can be given in several forms. We will use the form

$$L = K_1 e^{K_2/T}$$

$$\ln L = \ln K_1 + \frac{K_2}{T} \tag{4.53}$$

$$y = b + m\left(\frac{1}{T}\right)$$

We have again manipulated the formula so that its final form gives a straight-line plot for ln life versus $1/T$. Once the empirical constants in this equation have been found by experiment, we can predict life at the design temperature from accelerated test data of life at above-design temperatures. Since such data normally exhibits some scatter, good estimates of the empirical constants require the running of several specimens at each of several different temperatures. Statistically efficient plans for such testing have been devised, and we will touch on them shortly.

We stated earlier that accelerated testing used two different types

---

[1] K. J. Laidler, *Chemical Kinetics,* 2d ed., McGraw-Hill, New York, 1965, p. 50.

[2] Nelson, op. cit., pp. 75–85.

of models, one for the relation between average life and stress and another for the statistical scatter of life around the average life. We just finished discussing the two most common stress versus life models (inverse power and Arrhenius), and we now need to look at the life scatter models. Again we find that most problems can be treated with a proper choice from two or three widely used models. These are the exponential, Weibull, and lognormal distribution functions. We briefly encountered the lognormal model in Sec. 2.7 (metal fatigue), so here we will concentrate on the other two, which have not yet been treated. The *exponential distribution*, when used for product life studies where the independent variable is time t, corresponds to a failure rate (see Fig. 4.62) which is *constant* at all times. The probability density function is

$$f(t) \triangleq \lambda e^{-\lambda t} \tag{4.54}$$

where $\lambda \triangleq$ failure rate in failures per hour if t is in hours. The cumulative distribution $F(t)$ gives the percentage (decimal) failed at time t and is obtained by integrating Eq. (4.54) to get

$$F(t) = 1 - e^{-\lambda t} \tag{4.55}$$

The average (mean) life is $1/\lambda$, and the median life is $0.693/\lambda$. The standard deviation of this distribution turns out to be the same as its mean, $1/\lambda$. Figure 4.64 displays these facts graphically.

Many products have failure rates which are *not* approximately constant with time. We can then try to fit such data with a lognormal or Weibull distribution. The *Weibull distribution* probability density function used for most life testing is

$$f(t) \triangleq \left(\frac{\beta}{\alpha^\beta}\right)(t^{\beta-1})(e^{-(t/\alpha)^\beta}) \tag{4.56}$$

The cumulative distribution $F(t)$ gives the percentage (decimal) failing by age t:

$$F(t) = 1 - e^{-(t/\alpha)^\beta} \tag{4.57}$$

Parameters α (scale) and β (shape) are taken positive, and choice of their numerical values allows us to fit different kinds of real-world data. For $\beta = 1$ the Weibull distribution becomes the exponential distribution. The Weibull failure rate varies with time and can be adjusted in a versatile fashion to fit rates which increase or decrease with time.

$$\text{Failure rate} = \left(\frac{\beta}{\alpha}\right)\left(\frac{t}{\alpha}\right)^{\beta-1} \tag{4.58}$$

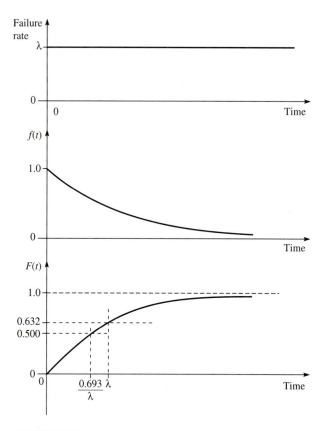

FIGURE 4.64
Exponential distribution.

Figure 4.65 illustrates these relations. The parameter α is sometimes called the *characteristic life,* and it is the life at which 63.2 percent of the failures will have occurred. If we are interested in the *median life* (50 percent failures), it is related to α and β by

$$\text{Median life} = \alpha(0.693)^{1/\beta} \tag{4.59}$$

We are now ready to show how all these concepts are actually used in an accelerated test. One must first choose the most appropriate life-stress model and the most appropriate life-distribution model and then apply them both to the measured data. Past experience (personal or vicarious) is a useful guide at this point, but if this is lacking, one simply tries the various combinations and finds out which fits the data best. We

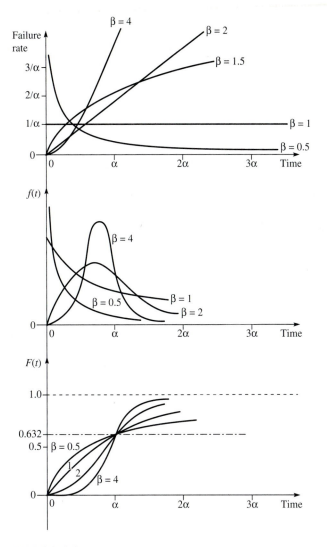

FIGURE 4.65
Weibull distribution.

will choose a specific combination, the Arrhenius life-stress model and the Weibull life-distribution model, for our example. The general procedure for any of the other combinations follows the same steps and methods, so one example is sufficient. Nelson (p. 113) discusses both graphical and analytical approaches and encourages the use of both, since they are complementary, even though many engineers are satisfied with just the graphical methods. Since our purpose here is an introductory overview, we will pursue only the graphical techniques; they clearly show the essence of what is involved. Special graph papers, not usually found in drafting supply houses, facilitate the procedures and are available from several sources.[1]

Just as we have seen earlier with fractional-factorial experiments, statisticians have devised efficient test plans for accelerated testing. Nelson devotes an entire chapter (6) to this subject, but we can quickly summarize some main points. Traditional test plans are not recommended by Nelson but are actually often used since the availability of more efficient ones may not be known to the experimenter. A traditional plan usually runs tests at three or four stress levels and uses an equal number of specimens at each stress level. As we have seen in earlier statistical planning, the reliability (accuracy) of our results improves as we use more specimens. Traditional plans require 25 to 50 percent more specimens than the best available plans, to achieve the same accuracy.

So-called optimal plans yield the most accurate estimates of life at the design stress. They run tests at only two levels, the highest and lowest of the chosen test range. Intermediate levels are not used. Of the total number of specimens to be tested, what portion should be allocated to the high level and what portion to the low? Nelson gives methods for making this decision. When the low stress is far from the design stress, then equal numbers of specimens are used at each level. If the low stress is very near the design stress (this, of course, does not accelerate testing very much), then almost all the specimens should be run at the low stress. For situations away from these two extremes, formulas are available to calculate the best allocation. In general, we will test more specimens at the low level. Nelson actually recommends optimal plans *only* when it is known that the data almost perfectly fit the assumed models. For the more common case of nonperfect models, he prefers plans which combine the best features of the optimal and traditional plans. These are briefly described as follows. Use three or four equally spaced stress levels. Allocate more specimens to the extremes of the test range and fewer to the middle. Also use more specimens at the lowest level, especially if this is close to the design stress. While these recommendations are not numerical, they are useful in practical test design.

For our numerical example we will again use a simulation approach (rather than real data) so that we can compare predictions with true

---

[1] TEAM Graph Papers, Box 25, Tamworth, NH 03886, phone (603) 323-8843.

values. Nelson's book, which is based on his industrial experience, gives many useful studies of real-world problems. As stated earlier, our example will assume the Arrhenius life-stress model and the Weibull life-distribution model. Let's first give the true Arrhenius model from which the data will be generated. Our data is made up, but we can give it a physical interpretation if we wish. Let's pretend that the life is the hours to failure for an adhesive-bonded joint in a structural member under a specified load. The stress causing failure is the temperature of the member, and we wish to estimate by accelerated testing the life at the nominal design temperature of 300 K. The relation between median life L and absolute temperature T is taken as

$$L = 1.5826 \, e^{3107.9/T} \qquad (4.60)$$

where L and K_1 are in hours and K_2 and T are in kelvins. These numbers were chosen to give a median life of 50,000 h at the design temperature of 300 K and 500 h at 540 K. The accelerated test plan will use 10 specimens at 540 K, 6 at 440 K, and 14 at 340 K. How were these numbers chosen? Deciding on the total number of specimens (30) requires a tradeoff between cost and accuracy and is, of course, helped in a real-world application by the previous experience of the company in similar situations. Choice of three levels of temperature and the allocation of specimens among these levels follow Nelson's general guidelines, which we gave above. The actual temperatures used are guided by knowledge of the material behavior and tradeoffs between accuracy and cost. If we choose our lowest temperature *close* to the design temperature of 300 K, then we will not need to extrapolate our results very far and they will be more accurate. However, low temperatures result in *long* lives, which defeats the purpose of accelerated testing.

We next need to use the Weibull distribution and the STATGRAPHICS random number generator to produce the simulated data. Parameter α can be calculated from our desired median life and Eq. (4.59) as soon as we decide on β. Since we are generating simulated data, we can use any β we wish; let's use 0.8 to get a failure rate that decreases with time. Using $\beta = 0.8$ and α values of 23,354., 2924.9, and 790.78, respectively, for temperatures of 340, 440, and 540 K, STATGRAPHICS produces the simulated data of Fig. 4.66. Since these are all rather small samples, we need to remember that small samples of any distribution will not have the characteristics of the total population. (This does *not* invalidate our data as a simulation of real-world behavior. Small samples of real-world data behave in exactly the same way). To check out this effect, I had STATGRAPHICS fit the best Weibull functions to these data and got the following results.

Run	Life, h $T = 340$	Life, h $T = 440$	Life, h $T = 540$
1	101,641.	12,730.	3,441.
2	37,008.	4,635.	1,253.
3	39,829.	4,988.	1,349.
4	30,873.	3,867.	1,045.
5	12,604.	1,579.	427.
6	9,418.	1.180.	319.
7	6,266.		212.
8	10,967.		371.
9	19,717.		667.
10	375.		12.
11	72,928.		
12	28,019.		
13	63,503.		
14	11,283.		
Total test hours	444,432.	28,978.	9,098.

FIGURE 4.66
Simulated data for accelerated test experiment.

Temperature, K	α	β
340	32,269.	1.04
440	5,308.	1.36
540	885.	0.94

These deviations from the numbers that we put into the random number generator are typical of small-sample behavior.

When we recall that 8760 h $= 1$ year, the test data at $T = 340$ indicate that most of these data would be impractical to obtain in a real-world experiment (101,641 h is almost 12 years). These long times to failure are, of course, a consequence of the assumed 50,000-h (5.7-year) median life at the design temperature. If we knew from past experience that such long lives could be expected, we would not plan our experiment with runs at such low temperatures. If we have no past experience, it might be best to run some preliminary experiments at high temperatures such as the 540 K data and extrapolate these data to lower temperatures, to help us decide on the lowest temperature at which to test specimens. The data at 340 K in Fig. 4.66 also make clear the *need* for some kind of accelerated testing when the design life is so long. Even though our data is partly unrealistic, we will still use it to show the graphical analysis methods since the methods are the same regardlesss of whether the data is economically realistic.

We now pretend that the data of Fig. 4.66 is the result of actual testing, and we prepare the two graphs on the special graph paper. Figure 4.67 is the Weibull plot, and Fig. 4.68 is the Arrhenius plot. While either of these two plots could be made first, I prefer to make the Weibull first, using the median-ranks plotting scheme of Eq. (2.48), since these are all rather small samples. On Weibull paper, perfect Weibull data plot as perfect straight lines; but, as usual, small samples cannot give this behavior, even if they have been picked from a perfect Weibull population, as our STATGRAPHICS random number generator does. The Weibull distribution has two parameters, and these can be estimated from the graph if the data is judged close enough to a straight line. One fits the best straight line by eye (least-squares calculations are usually not worth the effort) and then translates this line over to the upper left corner of the paper, where a scale for estimating β is found. Move the line until its right end passes through the crosshair symbol, and then read β from the vertical scale at the left. Rather than *finding* a β value this way, since our simulated data has a known beta, I chose to lay in a $\beta = 0.8$ line on the scale and then translate this line over to the three sets of data. You can now pretend that you *don't* know β and judge how badly you might miss the correct value by comparing the "correct" line with various ones you fit by eye to the data. While one can obviously fit lines of various slopes to this data, it is clear that lines with $\beta = 0.8$ *are* about as reasonable as any. The other Weibull parameter, α is quickly found by noting where the 63.2 percent level cuts across the data. (The graph paper highlights this level by marking it with another crosshair symbol, in the interior of the paper.) One can do this either before or after fitting the straight lines. If we use the straight lines shown, the α values for temperatures of 340, 440, and 540 K are, respectively, about 33,000, 6000, and 1000. You can compare these values with those used to generate the data, but remember that we can get *various* values of both α and β simply by shifting our best-fit lines.

We now turn to the Arrhenius plot of Fig. 4.68. The data can be plotted in various ways here. I decided to not show all the data points. Rather I transferred from the Weibull graph enough points to show the median (50 percent) line and lines for 25 and 75 percent. Again, one could estimate these numbers either from the fitted lines or from the numerical data values. In this case I used the numerical values in a

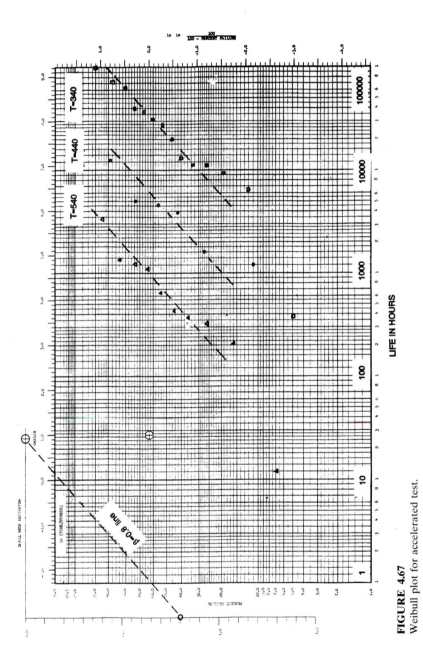

FIGURE 4.67
Weibull plot for accelerated test.

304

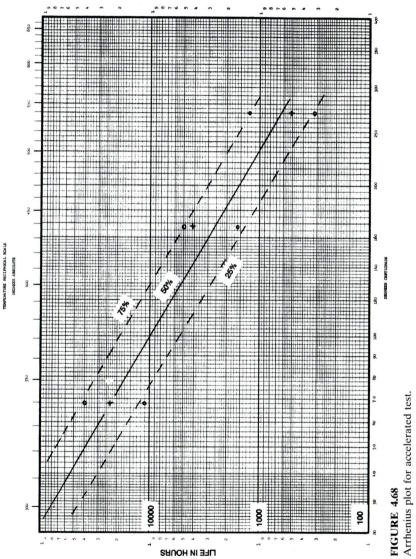

FIGURE 4.68
Arrhenius plot for accelerated test.

305

STATGRAPHICS routine which outputs the desired values. Using this approach, we get three points each for the median line and the other two lines. As usual, the three points for each line do *not* all fall in one line, so we again use judgment in deciding where to place these lines. Here we should make use of the fact that the $T = 340$ K points are based on a sample of 14 while the $T = 440$ K points number only 6 and $T = 540$ K points number 10. In placing the lines we thus give most weight to the 340 K data, next most to the 540 K data, and least to the 440 K data. For perfect data and large samples, the three lines are parallel. Using these guidelines, I drew the lines as shown. Note that the $T = 440$ K data are least reliable since the points for 50 and 75 percent are much closer to each other than they should be.

In a real-world problem, at this point we would have to decide whether all these graphs were close enough to the assumed models that we felt comfortable using the Arrhenius plot to predict the median life at the design temperature of 300 K. Since we have shown you what *perfect* (but small-sample) data might look like, you can use this to help you make the needed decision. In our present example the extrapolated median line gives a life of about 80,000 h. This deviation from the known true value of 50,000 h shows the approximate nature of these methods, even when we have perfect (but small-sample) data.

As a final calculation I had STATGRAPHICS generate a Weibull sample of 15 specimens, using $\beta = 0.8$ and a temperature of 300 K (median life = 50,000 h). This simulates an experiment we might run *instead* of the accelerated tests. That is, our real goal is an estimate of life at the design temperature, so why not run our tests there? Since we know that the lives will be long here, we decide to use a total of 15 specimens rather than the 30 used in the accelerated testing. Figure 4.69 shows this data set. Its median life is 94,877 h (even though we asked for 50,000 h), and the total test time is 1,770,920 h. If our test ovens can hold enough specimens to test them all at once, then the total test time is not the proper measure of time and expense. A better comparison would be made with the *longest* individual specimen life. For the accelerated test this is 101,641 h while for the test at design temperature it is 344,173 h. This gives an acceleration factor of about 3.4. We mentioned earlier that the testing at $T = 340$ K probably was not wise since this value was too close to 300 K, giving long test times. If we had been wiser and used, say, 380 K as the lowest test temperature, our acceleration factor would have been considerably better. In addition, testing at several temperatures rather than only one gives us useful information about the validity (or lack thereof) of the Arrhenius model for the physical process we are studying. This single example of accelerated testing should give you a good appreciation of its capabilities and the methods used to implement it.

Run	Life, h $T = 300$
1	344,173.
2	125,316.
3	134,868.
4	104,541.
5	42,679.
6	31,892.
7	21,216.
8	37,134.
9	66,762.
10	1,269.
11	246,946.
12	94,877.
13	215,031.
14	38,206.
15	266,005.
Total time	1,770,920.

FIGURE 4.69
Simulated data for temperature at design value.

We now want to give a brief discussion of *environmental stress screening,* the second type of accelerated testing application which we mentioned earlier. Recall that the main purpose here is to subject 100 percent of the output of an established and normally operating manufacturing process to some kind of overstress which does *not* damage good products but does force latent and incipient failures to occur before shipment to customers. It is perhaps most widely used with electronic, electromechanical, and electro-optical products. Although the terminology is not completely standardized, when the procedure is applied to a component (transistor, diode, integrated circuit, etc.) rather than a finished product, the term *burn-in* is often used.[1] The simplest forms of burn-in subject the component to, say, 168 h of 125°C operation. This

---

[1] *A Guideline to Component Burn-in Technology,* Reliability Inc., Houston, TX, phone (713) 492-0550.

weeds out a large percentage of faulty components before they get built into complex systems where they would cause expensive trouble later. Some applications require more complex forms of overstress, the most common here being temperature *cycling*.[1] Here, components or complete systems are subjected to temperature cycles with carefully selected rates of temperature change between a hot and a cold temperature. This has been found to be the most effective form of stress for many applications since it causes differential thermal expansion between component parts, precipitating failures in solder joints, glass-metal seals, wire bonds, etc. In some cases, combining mechanical random vibration with thermal cycling can encourage latent failures that would not be brought forth by either stress alone. This combination may also lead to a shorter testing time to accomplish the same level of screening. Recently, novel vibration shakers using pneumatic actuators and segmented shake tables to produce a vibration environment including all six degrees of translational and rotational freedom have been claimed to be particularly effective in such screening programs.[2]

We conclude this section on accelerated testing with a comment on possible future directions. Most present-day work uses stress-life relations (such as the Arrhenius and inverse power) which have little or no theoretical background. Of course, this does not prevent them from being effective, but a general trend in all engineering is to put procedures on a more scientific and rational basis whenever practical. Such studies are particularly valuable since they allow us to improve products at the *design* stage, rather than having to depend entirely on test results *after* design and manufacture. This is starting to happen in accelerated testing, and I wanted to reference at least one study of this type to give those interested some insight into the details. It comes from the electronics manufacturing area, one of the most important economically for any industrialized country. Specifically, it deals with the failure of *surface-mount* devices, a recent innovation in the technology of assembling electronic components to circuit boards. Environmental stress screening using temperature cycling is widely employed to enhance the reliability of such products, so we are interested in this stress-life relation. By using established techniques from materials science and mechanical engineering, equations have been derived using basic material properties and part dimensions to predict relations between temperature cycling parameters and fatigue

---

[1] *The Environmental Stress Screening Handbook,* Thermotron Industries, 291 Kollen Park Dr., Holland, MI, phone (616) 392-1491.

[2] Screening Systems Inc., 7 Argonaut, Laguna Hills, CA 92656. QualMark, 10218 Osceola Court, Westminster, CO 80030.

life.[1] This more scientific approach to accelerated testing is a fruitful area for future research.[2]

PROBLEMS

4.1. Assume that the modulus of elasticity of a certain class of materials is known from past experience to have a coefficient of variation ranging from 0.05 to 0.10. You need to measure the modulus of a new material of this class with an accuracy of 5 percent at a 95 percent confidence level. How many specimens should you plan to run? If this many specimens is too expensive, how much can you save by relaxing the confidence level to 90 percent?

4.2. Add a curve for 85 percent confidence to Fig. 4.3b.

4.3. In Fig. 4.2a, consider specimens 1 to 10, 11 to 20, 21 to 30, . . . , 91 to 100 as separate 10-item samples, and compute the average value for each such sample. Briefly comment on the significance of these results.

4.4. Using available software, repeat the study of Fig. 4.8 for R and C mean values each 2 percent high, standard deviations each 10 percent high.

4.5. Assume the cost of testing is $500. + 20n$, where n is the number of specimens. The total testing budget cannot exceed $1000. If you want to know the standard deviation within ± 25 percent, can you afford to require 95 confidence? Can you afford 90 percent confidence?

4.6. Modify Eq. (4.9) for the cases where the confidence interval is 5 and 20 percent of $\bar{x}_{A-B}$. If $s = 1.0$ and $\bar{x}_{A-B} = 5.0$, compare the sample sizes needed for confidence intervals of 5, 10, and 20 percent of $\bar{x}_{A-B}$.

4.7. Modify Eq. (4.9) for the case where you want a 99 percent confidence level.

4.8. In comparing two mean values which are very close to each other, explain the critical role played by the amount of scatter in the data.

4.9. Do the calculation needed to augment Fig. 4.14 with the results for $n = 35$.

4.10. Do the calculations suggested in the text after Eq. (4.22).

4.11. Carry out the derivations suggested in the text after Eq. (4.26).

4.12. Verify the sum of squares due to factor effects results given just after Eq. (4.37) for t, w, and m.

4.13. Using Fig. 4.37, devise a plan for an experiment with six factors, each at three levels, and randomize the run sequence. If you use multiple regression to analyze the data and include only linear terms in the six factors, what is the maximum number of two-factor interaction terms you can estimate? If you include both linear and squared terms in each factor, how many two-factor interactions can be estimated?

---

[1] W. Engelmaier, "Surface Mount Solder Joint Long-Term Reliability: Design, Testing, Prediction," *Soldering and Surface Mount Technology*, no. 1, pp. 14–22, February 1989; "Surface Mount Attachment Reliability of Clip-Leaded Ceramic Chip Carriers on FR-4 Circuit Boards," *International Electronics Packaging Society Journal*, vol. 9 (4): 3–11, January 1988.

[2] J. M. Hu, et al., "Role of Failure-Mechanism Identification in Accelerated Testing," *Journal of the IES* (Institute of Environmental Science), July/August 1993, pp. 39–44.

4.14. Is the six-factor, three-level plan of Prob. 4.13 the most frugal if we want to estimate *only* the six main-effect linear terms? Consult references to find the most frugal orthogonal plans for such an experiment. Repeat for a six-factor, two-level experiment.

4.15. Using available software, rerun the analysis leading to Eq. (4.41), but use noise terms twice the size of those in Fig. 4.44. Compare your results with those in the text.

4.16. Make up an expression like Eq. (4.45), except use *three* exponential terms (three time constants). Generate noisy data as in the text, and then analyze it using available nonlinear regression software.

4.17. What is the *minimum* number of runs that you could use with the data of Fig. 4.56 to fit a plane? Choose several such data sets from the nine points given, and compute the best-fit planes. Compare these results with Eq. (4.46), and discuss the significance of this study.

4.18. What features of Fig. 4.57a and b tell us that a plane is not the most correct shape for fitting this surface? Why do we still prefer to use a plane at this stage of the study? What simple graphs could you plot from the data of Fig. 4.56 to visualize the shape of the surface?

4.19. Using the data of Fig. 4.66, confirm the plotting of these data on Figs. 4.67 and 4.68. Try different visual straight-line fits of the data in Fig. 4.67 to see how much α and β might change. Would these different α and β values influence the graphs of Fig. 4.68? Explain.

4.20. Using available software, generate accelerated test data for the situation of Fig. 4.66, but use 380, 460, and 540 K as the test temperatures. The design temperature is still 300 K, so this test will be *more* accelerated than that of Fig. 4.66, an improvement suggested in the text. Analyze this new data and compare results with those of the text example.

4.21. Analyze the accelerated test data of Fig. 4.66, using the models below, and compare your results with those of the text model.
  a. Arrhenius model and lognormal distribution
  b. Arrhenius model and exponential distribution
  c. Inverse power law and Weibull distribution
  d. Inverse power law and lognormal distribution
  e. Inverse power law and exponential distribution

4.22. For the experiment of Fig. 4.40, suppose there are only two levels for each factor. If you analyze the data with multiple regression, what is the minimum number of runs needed if you don't require orthogonality? Make a table showing such a plan.

4.23. For the experiment of Fig. 4.40, suppose F has two levels, T has three levels, W has four levels, and M has five levels. What is the minimum number of runs needed for a multiple regression analysis if you don't require orthogonality? Make a table showing such a plan.

BIBLIOGRAPHY

1. R. L. Scheaffer and J. T. McClure, *Statistics for Engineers*, Duxbury Press, Boston, 1982.
2. C. Daniel and F. S. Wood, *Fitting Equations to Data*, Wiley, New York, 1980.

3. G. E. P. Box and N. R. Draper, *Empirical Model Building and Response Surfaces,* Wiley, New york, 1987.
4. M. S. Phadke, *Quality Engineering Using Robust Design,* Prentice-Hall, Englewood Cliffs, NJ, 1989.
5. N. Logothetis and H. P. Wynn, *Quality Through Design: Experimental Design, Off-Line Quality Control, and Taguchi's Contributions,* Clarendon Press, Oxford, 1989.
6. G. E. P. Box, W. G. Hunter, and J. S. Hunter, *Statistics for Experimenters,* Wiley, New York, 1978.
7. T. B. Barker, *Quality by Experimental Design,* Marcel Dekker, New York, 1985.
8. J. Sall, *SAS Regression Applications,* SAS Tech. Rep. A-102, SAS Institute, Box 8000, Cary, NC, 1981.
9. W. Nelson, *Accelerated Testing: Statistical Models, Test Plans, and Data Analyses,* Wiley, New York, 1990.
10. W. J. Diamond, *Practical Experiment Designs for Engineers and Scientists,* Wadsworth, Belmont, CA, 1981.
11. R. L. Mason, R. F. Gunst, and J. L. Hess, *Statistical Design and Analysis of Experiments,* Wiley, New York, 1989.
12. T. B. Barker, *Engineering Quality by Design: Interpreting the Taguchi Approach,* Marcel Dekker, New York, 1990.
13. F. W. Breyfogle, III, *Statistical Methods for Testing, Development, and Manufacturing,* Wiley, New York, 1992.
14. K. R. Bhote, *World Class Quality: Using Design of Experiments to Make It Happen,* Amacon, New York, 1991.

CHAPTER
5

PROJECT
PLANNING

5.1 OVERVIEW

The experiment plans described in Chap. 4 provide a conceptual framework for the experiment but do not address many other aspects of planning that are vital to the success of experimental projects. The plans of Chap. 4 are concerned mainly with rather theoretical and statistical issues, and we now want to address some of the more practical concerns associated with actually carrying out a project. Since this book is intended both as a textbook for engineering students and as a reference for practicing engineers, we will provide information useful to both groups. While many of these topics will apply equally to students and practitioners, student experimental work is subject to some unique constraints which we want to recognize. Figure 5.1 outlines the sequence of major steps involved in planning a comprehensive experiment. Simple experiments of course will not require consideration of all these questions; however, the outline can serve well as a checklist to aid in deciding which steps are necessary and which can be ignored or treated casually when one starts to plan any experiment. Although the arrows in Fig. 5.1 imply a definite time sequence for the various steps, the design of an experiment, like any design situation, involves an iterative process with feedback to earlier steps from information developed in later steps. Also, the scheme of Fig. 5.1 is certainly not the *only* way to organize the

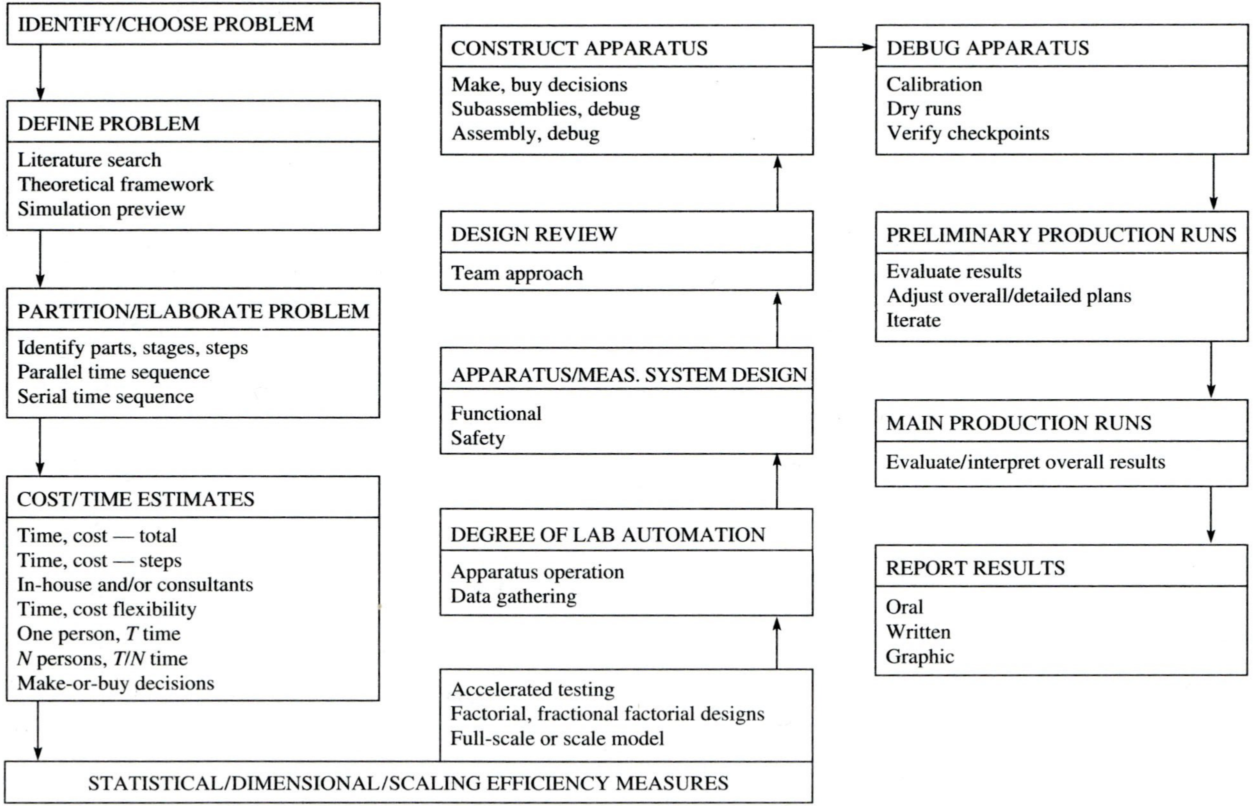

IDENTIFY/CHOOSE PROBLEM

DEFINE PROBLEM
Literature search
Theoretical framework
Simulation preview

PARTITION/ELABORATE PROBLEM
Identify parts, stages, steps
Parallel time sequence
Serial time sequence

COST/TIME ESTIMATES
Time, cost — total
Time, cost — steps
In-house and/or consultants
Time, cost flexibility
One person, T time
N persons, T/N time
Make-or-buy decisions

CONSTRUCT APPARATUS
Make, buy decisions
Subassemblies, debug
Assembly, debug

DESIGN REVIEW
Team approach

APPARATUS/MEAS. SYSTEM DESIGN
Functional
Safety

DEGREE OF LAB AUTOMATION
Apparatus operation
Data gathering

Accelerated testing
Factorial, fractional factorial designs
Full-scale or scale model

DEBUG APPARATUS
Calibration
Dry runs
Verify checkpoints

PRELIMINARY PRODUCTION RUNS
Evaluate results
Adjust overall/detailed plans
Iterate

MAIN PRODUCTION RUNS
Evaluate/interpret overall results

REPORT RESULTS
Oral
Written
Graphic

STATISTICAL/DIMENSIONAL/SCALING EFFICIENCY MEASURES

FIGURE 5.1
Flowchart for planning of experimental projects.

process of experiment planning, but it should serve most readers as a useful tool, especially if they have not used *any* orderly process for such work in the past. Some of the items in this process will require only a brief discussion, while others warrant their own section in this chapter or later chapters.

We begin our discussion by considering how one *chooses* a particular problem for experimental study. Here already, academic and industrial situations differ. Most industrial problems "choose themselves." That is, problems just naturally arise during the design, development, and manufacture of a product or service, and an experimental attack on the problem may seem most appropriate. Of course, in an industrial lab devoted to *basic* research, managers *do* need to carefully choose the areas they wish to pursue as foundations for future product development or problem solving. In academic environments, the choice of problem topics can be wider since no economic payoff need result. The *scope* of a project may, however, need to conform to the time constraints of an academic quarter or semester.

Once a topic has been chosen, it must be carefully defined, and then a literature search is usually carried out. Even though we are emphasizing experimental studies, it is usually wise to initiate some theoretical exploration of the problem also. Computer simulation using statistical (SAS, STATGRAPHICS, etc.), dynamic system (ACSL, SIMULINK, etc.), or other special-purpose or general-purpose tools can quickly give insights into system behavior, useful in experiment planning. Next we need to see whether we can break down (partition) the overall experiment into parts, stages, or steps. This is especially important in complex experiments, where it is easy to lose sight of essential concepts in a morass of detail. If more or less distinct steps can be identified, must these steps be taken "in series," or can we speed up the investigation by a "parallel" attack on several steps simultaneously? Can we use, in part or entirely, any of the thousands of standardized tests and codes developed and disseminated by industry groups or engineering societies?

It will also be desirable to develop estimates of the total time and cost and how these will be allocated to the various stages of the study. Does a tight time schedule and/or lack of in-house capabilities suggest the hiring of outside consultants or testing labs? Has management given us any idea as to allowable slippage in time scheduling or budget, and the priorities of this project with respect to others going on? Do we have the flexibility to assign our human resources to suit our needs? That is, can we use one person on a job and tolerate the time extension, or should we assign N persons and cut the time by a factor of N? Can we trade money for time by purchasing from "outside" suppliers the equipment items that *could* be made in-house?

We should also define several points along the way where we

consciously stop and *evaluate* our progress toward the goal and decide if we need to *change* our original plan and how. This includes the sometimes painful decision to acknowledge failure and terminate the project. Engineers, just as other mortals do, have difficulty with self-criticism and get into the habit of concluding every "final" report with strong suggestions for "further study" and, of course, the funding to pursue this. When such suggestions are mainly self-serving and not justified factually, they lead to a confusion between company (and perhaps even national) goals that can result in misallocation of resources.

We next need to consider whether certain measures which can enhance the efficiency of experiments are applicable to our current problem. Chapter 4 has already described in detail some factorial, fractional-factorial, and accelerated testing schemes, so we need not discuss these again. We do need to introduce methods of dimensional analysis and scale modeling since these can often be used to good advantage. A separate section of this chapter will describe these techniques.

For all except the simplest studies, we need to decide on the degree of automation that is appropriate for our experiment. This refers to the use of computer hardware and software, together with associated sensors and actuators, to replace or augment human supervision and/or operation with computer-controlled methods. We need to consider both the operation of the experimental apparatus and the gathering and processing of the data. While computer-controlled operation of the apparatus will often be too complex and expensive to justify for modest experiments, computerized data gathering and processing equipment is widely available and general-purpose, so it may be feasible for even brief experiments. Lab automation will be considered in more detail in Chap. 6.

While we have already devoted a chapter to measurement system design, it needs to be integrated with apparatus design, and this will be done in Chap. 6. Apparatus design is mainly concerned with meeting functional specifications, but we also need to carefully consider safety aspects, with respect to protecting both human operators and valuable equipment.

When the experiment planning reaches this point, it is appropriate to schedule a comprehensive design review before the design is finalized. Brief design reviews of critical parts of the design may be desirable even earlier in the overall process. It is important to involve in this comprehensive design review all individuals and groups who can contribute to discovery of faults in the design and can suggest improvements. Machine shop personnel can give valuable input concerning the ease of manufacture. In the industrial situation where the experiment may be designed by engineers but executed by technicians, it is vital to involve these individuals in an interactive way during the design review. This is particularly important when the engineer is a recent graduate of a curriculum which may have been almost entirely theoretical in focus. The practical viewpoint of experienced technicians can augment the engineer's contribution in many useful ways.

When the design review is finished, apparatus construction can logically proceed. Although make-or-buy decisions were part of the earlier cost/time estimating stage, they may again need to be addressed at this later stage. For complicated apparatus, preliminary testing and debugging of subassemblies may be desirable to discover problems before they become incorporated into the larger apparatus, where changes may be more difficult and expensive to make.

When the preliminary debugging of the construction phase is complete, the apparatus and measurement system are joined (partial joining may already have occurred earlier) for final overall debugging. In-place calibration, if possible, of all components of the measurement system should now be carried out. "Dry runs" intended to exercise the overall apparatus and measurement system rather than produce useful data are now appropriate. Any remaining flaws revealed at this stage need to be corrected.

We are now ready for so-called "production runs," i.e., runs which develop the information desired from the experiment. Early production runs may be considered preliminary in the sense that we are just beginning to develop some understanding of the subject of our investigation. This early information may *change* some of the goals and techniques that we previously thought appropriate, requiring some rethinking and redesigning of our plans and apparatus. Several iterations of this type of adjustment may be necessary as we develop more and more insight into the phenomena under study.

While the adjustments just mentioned may continue for the life of the study, in some experiments things do settle down at some point and the apparatus and method of use become quite stable. When and if this occurs, one can then speak about "main" production runs. Here operation is relatively routine and programmable, and large quantities of data can be quickly obtained. As data is gathered, it should be immediately subjected to ongoing evaluation and interpretation so that any trends detected can be used to steer the investigation in the most profitable directions. Such evaluation is also vital for correctly deciding when the study should be *halted.*

Figure 5.1 explicitly displays a reporting stage at the *end* of the overall process; however, interim reports of various types and scopes should be used earlier wherever appropriate. Since reports of one kind or another are the vital information links between members of the experimental team and management, the importance of the reports cannot be overestimated, and we later devote an entire chapter to this area.

The rest of this chapter will be devoted to discussion of those parts of Fig. 5.1 which do not require separate chapters but do warrant a more extended treatment than given above.

5.2 SELECTION OF PROBLEM FOR STUDY

Since we stated earlier that many industrial problems "select themselves," our brief discussion here will focus on the academic environment. Some guidelines for planning industrial research and development addressed to long-term questions rather than immediate and specific problems can be found in the literature.[1] In choosing the best problems in an academic environment, the major constraints are of course the available time and facilities and the previous preparation of the students. Since major efforts such as master's theses and Ph.D. dissertations will normally be dictated by the ongoing programs and interests of the faculty adviser, we will not address this area but instead focus on short-term projects typified by studies occupying all or part of a quarter-long or semester-long "project lab."

Since the faculty in charge will likely exert a deciding influence on the choice of topics, our advice is addressed perhaps more to them than to student readers of this book, although students may have some input into the process and thus benefit from understanding what is involved. Clearly, advice to faculty will be welcomed more by those who are teaching such a course for the first time than by seasoned veterans, thus our suggestions should be interpreted in that light. Our first and perhaps overriding criterion is *simplicity*. Topics which at first glance appear to offer no insurmountable challenges have a way of becoming frustratingly complex as they are pursued into actual detail. While much can be learned from "unsuccessful" projects, when students, due to crowded curricula, may encounter only one project lab experience in their entire student careers, we want them to feel some positive accomplishment. We thus recommend that instructors err on the side of simplicity and also try to think through the experiment in sufficient detail so as to anticipate any

---

[1] R. C. Parker, "The Art and Science of Selecting and Solving Research and Development Problems," *Proceedings of the Institute of Mechanical Engineers*, 185: 879–893, 1970–71, 64/71. L. W. Steele, "Selecting R&D Programs and Objectives," *Research-Technology Management*, March-April 1988, pp. 17–36. E. B. Roberts, "Managing Invention and Innovation," *Research-Technology Management*, January-February 1988, pp. 11–29. T. W. Jackson and J. M. Spurlock, *Research and Development Management*, Dow Jones-Irwin, Homewood, IL, 1966.

problems that might detract from the students' lab experience. Of course, this is a time-consuming exercise for the teacher but is generally well repaid in student achievement.

Our next suggestion is somewhat controversial since it goes against a fairly common current practice: Faculty use a project lab as an adjunct to their sponsored research programs, which are concerned mainly with the education of graduate students. Graduate research is often of a basic rather than applied nature, even at the master's degree level. Assuming that the project labs to which this text is addressed are part of an *undergraduate* program, I believe that they should have a different focus, more appropriate to the work that most bachelor of science degree holders will encounter in their industrial careers. This focus is on design, development, and manufacture rather than basic research. I would thus encourage the selection of project lab topics closely related to product and process development rather than the extension of basic engineering science. This again requires additional faculty time since it implies a broadening of interest and competence beyond the narrow specialization necessary in much basic research. To be more specific, I feel that one can find many excellent project lab topics by considering specific *products* or *processes* already in existence and looking into ways to better understand their behavior and improve their performance. When properly defined, this is *not* "routine testing" inappropriate for academic study but rather the essence of most experimental work in engineering practice. Most industrial companies these days are very willing to contribute sample components for student lab studies, and the company engineers may even be eager to help academics to appreciate practical problems. There is actually much to be learned from such informal contacts for faculty who are daily immersed in the academic environment, which may be entirely focused on basic research.

Some specific examples of product and/or process development topics suitable for project lab use may be helpful. Fluid power equipment often is a good choice since it involves an interesting mix of phenomena from several engineering science areas, such as thermodynamics, fluid mechanics, solid mechanics, and vibration. An apparently simple device such as a pneumatic cylinder displays a wealth of behaviors that are of interest to both designers and users of cylinders and that can be studied with fairly simple instrumentation. The friction forces due to seals and bearings at the piston and piston rod are significant in many applications but are not well documented or understood. These forces depend on many variables such as temperature, pressure, piston velocity, and degree of wear-in and are not the same from one cylinder to another of identical type. Many interesting and practically useful studies can be designed around these frictional effects. The dynamic behavior of these cylinders offers more areas for study. Time histories of displacement, velocity, and/or acceleration for sudden opening of supply valves and for different types of loads (masses, springs, dampers, etc.) can be studied. Economic studies involving air consumption for various defined tasks are possible. Comparison studies of cylinders from competitive manufacturers, using various performance and cost criteria, can be instructive. Another interesting area is the determination of failure modes and the associated safety factors. (For "dangerous" failures, faculty supervision of course must be especially careful.) Air cylinders are designed and marketed to be used under specific environmental conditions, usually with rather large safety factors. We might want to use a cylinder under extreme conditions, and we might be willing to accept a shorter life and/or degraded performance. Thus we might study the question: What is the highest pressure the cylinder can be operated with if we need only (say) 10 cycles of operation? For excessive air pressure, *how* does the cylinder actually fail? Similar studies involving excessive temperatures, dirty air supply, excessive side loads on the piston rod, and excessive moisture in the air supply might be of interest.

Another simple device permitting many experimental studies is a linear (translational) electromechanical solenoid. These devices are difficult to accurately analyze and usually are characterized by experimental testing. Some possible project lab studies include static force/displacement curves for a fixed current; force/current relations at various fixed displacements; time histories of displacement, velocity, and/or acceleration for sudden application of various fixed voltages; temperature/time histories for various repetitive duty cycles; and frictional effects. Again, studies to discover and document the various failure modes due to steady overvoltage, excessive cycles of operation at overload conditions, etc., are interesting and practical topics. We can also compare selected performance characteristics for several supposedly identical solenoids to document the effects of manufacturing variations.

I hope the examples just given will communicate the philosophy behind making such choices and allow you to extend this viewpoint to many other simple mass-produced devices available at no cost from their manufacturers. Many students welcome such practical adjuncts to their mostly theoretical curriculum and respond with enhanced motivation and effort.

5.3 PROBLEM DEFINITION

While the selection of a topic obviously includes a certain level of definition of the problem, it is most important to develop a thoughtful and detailed problem description early. "A problem well defined is a problem half solved." We should overcome our eagerness to get underway and consciously set aside sufficient time to thoroughly think through and discuss with colleagues what we *really* want to accomplish. In the relatively simple tasks appropriate to a student lab project, the real task may be fairly obvious, so this aspect of project planning applies most importantly to industrial projects. As students, however, we don't want to develop bad habits which will be hard to break later, so try to perform this task conscientiously even if your specific project seems not to require an in-depth study.

A brief example illustrating various aspects of problem definition may be helpful. Suppose you work for a manufacturer of electric motors whose market has traditionally been composed of users whose applications are characterized by mainly steady-state (constant speed and load) operation. Lately, technological developments have required these customers to deal with unsteady motor operating conditions such as frequent starting and stopping. Our sales engineers report to management that they are losing significant numbers of sales to a competitor who appears to be able to meet customers' needs with smaller and cheaper motors than our applications engineers are recommending for the same applications. A preliminary investigation shows that we are determining motor sizes based on the *average* horsepower requirement of the customer's load. This is not surprising since our past business has been almost entirely based on steady-state applications, where average horsepower *is* a proper criterion.

Management has decided that we need to study and probably revise our motor sizing procedures to enable us to compete more effectively in this changing market. It is well known in the motor business that the life of a motor is closely related to operating temperatures since motor insulation deteriorates much more rapidly at higher temperatures. Our chief engineer feels that our motor sizing methods based on average-load horsepower and with no detailed consideration of heating problems are probably too rough and conservative, allowing our competitors, who may be using a more scientific approach, to successfully specify smaller and cheaper motors. She wants you to undertake some experimental studies to develop company expertise in the area of motor heating, so that we can more effectively compete. To be specific, you are to set up tests to find the final temperatures attained by motors running under various steady loads and, for the highest loads, the life of these motors.

While we all have a self-preservation instinct to do what the boss asks us, most bosses prefer that their subordinates exercise some independent judgment; so we would be wise to consider the boss's initial problem definition with some care, to see if this is what we *really* need to do. Supervisors are often busy with several different projects and may have become more concerned with management than technical issues, so they rely on their technical specialists to fill in the details. In the present example, since our company has *no* background in thermal problems, the experiments suggested certainly would add some useful information to our general company database; however, a little further consideration shows that a better plan can probably be devised. The problem here is that our company has been long dominated by "steady-state thinking," whereas the current problem seems to be inherently unsteady or transient.

The principle behind our competitor's ability to specify smaller motors for transient applications is that a motor has both thermal resistance *and* thermal capacitance. When a high load of *short* duration occurs, the motor's thermal capacitance (energy storage) can absorb this heating load without causing damaging high temperatures. For unsteady motor duty cycles, periods of high load are followed by periods of low or no load, during which the heat absorbed can be dissipated before the next high-load portion occurs. To take advantage of this phenomenon in our motor sizing practices, we must learn how to relate actual transient duty cycles to peak (not average) temperatures. It may also be necessary to find out how motor life is related, not to steady temperatures, but to definite temperature/time cycles or noncyclic time histories.

Experiments to develop this type of information will necessarily be more complex than those originally suggested, but they are more closely related to our *real* problem and therefore have a greater chance of giving a good return on our investment. Whether we originate a problem ourselves or have it given to us by others, we should always reserve an initial period of time for thoughtful consideration of how our proposed plan relates to the *real* goals of the study.

5.4 LITERATURE SURVEYS

Whether we are considering the theoretical or experimental aspects of some project, once we have agreed upon a clear problem definition, a logical next step in many cases will be some kind of background literature survey. Anyone who has ever started such a survey will agree that, even for seemingly trivial topics, the amount of material to be reviewed quickly becomes overwhelming; so we must exercise extremely critical judgment as to where to look and when to stop. In the academic environment of students, the task may be even more difficult than for the industrial investigator. This is because most industrial problems are closely related to the past and present business of the company, and there may be extensive in-house files on earlier projects in a related field. The urgency of many industrial problems also encourages a willingness to terminate a survey earlier than one typically would in academia.

We need to acknowledge at the outset the hopelessness of ever getting a complete literature file on any topic. There is now such a proliferation of

1. Technical journals (*Engineering Index* reviews nearly 5000)
2. Conferences (*Engineering Index* includes about 2000 per year)
3. Government publications
4. Patents (75,000 new ones per year in the United States, more than 1 million worldwide)
5. Academic theses and dissertations
6. Manufacturers' catalogs and technical notes
7. Standards and recommended practices from engineering societies
8. Books, handbooks, and encyclopedias

that one hardly knows where to start. A major study[1] of engineers and how they use information shows that engineers consider the informal network of colleagues as the most important source of information. Abstracting and indexing services, such as *Engineering Index,* are of course a great aid in locating possible primary sources. Some of these are now available on CD-ROM, making search more efficient. The CD-ROM version of *Engineering Index* is called COMPENDEX*PLUS. A CD-ROM service is also available[2] for U.S. patents. Newly issued patents are indexed within 7 days of publication. Some other useful abstracting and/or indexing publications are

1. *Applied Science and Technology Index*
2. *Government Reports Annual Index*
3. *Energy Research Abstracts*
4. *Science Citation Index*
5. *Scientific and Technical Aerospace Abstracts*
6. *International Aerospace Abstracts*

The purposes of a literature search are several. A main purpose, is, of course, to find out whether the study you are about to undertake has

---

[1] H. L. Shuchman, *Information Transfer in Engineering,* The Futures Group, Washington, 1981.

[2] MicroPatent, Chadwyck-Healy, 1101 King St., Alexandria, VA 22314, phone (703) 683-4890.

already been done. We don't want to waste time and money repeating someone else's work, unless we question their results for some reason. Because of the vastness of the open literature and the inaccessibility of proprietary company files, of course we can never be *sure* that we have uncovered all the work that has been done. When our proposed project does *not* exactly duplicate earlier work, the literature survey makes us aware of what *has* been done in the general area. Critical reading of this literature often gives us valuable guidelines in planning our own project.

An advantage of the academic environment is the availability of expert technical librarians who can help one get started on a search most efficiently. Be sure to make use of such personnel when they are available. Our discussion of literature searching here is intentionally brief. Consult with your local librarian if one is available. If not, we provide some references for self-study.[1]

5.5 USE OF DIMENSIONAL ANALYSIS IN SCALE MODELING AND FOR IMPROVEMENT OF EXPERIMENTAL EFFICIENCY

In Chap. 4 we saw that significant improvements in efficiency (more information with less investment of time and money) were sometimes possible by use of factorial or fractional-factorial experiment plans. These replaced the intuitive one-factor-at-a-time approach with a more subtle one of varying many factors at once. Further gains in efficiency can sometimes be realized by application of dimensional principles. Since an introduction[2] to such techniques is often covered in fluid mechanics or heat-transfer courses taken by many of the readers of this book, we will use this section mainly to remind you that these methods are available and focus on how they can be used in planning more efficient experiments in *all* fields, not just fluid mechanics and heat transfer.

---

[1] L. J. Anthony (ed.), *Information Sources in Engineering,* 2d ed., Butterworth, London, 1985. K. Subramanyam, *Scientific and Technical Information Sources,* Marcel Dekker, New York, 1981. C. A. Erdmann, "Improving the Information-Gathering Skills of Engineering Students," *Engineering Education,* May/June 1990, pp. 456–460. D. S. Ingram and J. D. McCoy, "Engineering Students and the Library: Teaching the Technology of Library Research," *Engineering Education,* May/June 1990, pp. 461–462.

[2] R. W. Fox and A. T. McDonald, *Introduction to Fluid Mechanics,* Wiley, New York, 1985, chap. 7.

The dimensional methods we want to discuss depend on the identification of *dimensionless groups* pertinent to the problem under study. These groups are found in two basic ways. One method requires that the fundamental differential equations governing the physical phenomena be written and nondimensionalized. The other method requires only that we make a list of all the pertinent independent variables on which our dependent variable depends. If we can find the appropriate nondimensional groups, they can be used in two ways that involve experimentation. One class of applications relates to the use of scale models, rather than the full-scale (prototype) device. The other class of applications involves a reduction of the number of independent variables that need to be varied in the experiment, thus saving time and money.

Let's first look at the class of applications where our goal is to reduce the number of independent variables that need to be studied in order to explore the full range of behavior exhibited by some device or process. Actually, such reduction is usually beneficial to *both* theoretical and experimental studies and is thus quite common in engineering analysis (*Optimization studies,* in particular, benefit from reduction of variables. If we are trying to find a combination of parameter values which optimizes some performance criterion, the optimum is found much more quickly and easily if the number of parameters is small.) When we are able to at least formulate (but perhaps not easily solve) the theoretical equations describing the object of our study, such formulation will often directly reveal useful nondimensional groups of parameters that can be utilized in an experimental study of the phenomenon. No special tricks of dimensional analysis may be needed in such cases. When the phenomena are too complex to even formulate a proper set of equations, then the specialized techniques of dimensional analysis often provide the dimensionless groups we need.

Suppose we intend to carry out an experimental study but we *are* able to derive the theoretical equations and can use them to find dimensionless groups of parameters useful for improving the efficiency of our experiment. An electrically heated, single-tube heat exchanger[1] makes a good example (see Fig. 5.2). Using suitable simplifying assumptions, we can derive the partial differential equations relating the unknowns (fluid temperature and tube wall temperature) to the system parameters (density, specific heat, and cross-sectional area of both fluid and tube wall; fluid velocity; convection heat transfer coefficient; inside

[1] E. O. Doebelin, *System Modeling and Response: Theoretical and Experimental Approaches,* Wiley, New York, 1980, pp. 437–447.

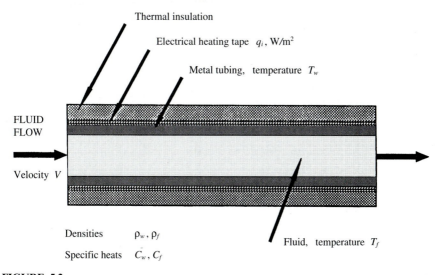

Densities ρ_w, ρ_f

Specific heats C_w, C_f

FIGURE 5.2
Heat exchanger.

and outside tube perimeters) and the system inputs (temperature of fluid as it enters the heat exchanger and instantaneous electric heating rate).

$$\rho_f c_f A_f \frac{\partial T_f}{\partial t} + \rho_f c_f A_f \frac{\partial T_f}{\partial x_a} = hP_i(T_w - T_f) \tag{5.1}$$

$$\rho_w c_w A_w \frac{\partial T_w}{\partial t} = q_i(t)P_o - hP_i(T_w - T_f) \tag{5.2}$$

These simultaneous equations contain 10 system parameters. In the reference, the objective was a theoretical (numerical) optimization study of a feedback control system in which the heat exchanger was the device controlled. To simplify this study, the equations were reformulated by using nondimensional groups of parameters, thereby reducing the number of *essential* parameters to *only two*. If the objective had been an experimental study of this class of heat exchangers, the same approach would have yielded a great simplification of the experiment. The reference may be consulted for all the details, but a sketch of the method is as follows. The actual space coordinate x_a (distance along the tube from the entrance) was made nondimensional by dividing by the tube length L, giving a new space coordinate $x \triangleq x_a/L$. Fluid residence time (time for a particle to pass through the tube) is defined by $t_r \triangleq L/V$. Using this residence time, we can define a new nondimensional time variable

$\tau \stackrel{\Delta}{=} t/t_r$. Time constants for a unit length of tube and fluid were defined by

$$\tau_f \stackrel{\Delta}{=} \frac{\rho_f c_f A_f}{h P_i} \qquad \tau_w \stackrel{\Delta}{=} \frac{\rho_w c_w A_w}{h P_i} \tag{5.3}$$

Definition of nondimensional temperatures θ_{wp} and θ_{fp} and a nondimensional electric heating rate H_p proportional to the actual heating rate q_i (see reference for details) leads to the final form of the equations.

$$\frac{\partial \theta_{fp}}{\partial \tau} + \frac{\partial \theta_{fp}}{\partial x} = \sigma(\theta_{wp} - \theta_{fp}) \tag{5.4}$$

$$\frac{\partial \theta_{wp}}{\partial \tau} = \frac{\lambda}{\sigma} H_p - \lambda(\theta_{wp} - \theta_{fp}) \tag{5.5}$$

These equations now contain *only two* parameters, λ and σ, where

$$\lambda \stackrel{\Delta}{=} \frac{t_r}{\tau_w} \qquad \sigma \stackrel{\Delta}{=} \frac{t_r}{\tau_f} \tag{5.6}$$

All possible variations of these two equations can be studied (either theoretically or experimentally) by exploring combinations of only the two parameters σ and λ, a much simpler task than for the original equation, which had 10 parameters. Furthermore, if we want to explore, say, a λ value of 3.0, we can now *choose* which of the seven *detailed* parameters (L, V, h, etc.) that go into lambda that are most convenient to actually change in our apparatus. Any combination of values which results in a λ of 3.0 will have the same effect on system behavior. We do, of course, need to be watchful for *interactions*. For example, an intentional and known change in velocity V may cause an unintentional (and perhaps unknown) change in the heat-transfer coefficient h, resulting in a λ change different from that expected for a simple change in V only.

Next we want to discuss the class of applications where theoretical equations are *not* available and systematic use of dimensional analysis techniques is required to identify the dimensionless groups useful in simplifying experimentation. As stated earlier, here we do not want to teach the details of how to find the dimensionless groups. Rather, we want to show how they enhance experimentation once they have been found. We do need, however, to review some of the essential features of dimensional analysis methods. Recall that one begins such a study by a careful consideration of the basic physical phenomena involved, in an attempt to identify *all* the independent variables that influence the dependent variable. While one may be able to later recover from a poor initial choice of variables, this aspect of the method puts a premium on experience and judgment, qualities not yet highly developed in most undergraduate engineering students. In some fields of application, however, these methods have been widely applied for many years, and considerable guidance is available from specialized texts.[1]

Once one has decided on the list of independent variables, the dimensionless groups can be found in a relatively routine manner. Originally, the dependent variable was some unknown function of all the independent variables. If we wanted to define experimentally this unknown function, our experiment would need to "exercise" all the independent variables in all their combinations. Once the dimensionless groups have been identified, the dependent variable can be expressed as some unknown function of these groups. Since the number of dimensionless groups to be "exercised" is generally much smaller than the number of original variables, experimentation to define this new unknown function is much more efficient. Langhaar[2] gives an example useful for our present purpose. We wish to study the friction torque T of a lubricated journal bearing which carries a radial load W and a moment loading M and which runs at constant angular speed N. The bearing has diameter D and length L, and there is a diametral clearance C between the shaft and the hole. It is conventional to work with bearing pressure $P \triangleq (W/LD)$ rather than W itself. Oil viscosity is μ, and we define our dependent variable as a friction coefficient $f \triangleq 2T/(WD)$.

Before dimensional analysis, we consider f to be some unknown function of seven variables:

$$f = f_1(P, M, L, D, C, N, \mu) \tag{5.7}$$

Langhaar shows that there are four dimensionless groups in this problem, so we can now represent the friction coefficient f as a new unknown function, but now there are only four independent variables, the four dimensionless groups.

$$f = f_2\left(\frac{\mu N}{P}, \frac{M}{PD^3}, \frac{L}{D}, \frac{C}{D}\right) \tag{5.8}$$

Let's see how our experimentation is facilitated by this new information. If we want to exercise the first dimensionless group in Eq. (5.8) while

---

[1] P. W. Bridgman, *Dimensional Analysis,* Yale University Press, New Haven, CT, 1931. H. L. Langhaar, *Dimensional Analysis and Theory of Models,* Wiley, New York, 1951. L. I. Sedov, *Similarity and Dimensional Methods in Mechanics,* Academic Press, New York, 1959.

[2] Op cit., p. 42.

holding the other three groups constant, we have a choice of varying the viscosity, the speed, the pressure, or any combination of these. As long as we explore the desired range of $(\mu N/P)$ values, it doesn't matter how that is achieved. For example, it might be most convenient to keep P and the viscosity constant and just vary the speed N. Once we have done this, there is *no* need to run further tests in which we would vary μ and/or P. Their effect on friction f has *already* been found when we varied N. Note that if we had decided to vary $\mu N/P$ by changing P, then since the other three groups must be kept constant, we would have had to make changes in M and/or D in order to keep $M/(PD^3)$ constant. If we chose to change D rather than M, then L and C would need to be changed to keep L/D and C/D constant.

You may recall (if you have studied dimensional analysis previously) that such analysis always begins with the analyst's making a list of all the physical variables which are *believed* to be involved in the phenomenon under study, but physical *laws* (such as Newton's law and Kirchhoff's laws) are *not* employed. This feature, which is both a strength and a weakness of the method, means that the list of variables chosen by a particular analyst may include exactly the correct variables but might also include some that are *not* pertinent or exclude some that *are* pertinent. When the list of variables is not exactly correct, one cannot assume that the *results* of the dimensional analysis will be correct either, and thus some precautions need to be taken. In our bearing example above, we might want to check whether $\mu N/P$ really behaves as a dimensionless group should. That is, instead of *assuming* that any values of μ, N, and P which give the same value for $\mu N/P$ are equivalent, we should run experiments using various combinations of these parameters and *check* whether these results verify the assumed behavior. This of course partially defeats the advantage of experimental economy which is a main feature of the method, but may nonetheless be necessary when we first try to develop dimensionless groups for a new phenomenon. Once the validity *has* been established, we can, of course, then take full advantage of this feature in future applications of the same phenomenon.

Let's use the bearing example above to give some details on how dimensionless groups are actually applied to experimentation. Bearings without the moment loading M were studied[1] by the National Bureau of Standards (now the National Institute of Standards and Technology). Since there was no moment loading in these tests, Eq. (5.8) has only the

---

[1] S. A. McKee and T. R. McKee, "Friction of Journal Bearings as Influenced by Clearance and Length," *Transactions of the American Society for Mechanical Engineers,* 51: 161–171, 1929.

three dimensionless groups $\mu N/P$, L/D, and C/D. This makes the friction coefficient a function of three "independent variables." Theoretical studies of such bearings indicate that the major effects should involve $\mu N/P$ and C/D and that L/D has a minor effect most noticeable when L/D is less than 0.7, a range rarely used in practical bearings. The simplest theories assume a bearing of *infinite* length to allow neglect of the oil flow in the axial direction. When L/D is greater than about 0.7, this "leakage" flow is small enough to cause little error in theory. For a given bearing, C/D and L/D are fixed, and we can display the effect of $\mu N/P$ on the friction coefficient conveniently by plotting a graph of friction coefficient f versus $\mu N/P$. We can then fix L/D at a chosen value and vary C/D to get a family of curves of f versus $\mu N/P$, showing the effect of C/D on friction. Finally, these families of curves can be obtained for different fixed values of L/D, to show the effect of the final dimensionless group. Figure 5.3 shows the general shape of these measured relations. If these graphical results are not themselves sufficient for bearing analysis and design, one could use multiple regression methods to fit a function to the tabular data. Since this could certainly be considered a "factorial" type of experiment, should we be using orthogonal arrays or other efficiency-improving methods that we discussed earlier, rather than the classical one-variable-at-a-time approach used by the referenced experimenter? Since multiple regression methods require assumption of the *form* of the functional relation, it might be best to use the classical graphical approach for the first experiments to get some feeling for the nature of suitable functions. Once graphs like Fig. 5.3 show us that the functions can be rather smooth and simple, *further* experiments might take advantage of multiple regression methods and orthogonal arrays.

The design and application of *scale models*[1] in engineering often employ dimensional methods and are thus appropriate for inclusion in this section. In the list of references just given, Schuring is especially useful. He lists over 400 references covering a wide range of model applications, and he introduces an unconventional approach to dimensional analysis. This approach utilizes physical *laws*, not just pertinent variables, in the formulation of the dimensionless groups, making the

[1] D. J. Schuring, *Scale Models in Engineering*, Pergamon, New York, 1977. Langhaar, op. cit. G. Murphy, *Similitude in Engineering*, Ronald Press, New York, 1950. J. Allen, *Scale Models in Hydraulic Engineering*, Longmans, Green, London, 1947. R. E. Johnstone and M. W. Thring, *Pilot Plants, Models, and Scale-up Methods in Chemical Engineering*, McGraw-Hill, New York, 1957. B. Hilston, *Basic Structural Behavior via Models*, Wiley, New York, 1972.

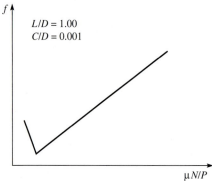

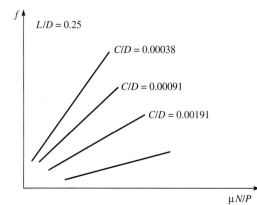

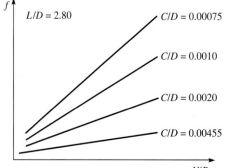

FIGURE 5.3
Relations among dimensionless groups for a bearing.

development more scientific and less arbitrary. A large part of the book is devoted to detailed discussion of specific case studies in modeling from a wide range of application areas, a few from Schuring's own work but mainly from the open literature. He analyzes and interprets each case study from a uniform point of view, giving critical insights into the methods used and the results obtained.

While computer simulation (mathematical modeling) has reduced the need in many cases for scale modeling, it continues to serve a useful function in many design projects. When the device under study is either very large or very small, experimentation on a scaled-down or scaled-up model may provide significant savings of time and money. Processes involving release or expenditure of large amounts of energy may be more safely and cheaply studied at a smaller scale. In the field of scale modeling, the full-scale device itself is usually called the *prototype*. When possible, we make our model geometrically similar to the prototype by requiring lengths l_m in the model to be related to lengths l_p in the prototype by

$$l_p = k_l l_m \tag{5.9}$$

where k_l is the length scale factor. Since the model is usually (but not always) smaller than the prototype, k_l is usually greater than 1.0. When we say that the model is, e.g., one-fifth the size of the prototype, we immediately must admit the need to often *deviate* from strict scaling of every detail. Most models are intended to be useful for only certain aspects of behavior, not *all* aspects. That is, when we are concerned about vibration problems of a rocket, a model of the rocket need be faithful to only vibration phenomena, not, say, heat-transfer phenomena. (We might build another model which addressed the heat transfer and not the vibration.) In a model strictly for vibration, we may want or need to relax perfect geometric scaling even for features which *are* pertinent to vibration. It is necessary only to duplicate the *essential* features of a phenomenon, not the minute scaled detail. For example, bending vibration models of rockets do not reproduce the complex internal details of the rocket engines. This would make the model extremely expensive and time-consuming to produce. Rather, the "engines" on the model are simplified to solid cylinders whose mass and moments of inertia adequately represent those of the prototype. Judgments as to when to relax perfect length scaling and when to enforce it obviously require both an understanding of the physical phenomena involved and lengthy experience in the particular industry.

Another fundamental decision in modeling relates to the *materials* to be used in the model. While we can choose our length ("size") scale factor at will, scaling relations for other aspects of the model may require impossible material properties. That is, we *cannot* require arbitrary relations between material properties; we must conform to the real materials available to us. If scaling requires for some reason that E/ρ for the model material be some specific number, a real material with that value may simply not exist. Schuring[1] shows, for example, that in structural problems if we use a model material with a modulus of elasticity E different from that of the prototype, then this model material must have the *same* Poisson's ratio as the prototype material, which is generally impossible. For this reason, many structural models relax the requirement to model Poisson's ratio and accept the errors so engendered. Fortunately, many such problems are little affected by Poisson's ratio, and usable results are obtained from model studies. When there is no advantage to using a different material for the model, we should use the same materials and avoid these scaling problems.

To give a simple illustration of some of the techniques used in relating model behavior to prototype behavior, let's now use as an example a structural vibration model employed during the development of the Saturn rocket. Figure 5.4 shows both the $\frac{1}{5}$-scale model and the full-size prototype.[2] Since the prototype is so big (160 ft long and 20 ft in diameter), the model itself is quite large (32 ft long and 4 ft in diameter). The large size of the model enjoys the advantage that most of the structural features can simply be directly scaled down without causing any manufacturing problems. It was also decided to use the same materials in the model as for the prototype. A rocket of this type has *many* potential vibration problems, including longituginal, torsional, and transverse (bending) modes of vibration. Since the structures are very efficient (high strength/weight ratio), the liquid fuel and oxidizer in the tanks make up a large percentage (about 90 percent) of the total weight. This large amount of liquid can give rise to another type of vibration, sloshing of the liquids in their tanks. While the same materials were used for the solid parts of the model and prototype, it was not necessary to fill the model tanks with the expensive and troublesome cryogenic fuel and oxidizer. Ordinary water was used, but care was taken that the inertial effect of the water (rocket mass per unit length) duplicated that of the actual liquids. The mass per unit length is the pertinent inertial property for the transverse (bending) vibrations which were to be the subject of study with this model. The sloshing vibrations were *not* under study in these tests, and thus accurate modeling of the sloshing phenomena was not critical.

Even though sloshing was not under study, the sloshing could *affect* the bending vibrations. It was found that the sloshing natural frequencies scaled as the square root of the length scale while the bending vibrations scaled directly as the length scale. Thus the *interaction* of these two

---

[1] Op. cit., p. 94.

[2] H. L. Runyan, H. G. Morgan, and J. S. Mixson, *Use of Dynamic Models in Launch-Vehicle Development*, NASA TMX 51680, 1964.

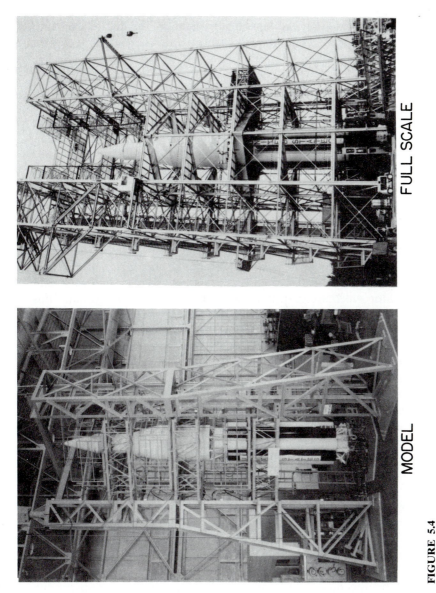

FULL SCALE

MODEL

FIGURE 5.4
Saturn rocket full-scale and $\frac{1}{5}$-scale models for vibration testing.

333

vibratory phenomena would *not* be the same in the model as in the prototype, creating a potential error source in the bending vibration measurements on the model. Fortunately the sloshing natural frequencies were quite different from the bending frequencies, tending to uncouple these two effects and minimize any errors in the bending frequencies measured on the model.

It was decided that a model useful for study of only the bending (transverse) vibrations would be constructed and tested. We mentioned earlier the difficulty in building a single model which is correct for all the physical phenomena occurring in a device. Actually, several different rocket models, each addressed to a particular problem area, were used in the overall development of the Saturn vehicle. For bending vibrations, designers are usually interested in getting accurate values for natural frequencies, mode shapes, and damping ratios. Scale models can give quite accurate values for frequencies and mode shapes, but since the damping in such structures is not something that is intentionally designed in, but just what "naturally" occurs, damping predictions from model studies are often inaccurate. The damping is due to several different effects (internal friction in stressed structural members, microscopic "scrubbing" in riveted and welded joints, viscous dissipation in gases and liquids), all of which are difficult to control and poorly understood. Thus the goal of this model was limited to prediction of natural frequencies and mode shapes at each of these frequencies. (Recall that a mode shape is just the dynamic deflection curve of the structure when it is in steady sinusoidal vibration at one of its natural frequencies.)

When we build a model and carefully scale all its dimensions to the prototype, then deflections, which are also lengths, will scale with the same factor, in our case a 5-to-1 factor. Think of the rocket (and the model) as a long, slim rod or shaft. When it is in transverse vibration at a natural frequency, each point along the length is oscillating sinusoidally at this same frequency, and all points are either in phase or 180° out of phase. The mode shape can be thought of as a strobe-light photograph snapped at the instant when all points are at maximum deflection. Since the length along the shaft (x axis of the curve) and the deflections at each station (y axis of the curve) are *both* scaled in the same 5-to-1 ratio, the mode shape curve we get from the model will accurately portray the prototype curve.

Next we need to find the relation between natural frequencies measured on the model and corresponding natural frequencies of the prototype. Since dimensions other than length are now involved, this relation will not generally be obvious (as in our mode-shape example) and we will need to carefully analyze the situation to find the correct scale factor relating frequencies for the model and the prototype. This kind of analysis is required whenever quantities with dimensions other than pure length are of interest to us, so it is needed in almost every model study. Now we want to use the rocket vibration example to explain three different methods used to establish the desired scale factors in any model study. Which of the three to use in a particular problem depends on the nature of the situation. The first method we discuss can be used if the differential equation governing the behavior of the prototype and model can actually be written and solved for the quantity of interest.

The simplest mathematical model (shear and rotatory inertia neglected) for beam bending vibrations leads to the differential equation[1]

$$\frac{\partial^2 y}{\partial t^2} + \frac{EI}{m}\frac{\partial^4 y}{\partial x^4} = 0 \tag{5.10}$$

where E is the modulus of elasticity, I is the cross-sectional area moment of inertia, and m is the mass per unit length. This equation holds for all the different types of end conditions to which the beam may be subjected. When the two ends are simply supported, the equation can be solved for the natural frequencies ω_n:

$$\omega_n = \frac{n^2\pi^2}{l^2}\sqrt{\frac{EI}{M/l}} \tag{5.11}$$

Here M is the total mass, l is the beam length, and $n = 1, 2, 3, \ldots$ for the various natural frequencies. Since the prototype and the model are *both* beams which obey Eq. (5.11), the equation can be applied to each in turn. Using subscript m for the model and p for the prototype, we can write

$$\omega_{n,m} = \frac{n^2\pi^2}{l_m^2}\sqrt{\frac{E_m I_m}{M_m/l_m}} \tag{5.12}$$

$$\omega_{n,p} = \frac{n^2\pi^2}{l_p^2}\sqrt{\frac{E_p I_p}{M_p/l_p}} \tag{5.13}$$

We can now make a list of all the scale factors relating the model and prototype quantities, starting with Eq. (5.9) for the length scale factor:

Length: $\qquad\qquad l_p = k_l l_m \tag{5.14}$

Elastic modulus: $\qquad E_p = k_E E_m \tag{5.15}$

Density: $\qquad\qquad \rho_p = k_\rho \rho_m \tag{5.16}$

Mass: $\qquad\qquad M_p = (\text{volume})(\text{density}) = (l_m k_l^3)(k_\rho \rho_m) \tag{5.17}$

Area moment of inertia: $\quad I_p = (\text{length})^4 = k_l^4 I_m \tag{5.18}$

---

[1] L. S. Jacobsen, and R. S. Ayre, *Engineering Vibrations*, McGraw-Hill, New York, 1958, p. 483.

Combining Eqs. (5.12) and (5.13) with these scale factors allows us to write

$$\frac{\omega_{n,p}}{\omega_{n,m}} = \left(\frac{l_m}{l_p}\right)^2 \sqrt{\frac{(E_p/E_m)(I_p/I_m)}{(M_p/M_m)/(l_m/l_p)}} = \frac{l_m}{l_p} = \frac{1}{k_l} \tag{5.19}$$

where we have used the same material for the model and the prototype, making the density and modulus scale factors equal to 1.0. This result tells us that for *any* structural model for fixed-end-beam bending vibrations, if we use the same material for the model as for the prototype, then natural frequencies measured on the model must be divided by the length scale factor k_l when the frequencies of the full-scale structure are predicted.

In the case of the Saturn rocket model, the length scale factor was 5.0, so frequencies of the actual rocket will be 5 times slower than those measured on the model. Returning to the NASA reference, we see that the lowest mode of bending vibration measured on the model was 14.0 Hz, which means that the prototype should have 2.80 Hz. When the full-scale structure was finally built and vibration-tested, the lowest frequency measured was 2.83 Hz, showing excellent agreement with the model tests. The model-predicted lowest frequency also coincided closely with a theoretical calculation. Theoretical predictions for the *second* mode of bending vibrations, however, overestimated the model-measured value by about 10 percent. Careful instrumentation of the model revealed that in this vibration mode the cluster of eight fuel tanks around the large central tank exhibited motions *perpendicular* to the direction of the applied force from the vibration shaker which is exciting the vibrations. The simple theory in use at this time had assumed the "beam" to vibrate in one direction only, causing this discrepancy. Learning from the model testing, the engineers modified the theory to account for this other degree of freedom, which led to more reliable theoretical predictions later in the project.

This first scaling method that we just described requires rather detailed knowledge of the system's behavior, in fact an actual theoretical solution for the quantity of interest, in our case the natural frequency. Often such complete information is unavailable, yet we still wish to pursue model testing. Let's next suppose that the differential equation governing the phenomenon *is* known but the solution is *not* known. This situation is clearly less restrictive than that encountered in our first method, making it more generally applicable. Let's rewrite Eq. (5.10) as it would apply to the prototype, also using $\rho A = M/l$, where $A \triangleq$ cross-sectional area:

$$\frac{\partial^2 y_p}{\partial t_p^2} + \frac{E_p I_p}{\rho_p A_p} \frac{\partial^4 y_p}{\partial x_p^4} = 0 \tag{5.20}$$

This equation can also be written for the model:

$$\frac{\partial y_m^2}{\partial t_m^2} + \frac{E_m I_m}{\rho_m A_m} \frac{\partial^4 y_m}{\partial x_m^4} = 0 \tag{5.21}$$

Note that we now allow *all* quantities, including the time variable, to have different values in the model and in the prototype. Again using k's for the scale factors, including a time-scale factor such that $t_p = k_t t_m$, we can rewrite Eq. (5.20) as follows:

$$\frac{k_l}{k_t^2} \frac{\partial^2 y_m}{\partial t_m^2} + \frac{(k_E E_m)(k_l^4 I_m)}{(k_\rho \rho_m)(k_l^2 A_m)} \frac{k_l}{k_l^4} \frac{\partial^4 y_m}{\partial x_m^4} = 0 \tag{5.22}$$

If the solution of this equation is to agree with that of Eq. (5.20), then the coefficients in one equation must be some constant times the respective coefficients in the other, since the right-hand side is zero in both. This gives

$$\frac{k_l}{k_t^2} = \text{constant} = \frac{k_E}{k_\rho k_l} \tag{5.23}$$

which becomes for identical materials

$$k_t = k_l \tag{5.24}$$

Applying this result to our model rocket example, we see that the time scale of model vibrations will be faster than that of the prototype since $t_p = 5.0 t_m$. That is, an event which happens at time $= 1$ sec in the model will occur at time $= 5$ sec in the prototype; thus model frequencies will be 5 times those of the prototype, just as we saw earlier by another method. Because this second method did *not* require that we assume specific end conditions for the beam, it is more desirable since it shows that the end conditions don't affect the scaling relations. The rocket example, in fact, has end conditions in flight that are more nearly *free* ends, not fixed ends. (Our earlier method did not analyze free ends because the frequency solution for this case is not available as a formula but rather requires a numerical solution.)

Our third and final method for establishing model scale factors uses the standard methods of dimensional analysis. As is typical of these methods, they require the least knowledge of the detailed behavior of the system under study, but sometimes give false results due to improper selection of the factors felt to be pertinent. For our beam vibration problem, we assume that the variables which affect the natural frequencies are E, ρ, and some characteristic length L_c of the vibrating object. Note that we don't even mention beams, our analysis holds for *any* vibrating structure.[1] Using a force, length, time (F, L, T) unit system, we first note that the dimensions of the dependent and independent variables are, respectively, $1/T$ for frequency, L for length, F/L^2 for elastic modulus, and FT^2/L^4 for density. Using standard procedures of dimensional analysis, we can write

$$\omega^a L_c^b E^c \rho^d = F^0 L^0 T^0$$
$$b - 2c - 4d = 0$$
$$-a + 2d = 0 \tag{5.25}$$
$$c + d = 0$$

The three equations in the four unknowns a, b, c, d allow us to freely choose an additional relation, say, $a = b = 1.0$. We can then solve for the other unknowns as $d = 0.5$ and $c = -0.5$. In this simple problem there results only one dimensionless group, which is found to be

$$\omega L_c \sqrt{\frac{\rho}{E}} \tag{5.26}$$

This dimensionless group must have the same value for the model and for the prototype.

$$\omega_m L_{c,m} \sqrt{\frac{\rho_m}{E_m}} = \omega_p L_{c,p} \sqrt{\frac{\rho_p}{E_p}}$$

$$\frac{\omega_p}{\omega_m} = \frac{L_{c,m}}{L_{c,p}} \sqrt{\frac{\rho_m/\rho_p}{E_m/E_p}} = \frac{1}{k_l} \sqrt{\frac{k_E}{k_\rho}} \tag{5.27}$$

We again see our usual result; for the prototype and model of the same material, prototype frequencies are $1/k_l$ times the frequencies measured on the model. This completes our brief survey of dimensional methods and scale modeling.

PROBLEMS

5.1. Prepare a book review of about 1000 words for L. J. Anthony, *Information Sources in Engineering*, 2d ed., Butterworth, London, 1985.

5.2. Repeat Prob. 5.1 for K. Subramanyam, *Scientific and Technical Information Sources*, Marcel Dekker, New York, 1981.

5.3. Prepare a literature survey on the following topics, limiting your library time to 2 hours.

---

[1] Langhaar, op. cit., p. 95.

a. Heating problems in electric motors
b. Testing machines and methods for studying friction coefficients of materials
c. Operating principles of quadrupole partial-pressure analyzers and their application in vacuum systems
d. Solar energy systems for large-scale electric power generation
e. Application of Taguchi methods in electronics manufacture
f. Diamond turning for optics manufacture
g. Piezoelectric actuators for control of micromotions
h. Sensors used in automobile control systems

5.4. All parts of this problem can be solved by using information from Chap. 5 alone. If you are interested in additional background, such as the derivation of the dimensionless groups, consult the references given.

a. The drag force F on a sphere is related to a single dimensionless group $\rho DV/\mu$ by the relation

$$\frac{F}{\rho V^2 D^2} = f\left(\frac{\rho DV}{\mu}\right)$$

where D is the sphere diameter, V is fluid velocity, ρ is fluid density, μ is fluid viscosity, and f is an unknown function. Describe the simplest experiment we could run to find the nature of the function f. Suppose we want to use a scale-model sphere of diameter 1 ft, tested in water at 60°F, to predict the drag force on a full-scale prototype sphere of diameter 10 ft subjected to 60°F air flowing at 50 ft/sec. What velocity should we use in our model test? How is force measured on the model related to force predicted for the prototype? (Langhaar, op. cit., p. 15.)

b. If fluid viscosity is assumed unimportant, the thrust force F of an ocean liner's propeller is related to two dimensionless groups $V^2/(gD)$ and $V/(nD)$ as follows:

$$\frac{F}{\rho V^2 D^2} = f\left(\frac{V^2}{gD}, \frac{V}{nD}\right)$$

Here, V is the ship velocity, D is the propeller diameter, g is the acceleration of gravity, ρ is the water density, n is the propeller's rotational speed, and f is an unknown function. Describe the simplest experiment you can that allows us to determine the nature of the function f. Such full-scale experiments are quite complex and expensive, so scale-model testing is common. Suppose the prototype has $D = 20$ ft, $V = 25$ ft/sec, and $n = 2$ rev/sec. If a $\frac{1}{10}$-scale model is tested in seawater, what values of V and n should be used? How is F measured on the model related to F predicted for the prototype?

If viscosity μ *were* important, a third dimensionless group, the Reynolds number $\rho DV/\mu$, would need to be included as an independent variable in the function f. If it is impractical to use any fluid other than water, show that now it is *impossible* to study the effects of each dimensionless group separately; i.e., hold two constant and vary the third. Once the prototype is built, suppose money becomes available for some full-scale testing. Explain how we might use such tests to check whether viscosity really had an effect. (Langhaar, op. cit., p. 65.)

c. Underwater detonation of explosives causes propagation of a spherical shock wave. The shock wave pressure p at any instant and at a radius R from the point of origin is related to the initial gas pressure p_0 and mass m of the explosive material, and density ρ and bulk modulus E of the water by an unknown function f of two dimensionless groups p_0/E and $m/(\rho R^3)$:

$$\frac{p}{p_0} = f\left(\frac{p_0}{E}, \frac{m}{\rho R^3}\right)$$

If the liquid in which the explosive is immersed must be water, describe experiments we can run to find the nature of the function f. We can use different explosives which each have their own known value of p_0 (TNT, for example, has a p_0 of about 2 million psi).

Scale "model" testing here refers to studies of explosive charges (m values) smaller than in the prototype device. If 1 lb of TNT causes a peak (timewise) shock pressure of 2200 psi at $R = 7.5$ ft, predict the R value at which a full-scale ($m = 1000$ lb of TNT) device will have this same pressure. (Langhaar, op. cit., p. 71.)

d. For elastic structures with a specified shape (cantilever beam, for instance) and a specified type of loading (concentrated load at midpoint plus an end moment, for example), any component σ of stress or any component u of deflection is related to three dimensionless groups $F/(EL^2)$, $M/(FL)$, and v through unknown functions f_1 and f_2:

$$\sigma \frac{L^2}{F} = f_1\left(\frac{F}{EL^2}, \frac{M}{FL}, v\right)$$

$$u \frac{EL}{F} = f_2\left(\frac{F}{EL^2}, \frac{M}{FL}, v\right)$$

Here F is a representative force, M is a representative moment, L is a representative length, E is the modulus of elasticity, and v is Poisson's ratio. Describe the simplest experiments we can run to determine the nature of functions f_1 and f_2.

For scale models, must the model have the same E value as the prototype? Why? What about Poisson's ratio? Develop useful relations for all the scale factors.

Using your background in strength of materials, list some simple structures and structural elements for which Poisson's ratio is important. List some for which it is not important. (Langhaar, op. cit., p. 80.)

CHAPTER
6

APPARATUS DESIGN AND CONSTRUCTION

6.1 DESIGN PRINCIPLES AND PLANNING

The design of apparatus for carrying out experiments is, of course, part of the larger area of engineering design of products, processes, and services. Unfortunately, this larger area of design has recently been recognized[1] as being in need of radical improvement, both in industrial practice and in engineering education, in the United States. This subject is a complex one which requires a thorough treatment (as in the above reference) for real understanding, so we will not be able to pursue it in any detail here. We do, however, want to incorporate as much as we can of these new approaches into our presentation of the subject of apparatus design. In fact, in earlier chapters we tried to tie in the treatment of experimentation to this important topic wherever possible. Much of Chap. 4, for

---

[1] *Improving Engineering Design*: *Designing for Competitive Advantage,* Committee on Engineering Design Theory and Methodology, Manufacturing Studies Board, National Research Council, National Academy Press, 2101 Constitution Ave., Washington, DC 20418, 1991.

341

example, is focused on the design of experiments aimed at improving products and manufacturing processes, vital ingredients of the latest approaches to design.

Designing an experimental apparatus shares many common features with product design, although of course we immediately recognize that almost every such apparatus is a one-of-a-kind, rather than a mass-produced, item. Some "products," however, are *also* one of a kind. Special-purpose machining and assembly equipment, for example, may often be designed and produced in quantities of one and is thus closely allied with experimental-apparatus design. In discussing apparatus design, we will thus be able to follow quite closely the patterns recommended in the new approaches to product design. We can thus present an up-to-date method for apparatus design and simultaneously improve and extend the *general* design competence of those readers who may not have previously encountered the recent advancements. Perhaps we should mention at this point that the "new" design approach which we are referring to is *not* the computer-aided design (CAD, CAE, CAD/CAM, etc.) with which most readers *are* familiar and which largely concentrates on "automating" the design of individual *components,* such as gears or linkages. Rather, the new design approach is a systematic way of approaching the design of *complete* machines and systems of *any* kind. Its importance lies in its *philosophy,* not in the details of its implementation, which will, of course, use computers wherever that use makes sense.

Since we have only a brief chapter available, our treatment will need to be quite efficient and condensed, but many useful viewpoints can be presented in sufficient detail to be self-contained. When this is not possible, references will be cited where the interested reader may find a more in-depth treatment. A good starting point for our discussion is a flowchart showing the major steps in apparatus (product) design and their interrelationships. Such charts, in one form or another, are presented in most of the recent books[1] devoted to explaining the more systematic design approaches now being recommended. Our Fig. 6.1 is a blend and condensation of these various descriptions. We see that, after an initial stage devoted to a clarification of the task, the design process is divided into three major phases: *conceptual design, substantive design,* and *detail design.* (The English translation of Pahl and Beitz uses the term *embodiment design* where I have used *substantive design.*)

Since the words *conceptual, embodiment, substantive,* and *detail* all

---

[1] G. Pahl and W. Beitz, *Engineering Design: A Systematic Approach,* Springer-Verlag, New York, 1988 (original German edition, 1977). N. Suh, *The Principles of Design,* Oxford, New York, 1990.

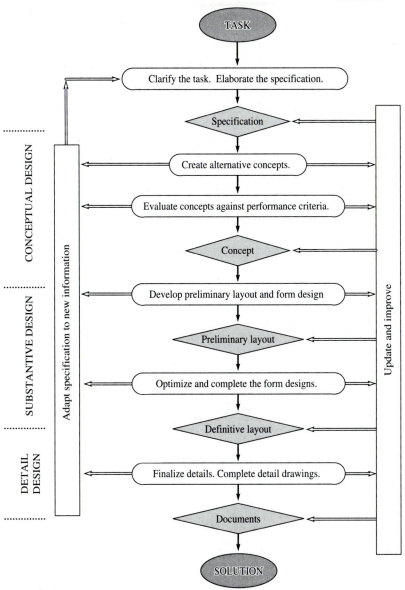

FIGURE 6.1
Flowchart for the product design process.

have various possible meanings, we need an example here to clarify their intended meaning in the context of design. If our design task were to move a 50-kg mass along an already existing straight guiderail 2 ft in

5 sec, the *conceptual* design phase would require us to think of "all" the different ways we could accomplish this basic task. Some possibilities would be human force, gravity force, fluid force, magnetic force, etc. If some of these initial concepts could be shown to be ridiculous, we would throw them out at this point; otherwise we include them all as we proceed further in the design process. Moving now to the *substantive* design phase, we need to embody our concept in an actual physical device which can be analyzed for function and cost. For our example, we would consider specific devices such as springs, electric motors of various types, pneumatic cylinders, hydraulic cylinders, and rotary fluid actuators with some mechanism to convert rotary to straight-line motion. Each of these different approaches must be carefully analyzed, improved, and adapted to our specific needs and then evaluated critically against performance and cost specifications. Often when we reach this stage, we will have learned so much about the general problem that we may return to the conceptual stage and add some new concepts to our initial list. This feedback from one stage of the design process to an earlier or later one is very common.

In some problems, after "completing" the substantive design phase, *one* of the alternative designs stands out as clearly superior, and we reject the others. More often, *several* designs remain as possibilities, and we must move on to the *detail* design stage in order to make a valid decision. For our example, if a pneumatic cylinder has survived the conceptual and substantive phases of design, at the detail stage we must consider aspects of the cylinder that were of no consequence earlier but are now necessary to give a description of the cylinder adequate for manufacture. That is, in substantive design we *assume* that the cylinder will not explode from its internal pressure and concentrate on what pressure and diameter are needed to meet the motion specification. In detail design we must compute the cylinder wall thickness, head-bolt size, etc., to ensure that an explosion will not actually occur. As such details are dealt with, we integrate considerations of manufacture, assembly, test, life-cycle cost, reliability, safety, quality control, marketing, etc. If several alternative concepts have survived to this stage, an intelligent choice among them can be made only when *all* such factors are considered. For one-of-a-kind designs, especially of experimental apparatus, many of these factors will of course be given little weight relative to simple technical functioning; however, it is generally best to at least *start* with a rather complete list of considerations to ensure that nothing important to the specific apparatus has been overlooked. It is usually quite easy and quick to eliminate those factors which are obviously not pertinent.

Having given a quick overview of the design process, we now discuss the steps in Fig. 6.1 in a little greater detail. Note that there are five distinct packets of information (shown in the diamond-shaped areas) that need to be produced in sequence: *specification, concept, preliminary layout, definitive layout,* and *documents.* Between these, in the ovals, are the design functions which must be carried out to move from each preceding level of information to the next. The sidebars (adapt specification to new information, and update and improve) display the iterative or feedback nature of the entire process, in which information developed in any step can influence either earlier or later steps.

Upon stating the *task,* to produce the *specification,* we must analyze and clarify the initial task statement since such statements often are somewhat vague. A list of the types of information needed in a specification is useful for organizing the process of task clarification. Pahl and Beitz[1] suggest the following checklist:

Geometry: Size, height, breadth, length, diameter, space requirement, number, arrangement, connection, extension.

Kinematics: Type of motion, direction of motion, velocity, acceleration

Forces: Direction of force, magnitude of force, frequency, weight, load, deformation, stiffness, elasticity, inertial forces, resonance

Energy: Output, efficiency, loss, friction, ventilation, state, pressure, temperature, heating, cooling, supply, storage, capacity, conversion

Material: Flow and transport of materials, physical and chemical properties of the initial and final products, auxiliary materials, prescribed materials (food regulations, etc.)

Signals: Inputs and outputs, form, display, control equipment

Safety: Direct protection systems, operational and environmental safety

Ergonomics: Human-machine relationships, type of operation, operating height, clearness of layout, sitting comfort, lighting, shape compatibility

Production: Factory limitations, maximum possible dimensions, preferred production methods, means of production, achievable quality and tolerances, wastage

Quality control: Possibilities of measuring and testing, application of special regulations and standards

Assembly: Special regulations, installation, siting, foundations

---

[1] Ibid., p. 54.

Transport: Limitations due to lifting gear, clearance, means of transport (height and weight), nature and conditions of dispatch

Operation: Quietness, wear, special uses, marketing area, destination (for example, sulfurous atmosphere, tropical conditions)

Maintenance: Servicing intervals (if any), inspection, exchange and repair, painting, cleaning

Cost: Maximum permissible manufacturing costs, cost of tools, investment and depreciation

Schedules: End date of development, project planning and control, delivery date

This checklist is intended not to be exhaustive but rather to remind one of the *kinds* of considerations that need to be examined when one is writing out a detailed specification. Each specific application will have its own peculiarities in addition to the more routine items shown in the above list. Again, when the task is the design of an experimental apparatus rather than a mass-produced "product," many of the listed considerations have little import and may be quickly passed over.

Once the specification has been thoughtfully developed, we can proceed to the first phase of actual design, the *conceptual* phase. As its name implies, here we are concerned with very basic concepts rather than their physical implementation. In fact one should strenuously *avoid,* at this stage, thinking of specific physical hardware and concentrate instead on strictly *functional* requirements. That is, we are to formulate in *abstract,* rather than concrete, terms what essential functions the apparatus must perform in order to meet the specifications. This abstraction is necessary if we are to avoid influencing the later steps with conscious or subconscious prejudices in our minds. Pahl and Beitz[1] again provide a useful checklist to help us move from the specification to one or more concepts which can then be developed in later steps:

Abstract to identify the *essential* functions.

Establish function *structures* (overall and subfunctions).

Search for solution principles to fulfill the subfunctions.

Combine solution principles to fulfill the overall function.

Select suitable combinations.

---

[1] Ibid., p. 58.

Firm up into concept variants.

Evaluate concept variants (technical and economic criteria).

The most difficult aspect of this phase of the design process is the just-mentioned need to define the required functions in an abstract way. Our natural instinct is to immediately draw on our past experiences in the general area, and these will invariably come to mind in very specific, rather than abstract, terms. Nevertheless, it is *vital* that we force ourselves to avoid this natural inclination since an unbiased viewpoint is essential at this stage of design. If this is *not* done, we may completely overlook many unconventional, but potentially profitable, design concepts. For example, in the area of building construction using concrete, many different variations of on-site machinery for mixing concrete were developed before somebody had the "bright idea" of mixing concrete *off*-site and delivering it in specially designed trucks. In this example, the designer needed to design *not* a new machine for mixing concrete but rather the best method of providing mixed concrete at the building site, which is the *basic* function that was to be provided.

Except for very simple machines or apparatus, the basic function will exhibit a *function structure.* That is, the overall function is a combination of *subfunctions,* and the identification of these subfunctions facilitates the search for solutions. Organization of the subfunctions into a clearly defined structure further clarifies this search. Since the same basic subfunctions may appear over and over in different combinations for different applications, systematic consideration of function structure in all design problems builds the designer's capacity for rapid innovation. "Logical" considerations are a useful tool in establishing function structures. That is, there may be an AND, OR, NOT, NOR, NAND, or other logic relation between subfunctions, and these links are part of the function structure. For example, in a tensile testing machine, the function "clamp specimen" must occur *before* the function "apply load"; "measure force," "measure deflection" must come *after* "apply force." Detailed examination of the specification will reveal which subfunctions exhibit logic relations. Physical considerations involving the flow of *material, energy, and signals* are also helpful in identifying subfunction structure since a small number of basic subfunctions have been found to cover most situations: *changing, varying, connecting, channeling,* and *storing.* For example, electric energy may be *changed* into mechanical, material dimensions may be *varied* in a rolling mill, a feedback signal may be *connected* with a desired value signal to compare them, material on a conveyor belt may be *channeled* to one of several alternate paths, and signals may be *stored* in a digital memory.

Once the overall function and the subfunction structure have been clarified, one can begin searching systematically for *solution principles* to fulfill the subfunctions. We have space here to list only a few approaches to this task that Pahl and Beitz[1] explain at length:

1. Literature search	**6.** Brainstorming
2. Natural systems	**7.** Delphi method
3. Known technical systems	**8.** Physical processes
4. Analogies	**9.** Classification schemes
5. Experimental studies	**10.** Design catalogs

Having identified one or more solution principles for each subfunction, we next need to form suitable *combinations* of these that will satisfy the overall function. Then these competing overall solutions must be evaluated, and the "best" one or several selected. Evaluation of alternative conceptual designs requires formulation of performance criteria suitable for this stage of design (recall that substantive design and detail design have not yet been carried out). We quote from Pahl and Beitz:

> For the systematic approach the solution field should be as wide as possible. By paying regard to all possible classifying criteria and characteristics, the designer is often led to a large number of possible solutions. This profusion constitutes the strength and also the weakness of the systematic approach. The very great, theoretically admissible but practically unattainable, number of solutions must be reduced at the earliest possible moment. On the other hand, care must be taken to not eliminate valuable solution principles, because often it is only in their combination with others that an advantageous overall solution will emerge. While there is no absolutely safe procedure, the use of a systematic and verifiable selection procedure greatly facilitates the choice of promising solutions from a wealth of proposals.[2]

A mere listing of key concepts useful in the evaluation and selection procedure must again suffice; more detail may be found in the reference.[3] Two major steps are involved in the process: *elimination* of totally unsuitable proposals and *preference* ranking for those that remain. Criteria for elimination should be applied at every step of the conceptual design process. This avoids spending time and money needlessly carrying forward unworkable schemes. Only such proposals should be

---

[1] Ibid., pp. 82–108.

[2] Ibid., p. 112.

[3] Ibid., pp. 112–139.

pursued as

1. Are compatible with the overall task and/or one another
2. Fulfill the demands of the specification
3. Are realizable in respect of performance, layout, etc.
4. Are expected to be within permissible costs

We have given above an outline of the conceptual phase of the design process. Even though this is a necessarily brief treatment of a complex subject, it should be helpful in organizing your thoughts as you begin a practical project. An in-depth discussion is available in Pahl and Beitz, and in particular I can recommend a detailed practical example found there.[1] Fortunately this example *is* an experimental testing apparatus rather than a mass-produced product, so it meets our needs very nicely.

After the conceptual phase, design next proceeds into the *substantive* phase (called *embodiment* design in the English translation of Pahl and Beitz). Whereas conceptual design has as its output one or more somewhat abstract concepts, substantive design produces (see Fig. 6.1) a "definitive layout" drawing. This drawing (or drawings), with associated documents, defines the design in sufficient detail to allow the next (and last) stage—detail design—to proceed. Again there are a number of distinct steps and functions (not all shown in Fig. 6.1, which is condensed) involved in this substantive design phase. Preliminary layout drawings are used to determine the general arrangement of parts and their spatial compatibility. Shapes and materials for components are selected as are production procedures, with a view to both technical and economic requirements. Substantive design involves a close interplay between synthesis and analysis as components are initially proposed and then analyzed for conformance to specifications. Size-determining, arrangement-determining, and material-determining requirements play a crucial role. Detection and correction of design flaws and errors must be given top priority. Designs should exhibit clarity of function, simplicity, and safety.

Although we gave a brief overview of conceptual and substantive design since some generally useful rules and principles have been formulated, *detail design,* the last stage, is so peculiar to each specific application area that it is not presented in general design texts such as that by Pahl and Beitz. Also, this phase of design is most likely to have been presented in some depth in academic curriculums in the various specialties, so we won't pursue it here.

---

[1] Ibid., p. 139.

6.2   STANDARD TESTS, CODES, AND RECOMMENDED PROCEDURES

Having just given a brief outline of design methodology, we now pursue more specific and concrete aids for apparatus design. Literature searching has already been discussed in a general way, but we now focus on a particular type of literature that can be very useful in many apparatus design projects. We refer to the thousands of "standard practices" produced and disseminated by various engineering societies and committees. These documents are a vast storehouse of accumulated engineering lore which has been critically sifted and evaluated by knowledgeable practitioners in various areas. Since they usually are consensus documents arrived at by agreement among several or many experts, standard practices may often be more credible than individual opinions expressed in books or papers. Of course, they may also *suffer* the fault of sometimes being "behind the times" since getting agreement within a committee can delay judgment.

I personally consider such literature to be highly useful, but many engineering students are hardly aware of its existence. This is partly due to the emphasis (*over*emphasis, in my opinion) on science and analysis, rather than practical design, in many curriculums. Standardized tests are considered uncreative and cookbooky. Although one would certainly not want to build a curriculum around them, these information sources are valuable in the practice of engineering, and we need to have some appreciation of what is available here.

From the many such documents available[1] I have selected the *Annual Book of Standards,* published by the American Society for Testing and Materials (ASTM), as a vehicle for discussing the nature of this whole class of materials. This "book" is actually 68 individual volumes which are updated annually, many volumes having over 1000 pages. The 68 volumes are organized by subject area into 16 sections, each section having one or more volumes. To begin to comprehend this vast collection of information, let's just list the titles of the 16 sections.

1. Iron and steel products, seven volumes

2. Nonferrous metal products, five volumes

3. Metals test methods and analytical procedures, six volumes

4. Construction, nine volumes

---

[1] P. Ricci, *Standards: A Resource and Guide for Identification, Selection, and Acquisition,* Bernards and Co., St. Paul, MN, 1989. L. R. Musser, "Identifying Standards, How to Find Out If a Standard Exists," *ASTM Standardization News,* April 1989, pp. 44–47.

5. Petroleum products, lubricants, and fossil fuels, five volumes

6. Paints, related coatings, and aromatics, three volumes

7. Textiles, two volumes

8. Plastics, four volumes

9. Rubber, two volumes

10. Electrical insulation and electronics, five volumes

11. Water and environmental technology, four volumes

12. Nuclear, solar, and geothermal energy, two volumes

13. Medical devices and services, one volume

14. General methods and instrumentation, three volumes

15. General products, chemical specialties, and end-use products, nine volumes

16. Index, one volume

While I usually go to the library in search of a specific book or journal, I also periodically set aside some time for "technical browsing" and have found many interesting and useful books and papers in this way. Card catalogs or computerized library systems are, in my experience, not entirely reliable in placing documents into categories that I would consider most correct. This means that if you search under a set of keywords that *you* think covers the topic of interest, you may miss some important documents. Also, information pertinent to a project is sometimes found in sources that at first glance would appear to be far afield. While technical browsing may be a luxury more easily afforded by academics, I have personally found it so valuable that I encourage its practice by all engineers, to the extent that their circumstances allow. With regard to our present topic, let me recommend to those unfamiliar with standards documents that they, right now, set aside 1 hour of library time for a browsing of the ASTM volumes.

Although the bare listing of titles of the ASTM sections given above provides a broad overview of their scope, we need to delve a little more deeply to get a useful appreciation of their content. To do this, we explore the content of a typical volume and then finally describe in considerable detail a specific standard. We have chosen volume 14.02 of section 14 for this illustrative treatment.[1] Its title is *General Test Methods, Nonmetal*; *Laboratory Apparatus*; *Statistical Methods, Appearance of*

---

[1] *ASTM 1987 Annual Book of Standards,* ASTM, 1916 Race St., Philadelphia, PA 19103, phone (215) 299–5400.

Materials; *Durability of Nonmetallic Materials*. Pages viii and ix of this volume give an alphabetical listing of the more than 200 subject areas treated in the complete set of 16 volumes. A brief sampling of these subjects should suffice to illustrate the coverage.

Acoustics, environmental
Activated carbon
Adhesives
. . .

Amusement rides and devices
Analytical atomic spectroscopy
Anesthetic and respiratory equipment
Appearance of materials
. . .

Business copy products
. . .

Ceramic materials
Chemical analysis of metals
. . .

Computerized systems
. . .

Consumer products
. . .

Durability of nonmetallic materials
. . .

Electronics
Emission spectroscopy
. . .

Fatigue
. . .

Fire standards
. . .

Forensic sciences
. . .

Geothermal resources and energy
...

Hazardous substances and oil spill response
...

Medical and surgical materials and devices
...

Occupational health and safety
...

Packaging
...

Pressure vessel plate and forgings
Products liability litigation, technical aspects of
...

Robotics
...

Sports equipment and facilities
...

Surface analysis
...

Temperature measurement
Textiles
...

Thermal insulation
...

Tires
Traveled surface characteristics
...

Wear and erosion
Wood

In volume 14.02, pages xv to xx list the titles of 189 standards included in this volume. Under "Appearance of Materials" some typical standards titles are:

Absolute calibration of reflectance standards
Evaluating change in color with a gray scale
Description of conditions for photographing specimens

Under "Methods of Testing":

Maintaining constant relative humidity by means of aqueous solutions

Performing accelerated outdoor weathering of nonmetallic materials using concentrated natural sunlight

Safe use of oxygen combustion bombs

Under "Statistical Methods":

Dealing with outlying observations

Probability sampling of materials

Under "Hazard Potential of Chemicals":

Pressure and rate of pressure rise for dust explosions

Autoignition temperature of liquid chemicals

Under "Apparatus for Testing":

Apparatus for the determination of water by distillation

Laboratory weights and precision mass standards

Calibration of volumetric ware

Under "Technical Aspects of Product Liability Litigation":

Examining and testing items that are or may become involved in product liability litigation

Reporting opinions of technical experts

Under "Filtration":

Gas flow resistance testing of filtration media

Determining the performance of filter media

Under "Durability of Nonmetallic Materials":

Determining resistance of plastics to bacteria

Under "Compatibility and Sensitivity of Materials in Oxygen-Enriched Atmospheres:

Designing systems for oxygen service
Ignition sensitivity of materials to gaseous fluid impact

To get a proper appreciation of the utility of such standards documents, we next examine a specific one—F 778–82 from volume 14.02—in some detail. This standard is entitled *Gas Flow Resistance Testing of Filtration Media* and appears on pages 907 to 923. Rather than reproduce the entire standard here, it is more effective to outline its contents and comment on its most significant features as they relate to the *general* utility of such documents, not the specific subject of filtration. The standard is organized into 19 numbered sections, as follows:

1. Scope	**11.** Requirements for method A
2. Referenced documents	
3. Summary of methods	**12.** Test apparatus
4. Significance and use	**13.** Procedure
5. Terminology	**14.** Requirements for method B
6. General requirements	
7. Sampling	**15.** Test apparatus
8. Number of specimens	**16.** Procedure
9. Specimen conditioning	**17.** Calculation
10. Specimen dimensional measurement	**18.** Report
	19. Precision and accuracy

The *scope* of the standard defines what is included and what is not. Here the filtered medium is gaseous (rather than liquid), and the filter media are of flat, pleated, or bulk form. It is not unusual for a standard to refer to other standards which apply to portions of the procedure being described. This referencing is desirable for compactness, but has the defect that the document is not self-contained and thus the reader must have ready access to all the referenced documents, not just the one originally consulted. In our current example, section 2 lists 11 pertinent ASTM standards and one ASME standard, making clear the need to maintain rather complete sets of standards for the fields in which one regularly works. The *summary of methods* briefly describes the general nature of the testing methods; complete details are given later in the "procedure" section. In our case, two alternative methods called A and B are described and instructions given on when to use each.

The *significance and use* of the standard describe its practical importance and various ways it might be applied. The airflow resistance of a filter characterizes its pressure/flow relation and could be used for production quality control, product development studies, or basic research. Many areas of engineering practice, in addition to using commonly accepted general terminology, will also have some specialized jargon of their own, perhaps unfamiliar to those outside the immediate field. A section on *terminology* may thus be desirable to foster accurate communication among the personnel involved. In our filter standard, the definition of airflow resistance, for example is different from the more general concept of fluid resistance. Most engineers consider the fluid resistance of a device to be given by the ratio (Δpressure drop)/(Δvolume flow rate), i.e., the local slope of a curve of pressure drop versus flow rate. Definition 5.2.2 of our filter standard states:

> Airflow resistance, ΔP—pressure drop or pressure differential across a test specimen of filter medium at a specified air face velocity or mass flow rate

Note that this specialized definition of fluid resistance does not even have the same units as the more general definition. Also it uses another term, *face velocity,* whose meaning may not be obvious, and thus *this* term is also defined in this section, as are 11 other quantities.

In this standard, *general requirements* has two subsections, instrument accuracy and test apparatus environment. The required accuracy of both pressure-drop and flow rate instruments is stated as ±3 percent, and it is required that the instruments actually be checked against suitable calibration standards. Test environment requirements are concerned mainly with possible variations in air viscosity which can introduce errors. The effects of temperature, pressure, and humidity on viscosity are explained, and control measures are recommended. Section 7 on *sampling* is so brief that we give it in its entirety:

> The sample to be tested as a flat media, pleated media, or bulk media should be obtained under the guidance of the particular standard or specification covering the generic material or as agreed upon between purchaser and seller.

(The use of the plural *media* rather than the singular *medium* might be challenged here by grammarians.)

Section 8, *number of specimens,* gives a detailed discussion of statistical tools useful for choosing an appropriate number of specimens to determine the average resistance of the filter material. If the lab has previous experience which has produced a reliable estimate of the coefficient of variation C_v (given in percent) and if the allowable variation of the test results (expressed as a percent of the average) is $A,$ then for a 95 percent confidence level, the number of specimens n is given by $n = 3.842 C_v^2 / A^2.$ Procedures and formulas for other conditions are also given. Section 9, *conditioning of test specimens,* explains that since many filter media undergo physical changes with temperature and moisture, the media should usually be exposed to a standard conditioning environment for a period of time before being tested for flow resistance. This environment is then described for the various types of media within the scope of the standard. It is also stated that if such conditioning is deemed inappropriate and thus not used for a specific test program, this must be made clear in the report. Section 10, *dimensional measurement of specimens,* refers the reader to pertinent standards listed in section 2, for details on proper methods of making the thickness and area measurements needed in calculating flow resistance.

Sections 11 to 13 give details on using method A, while Sections 14 to 16 cover method B. Method A uses a simpler apparatus and procedure and should be used whenever possible. Method B employs a "guarding" principle that is necessary for certain types of media which exhibit significant leakage of airflow out the edges due to high pressure drop or other reasons. The guarding feature reduces the leakage effect and thus gives more accurate measurements of the flow resistance. Guarding techniques are useful in many kinds of apparatus to reduce leakage effects of a thermal, fluid, or electrical nature. We treat this general apparatus design principle more completely later in this chapter. Figure 6.2a shows schematically the apparatus recommended for method A while Fig. 6.2b gives a variety of possible methods of clamping the filter medium for test purposes. These are both good examples of the general usefulness of standards in apparatus design. We are made aware of a successful apparatus layout and benefit as well from many investigators' past experience with the details of clamping schemes which can be adapted to different situations. For media with moderate pressure drops, 10. inHg or less, the suction type of system shown in Fig. 6.2a is recommended, while a blow-through system works better with higher pressure drops. Again, these kinds of experience-based guidelines save us a lot of time and effort when we need to build or operate similar apparatus. Figure 6.3 shows the more complex apparatus associated with method B.

Sections 11 and 12 give detailed instructions for carrying out method A. Here we give only a few excerpts to make clear the nature of the useful information available. Specimen-mounting techniques should be selected to eliminate edge leakage as much as possible without, however, deforming the filter medium to the extent that flow resistance is affected. To check a tentatively selected clamping scheme for edge leakage, we are told to make up and test a specimen in which the edges have been *sealed,* using a spacer bar of equivalent thickness to the specimen. If this sealed-edge specimen has the same flow resistance as an unsealed one, then the edge leakage of this clamping scheme is

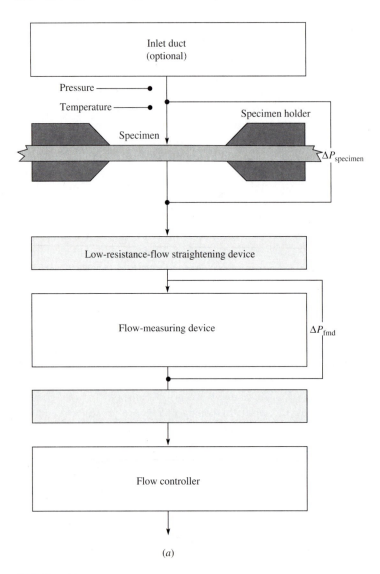

(a)

FIGURE 6.2
Apparatus diagrams from ASTM F 778-82.

considered negligible, and we can continue testing with the simpler and cheaper unsealed specimens. The specimen area for flat paper and paperlike media should be the standard value of 5.94 in^2 or larger. Specimen shapes can be rectangular or circular, but rectangular specimens must have a length-to-width ratio of 2 or less. Pressure measurement taps must be flush with the duct walls and sufficiently distant from the

AIRFLOW

AIRFLOW

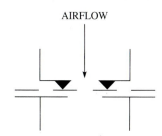

A1.1 Sharp edge—top clamp

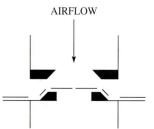

A1.5 Weight ring to produce
known tension

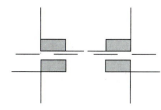

A1.2 Soft gaskets—top and bottom
clamp or chamber

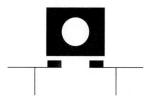

A1.6 Thick rigid specimen—impregnate
with barrier-type material except
for test area

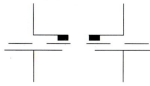

A1.3 Pneumatic actuated top clamp
with low-surface-area sealing

A1.7 Special holder to fit exact
thickness of specimen

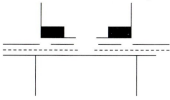

A1.4 Support grid of known air
flow resistance and
construction

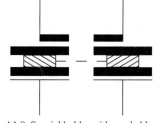

A1.8 Special holder with crushable
gasket to conform to specimen
surface irregularities

(*b*)

FIGURE 6.2
(Continued).

specimen that localized flow irregularities caused by the specimen and its
holder have disappeared. The air density at the specimen, obtained from
temperature and pressure measurements (humidity effects have been
shown negligible and may be ignored), must be calculated and reported.

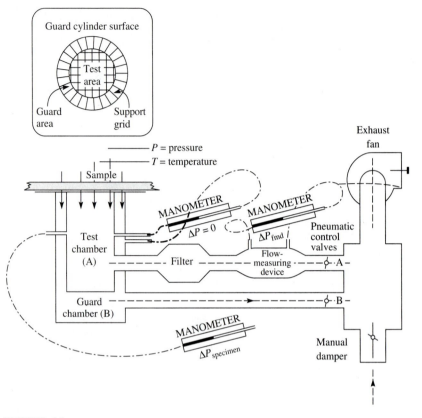

FIGURE 6.3
Guarded-cylinder apparatus for filter testing.

If the filter medium must be supported by some kind of gridwork, a pressure-drop test of the gridwork *alone* must be run to prove that the effect of the gridwork is negligible or else corrected for. Flow rate measurements using orifices and nozzles, the most common types of flowmeters used in such tests, must be corrected if filter test conditions are different from the meters' calibration conditions. Methods for these corrections are given. This nuisance can sometimes be avoided by use of a laminar-flow element[1] instead of an orifice meter. Details on this simplification are given in the standard. This kind of helpful hint is

---

[1] E. O. Doebelin, *Measurement Systems,* 4th ed., McGraw-Hill, New York, 1990, p. 573.

typical of the useful information we often find in standards documents and of which we might otherwise be unaware.

Before formal testing is begun, an overall check of the apparatus, procedures, and calculation methods can be done by substituting for the filter medium a standard test plate whose flow resistance is independently known. We are cautioned that the flow resistance of some filter media depends on which of the two surfaces is facing into the flow. Unless an accepted standard direction has been noted, we should test in both directions to discover any such effects which may exist. For the guarded-cylinder apparatus of method B (Fig. 6.3), the specimen face area must be at least 38.75 in^2, the recommended guard area is 3 times the measuring area, and their ratio is never to be less than 1 to 1. For the guard scheme to effectively prevent edge leakage, the guard area flow rate is increased until the pressure difference measured between the guard and measuring chambers downstream of the sample is zero. When this pressure difference is zero, the pressure gradient (in the filter medium) perpendicular to the filter thickness will also be minimized, reducing the edge leakage from the measured area of the filter to a negligible value and thus improving the accuracy of the measured flow rate. In addition to the specific items just mentioned, sections 11 to 16 also give a complete and detailed step-by-step explanation of how to carry out the test.

Section 17, *calculations,* gives all the formulas needed to make the required calculations and discusses when and how they are to be applied. Here we present only a sampling of these instructions to make clear their utility, not just for this standard, but also in general. The flow behavior of filter media can be classified according to how the flow resistance varies with the flow rate (really, the Reynolds number), and this classification influences the calculations. Three cases that occur in practice are identified in the standard, and directions are given on how to deal with each. Case I applies to media for which the flow resistance is known to be nearly linear with the flow velocity. Here test data are to be reported as single- or multiple-data-point ordered pairs of $(\Delta P, V)$, where V is the filter face velocity. If the pressure drop is greater than 5 percent of the inlet absolute pressure, then the air density changes significantly across the filter and must be taken into account. The standard way of doing this is to calculate and report the data, not as $(\Delta P, V)$ pairs but rather as $(\sigma \Delta P, G)$ pairs, where σ is the ratio of the measured air density to a standard value of 1.201 kg/m^3 and G is the filter mass flow rate. Case II procedures should be used when the variation of filter resistance is known to be *nonlinear* with the flow velocity. Here the data are to be reported as $(\sigma \Delta P, G)$ pairs, the same as for case I with high pressure drops. Case III applies to the situation where we are not sure how the resistance varies with the flow rate, e.g., when we are testing a new material for the first time. Here we are required to run tests at 25, 50, 75, 100, 150, and 200 percent of rated flow. These data are used to plot a graph from which one can deduce whether to use case I methods or case II methods in reporting the data. Again, complete instructions on how to do this are supplied.

Section 18 on *reporting* gives a detailed listing of all the results that must appear in a report of this type. It does not give any information on *general* report-writing practices such as organization and grammar, since the reader is assumed to be proficient in this area. The last section, 19, is so brief that we cite it all:

> 19.   Precision and accuracy
> 19.1   Precision. The precision of these methods for testing air-flow resistance is being established.
> 19.2   Accuracy. No justifiable statement can be made on the accuracy of measuring air-flow resistance, since the true value of the property cannot be established for most filter materials or specimens.

These statements may appear unsatisfactory, but they reflect the state of the art at the time the standard was written. For the neophyte filter tester, these rather negative statements are actually very useful since we are made aware of problems that would otherwise not be obvious. If you browse the ASTM volumes as we suggested earlier, you will find that some standards give quite specific and numerical statements on accuracy and precision. For example, ASTM Standard E 831-86 (*Linear Thermal Expansion of Solid Materials by Thermomechanical Analysis*) devotes almost two pages to this subject, giving many numerical results. Such complete and reliable data on accuracy and precision are often the result of round-robin interlaboratory studies.[1] The term *round-robin* is used to describe a series of tests where several (perhaps six to eight) independent labs all agree to test the same specimens, using the same methods, to see how well or badly the results agree. If the specimens are not significantly altered or damaged by the testing, the specimens are sent from one lab to the next, so that all labs use the same ones. Since such a procedure exercises most of the extraneous effects which *cause* variability in results (things such as different operators, different lab environments, and different methods of apparatus construction), it is very effective in producing good estimates of accuracy and precision. Since it is quite costly and time-consuming, this procedure is used only for tests that are sufficiently important. While there may be occasional embarrassment

---

[1] W. J. Youden, "Ranking Laboratories by Round-Robin Tests," in H. H. Ku (ed.), *Precision Measurement and Calibration,* NBS Special Publication 300, vol. 1, 1963, pp. 165–169. (NBS is now NIST.)

when a single lab disagrees with all the others, even this experience has a positive aspect, and we thus encourage such cooperation among testing labs and companies as a contribution to the common good.

Somewhat different from the type of standard we have just discussed, but also useful, are the *standard reference materials* available from the National Institute of Standards and Technology (NIST). These are described in detail in their yearly catalog, and we discuss briefly the contents of this 161-page document.[1] An overview of the utility of these materials is nicely given in the accepted definition of a *reference material*—a material or substance one or more properties of which are sufficiently well established to be used for the calibration of an apparatus, the assessment of a measurement method, or for assigning values to materials. A few examples from the catalog will make this definition more concrete. Samples of many different metals and alloys, with accurately known chemical composition, are available for use in validating the analysis facilities used in a factory or lab for quality control, process development, or basic research. That is, if you are concerned about the possible inaccuracy of your local lab's analysis, have the lab analyze one of NIST's reference materials but don't tell the lab what the material is. You can then compare your lab's results with the known analysis to see whether there is a problem. Standard reference gases and liquids are also available, as are chemical, nuclear, biological, medical, food, agricultural, and environmental materials. Line width and depth standards for microcircuit manufacturing, thermal conductivity and thermal expansion, optical properties, magnetic properties, wear and corrosion, and particle size; reference materials and samples for all these and more can be obtained.

Our discussion of standards, codes, and recommended procedures was intended to show how these could be used in two major ways. First and most obvious, we can use them *directly* when we need to perform a quality control, purchase acceptance, or research and development test that is well established and acceptable to workers in the field. Second, if we are new to an area and need to design, construct, and operate an apparatus for an experiment which has at least some, and perhaps many, novel aspects, then a careful study of standards pertinent to the various features of the device can be an excellent starting point for our apparatus design project. Since we gave only one detailed example of apparatus from the standards literature, Fig. 6.4 shows (with no further

---

[1] *Standard Reference Materials Catalog, 1990–91*, NIST Special Publication 260, NIST Standard Reference Materials Program, Bldg. 202, Room 204, Gaithersburg, MD 20899, phone (301) 975–6776.

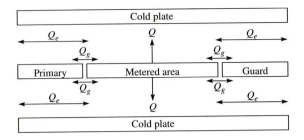

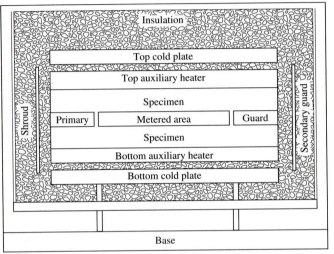

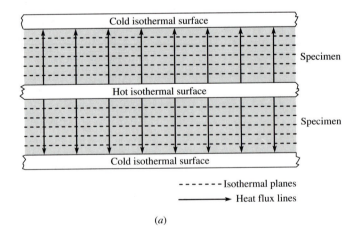

- - - - - - Isothermal planes

⟶ Heat flux lines

(a)

FIGURE 6.4
(a) ASTM C 177-85 Guarded hot-plate heat flux apparatus. (b) ASTM C 236-87 Building thermal performance using guarded hot-box. (c) ASTM C 835-82 Total hemispherical emittance of surfaces. (d) ASTM C 384-85 Acoustical impedance and absorbtion apparatus. (e) ASTM E 756-83 Measuring vibration-damping properties of materials. (f) ASTM C 635-86 Load-carrying capacity of suspended ceilings. (*Copyright ASTM. Reprinted with permission.*)

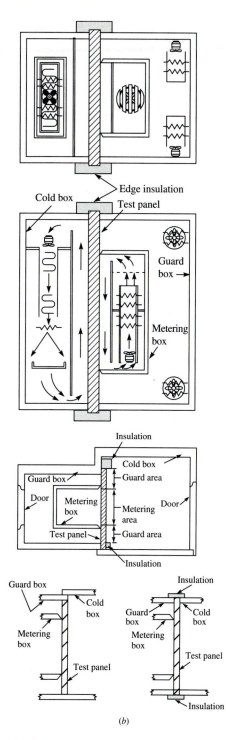

FIGURE 6.4
(Continued).

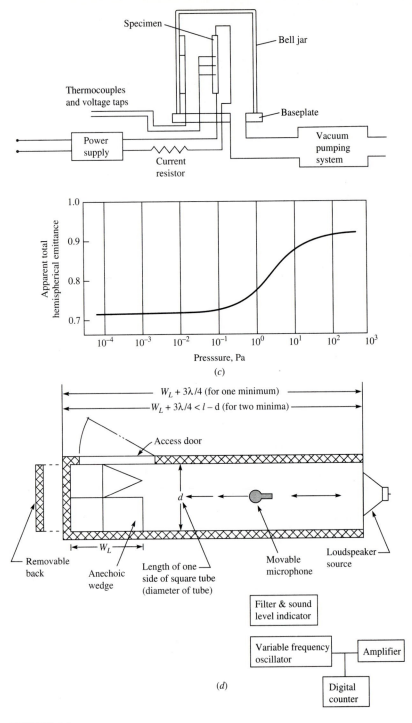

FIGURE 6.4
(Continued).

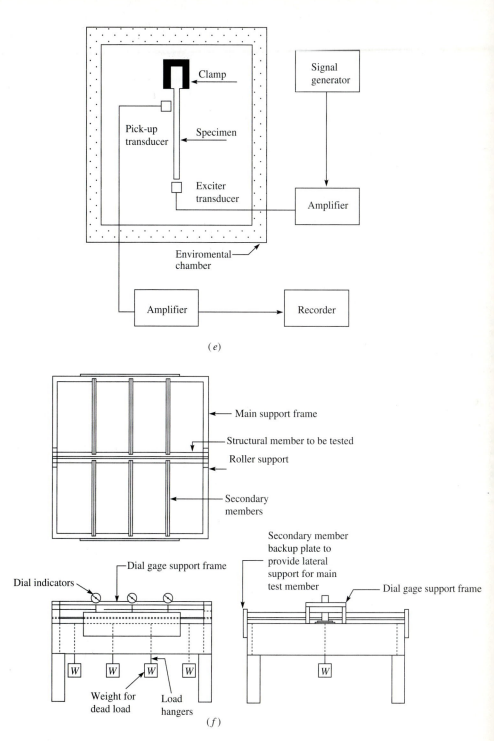

FIGURE 6.4
(Continued).

explanation) several more apparatus illustrations from the ASTM volumes, to showcase the diversity of information available there. Finally, be sure to recall that the ASTM material that we focused on here is only a small sample of the worldwide standards literature available.

6.3   READY-MADE APPARATUS, SUBSYSTEMS, AND COMPONENTS

An important part of the art and science of design, whether for mass-produced products or experimental apparatus, is the need for the designer to be familiar with existing devices which might provide all or part of the functions needed. This familiarity can lead to purchase of such parts or can serve as a source of *ideas* for a part that we will make ourselves and which can *combine* and *adapt* features from existing devices in optimum fashion for our specific needs. With the rich and continually changing technology available to us today, the task of keeping up with all the new developments which might be useful in our personal areas of work is indeed formidable. Many writers on design comment that the best designers seem to be "interested in everything" and are continually foraging for new ideas, even in fields which appear far removed from immediate needs. Since formal engineering education concentrates so much on developing competence in analysis, and leaves little curricular time for fostering familiarity with existing hardware, most recent graduates have not been given much opportunity to start accumulating this kind of background and may not even be aware of its importance in the practice of design. This section is thus a modest effort to do some consciousness raising with respect to the topic.

The ultimate example of the use of ready-made apparatus is found in the area of standardized testing of mass-produced products. Here, because the test method has been agreed upon by all involved and there are large numbers of manufacturers and products, an economically viable market for testing machines exists, and entrepreneurs will compete to fill this need. These vendors of testing machines can be located in various ways, the most systematic of which is the use of *buyer's guides.* Such guides are also useful for components and subsystems, not just complete testing machines. Most such buyer's guides are associated with trade magazines for the various technical areas, so you should become familiar with those that are pertinent to your area of work. Note that these are *trade* magazines, *not* the refereed technical journals for which academics write papers and with which students get somewhat familiar in school. Such buyer's guides are generally published once a year, perhaps as a special issue of the magazine. There are too many of them to list here, but to help those totally unfamiliar get started at the library, we name a few. *Machine Design, Product Design, R&D, Hydraulics and*

Pneumatics, Lasers and Optronics, Motion, and *Electronic Design.* In addition to companies which produce only one or a few kinds of testing machines, there are others that manufacture and/or distribute a wide variety. One supplier[1] lists hundreds of machines and includes tests in the following categories:

Laboratory automation	Burst	Abrasion
Compression set	Cure meters	Aging
Specific gravity	Cutters	Density
Drop	Dynamometers	Dies
Printing	Flex, fold, bend	Coating
Freeness, pulp	Flammability	Friction
Impact	Melt index, gel	Hardness
Moisture	Molds	Thickness
Plasticity	Resilience	Permeability
Optical properties	Pulp/fiber classifiers	Presses
Pulp beating	Sheet making	Stiffness
Tack, adhesion	Tensile, torsion, twist	Tear
Viscometers	Relative	Vibration
Bending, low	humidity/temperature	
temperature		

Some of these machines are general-purpose, but others are designed to specifically perform *standard* tests according to accepted industry or society codes, such as those of ASTM (American Society for Testing and Materials), TAPPI (Technical Association of the Pulp and Paper Industry), and ISO (International Standards Organization). Note again that we list *specific* companies and organizations to give the reader the most concrete and practical information, but this should not be taken to mean that these are the *only* worthy groups or that we personally recommend those named as superior to their unnamed competitors. For those who are new to a field, one specific contact or reference, when pursued, will often serve to open up the whole general area.

While the purchase of a ready-made machine or apparatus may sometimes be the most cost-effective means of obtaining a needed piece of experimental equipment, many times our needs are too specific to

---

[1] Testing Machines Inc., 400 Bayview Ave., Amityville, NY 11701, phone (516) 842–5400.

make this approach feasible. Even then, however, examination (not purchase) of the standardized machines may give us some excellent guidelines and ideas which we can use in the design of our own device. Pursuing the make-or-buy question at this next level of detail again is aided by familiarity with commercially available hardware. That is, instead of needing to be familiar with sources for entire machines or systems, we now need to know what kinds of general-purpose *subsystems* are on the market. By subsystems we refer to classes of equipment such as motion-control devices, optical benches and accessories, heating and cooling equipment, vacuum systems, measuring instruments and data processing devices, lab computers, software and interfacing systems, robots, electric and fluid power supplies, and control equipment.

The buyer's guides already mentioned are again useful in locating sources for this level of hardware. We also want to provide some additional information on certain classes of equipment which are required quite frequently and for which we can supply some guidance of a general nature.

6.3.1 Motion-Control Systems

Many experimental setups require that we *move* certain parts relative to others, quickly, accurately, and conveniently. Some examples are moving wind-tunnel models to different locations and into different angular attitudes, moving optical devices (lenses, mirrors, lasers, optical fibers, etc.) in an optical system, moving a hot-wire anemometer probe to a new location in a flow system, and manipulating specimens in an automated testing machine. We briefly survey the standard equipment available and list some specific sources to help you get started on an apparatus design with this type of requirement. Again, we may actually purchase such ready-made equipment or just get familiar with what is available so that we can steal some ideas for our own design. It is extremely helpful, for instance, to know what are the *best* specifications that are currently available. If we require something 2 orders of magnitude better than the best we could buy, then we know that we may have a rather difficult design problem to solve. This does not always mean that we should give up; there simply may not be a large enough market for such high-quality devices to make their routine production for sale commercially viable. Also, we may be able to come up with some *new approaches* that make the needed advances in performance feasible.

Let's now discuss the types of standard motion-control systems that are commercially available from one or more suppliers. First we should try to do a little classifying to organize our thinking. We need to keep in mind that while *motion* is our object, we need to also consider how much *force* is required in our application. *Manually executed* motions are usually the simplest, cheapest, and most reliable and thus should be considered before more complex methods. For translational motions, ready-made slides in various sizes and degrees of precision are available from several manufacturers.[1] These slides can be positioned by hand or by some kind of drive system. Fixed or adjustable mechanical stops can be used to reposition to a selected location. Screw actuation via manual knobs or motor drives is common, and calibrated micrometer-type screws provide both a positioning function and a measuring function. Differential screws[2] can be used to get high resolution without using extremely fine-pitch threads, which are weak. Backlash and lost motion can be reduced by using spring- or magnetic-force preloading schemes of one kind or another.

The repeatability of location can be improved by always approaching the final point from the *same direction*. This trick is used both in simple hand actuation and in some sophisticated computer-controlled electric drives. Slide bearings range from simple plain bearings, to linear ball bearings and crossed-roller bearings, and finally to hydrostatic air bearings.[3] Air bearings are essentially frictionless and provide the smoothest and most precise motion, but they require an air supply, are more costly, and are less stiff than the rolling-contact bearings. Manual control of very small motions can be accomplished with flexural microactuators[4] or hydraulic micromanipulators.[5] A flexural actuator of 1000-nm total travel has a resolution of 1 nm, while one of 10-nm total travel can resolve 0.01 nm. These devices have a micrometer screw input knob of "normal" size but produce translational output motions several thousand or million times smaller by using the elastic lever principle. Hydraulic micromanipulators operate on a simple liquid displacement principle whereby "large" motions of a small-diameter input piston produce much smaller motions of a large-diameter output piston. The input and output cylinders are connected with flexible plastic tubing and

---

[1] Velmex Inc., Box 38, E. Bloomfield, New York 14443, phone (800) 642-6446. Daedal, Box 500, Harrison City, PA 15636, phone (412) 744-4551.

[2] H. A. Rothbart (ed.), *Mechanical Design and Systems Handbook,* McGraw-Hill, New York, 1964, p. 26-5.

[3] New England Affiliated Technologies, 620 Essex St., Lawrence, MA 01841, phone (800) 227-1066. Dover Instrument Corp., 200 Flanders Rd., PO Box 200, Westboro, MA 01581, phone (508) 366-1456.

[4] A. E. Hathaway Inc., 595 E. Colorado Blvd., Suite 400, Pasadena, CA 91101, phone (818) 795–0781.

[5] Narashiga USA Inc., 1 Plaza Rd., Greenvale, NY 11548, phone (516) 621-4588. Dr. Lutz Pickelmann, Piezomechanik-Optik, Riedgaustrasse 11, D-8000, Munchen 80, Germany.

can thus be remote from each other. For a unit with 0.4-mm total range, the resolution is about 50 nm, but the temperature stability is about 1 μm/C°. In general, translational actuators are uniaxial devices but can be "stacked" in various ways to achieve multiaxis positioning capability, if that is needed.

If manual actuation is not acceptable, various forms of powered drives can be considered. Electric motors can be used for a wide variety of applications and are extremely common. Fluid power devices (hydraulic and pneumatic cylinders and motors) are another possibility. Strictly "mechanical" approaches (cams, Geneva drives, linkages, belts, chains, traction drives, power springs, falling weights, etc.) should not be overlooked. Although it is relatively rare, thermal actuation using, for example, a piston and cylinder packed with certain waxlike materials and electrically heated to cause expansion is another possibility.

Electric positioners are by far the most common and are available in several different forms from many manufacturers, or they can be designed and built in-house. Perhaps the simplest approach is to use *stepping motors* in an open-loop configuration. For motion-control systems in general, open-loop operation means that we do *not* use a feedback system. In a feedback system, the commanded motion is continuously compared with a measured value of the actual motion, and corrections are applied based on this error. Feedback (also called *closed-loop*) systems can be more accurate than open-loop ones but are more complex and require careful design[1] to prevent instability. Stepping motors (and associated driver electronics) respond to low-power electric input pulses with corresponding incremental steps of mechanical rotation. Typically, one input pulse causes an output rotation of 1.8°, giving 200 pulses per revolution of the motor shaft. If translational (rather than rotary) motion is desired, the motor shaft can be coupled to a screw-and-nut mechanism. If, for example, we use a 5 threads per inch leadscrew, one step of motor motion will result in 0.001 in of slide translation, and nominally we could position the load to the nearest 0.001 in. If no feedback (measurement of *actual* load position) is used, we must be careful to not apply our electric pulses at too high a rate, since then some electric pulses might *not* get converted to mechanical motion and the load position would be in error without our knowing it. Such errors can also occur if there are excessive *load torques* (beyond what the motor can supply without skipping pulses) acting on the positioned load. Stepping motors are fairly easy to control from digital computers, which

---

[1] E. O. Doebelin, *Control System Principles and Design*, Wiley, New York, 1985.

today are often already present in a laboratory apparatus for other purposes. The basic computer needs some kind of interface card which can produce a distinct number of pulses at a specified rate and also provides a logic (on/off) signal to reverse the motor when that is desired. For the simpler applications, the experimenter may be able to successfully put together a system of this type.

For more demanding requirements, or if we want to "trade time for money," we can purchase ready-made step motor systems from many vendors. We now describe a representative product of this type,[1] the Nanomover system. This is an open-loop step motor system which employs the technique called *microstepping* to achieve a resolution of ± 50 nm with a 400 step-per-revolution motor and leadscrew of pitch 2 threads per millimeter. Microstepping uses electronic methods (precise motor current stepping) to provide 10,000 steps per revolution from a 400-step motor. Repeatability of ± 100 nm is achieved by always approaching the final point from the same direction. Accuracy, which is generally less important than repeatability since systematic errors can often be compensated, is ± 1000 nm. If higher accuracy is needed, the individual system can be factory-calibrated (once) against an interferometer to document the systematic errors in motor and leadscrew. Corrections for these errors are then programmed into the system software. The travel range is 25 mm, maximum velocity is 2.5 mm/sec, maximum acceleration is 1.25 m/sec^2, and maximum load is 10 kg.

When load-positioning requirements involve high power and speed, *dc servomotor* systems may be preferable to step motor types, although one can find examples of both kinds of systems at the same power level.[2] Using closed-loop principles, dc servomotor systems combine a brushless or brush-type motor, digital control electronics, an analog power amplifier, and a motion sensor (often a digital position encoder) into a feedback loop. The performance characteristics can be quite similar to those quoted above for step motor systems. For high-speed rotary positioning of light loads, such as optical system mirrors, *galvanometer*[3] systems may be appropriate. These are essentially dc servomotor systems in which the motor is a galvanometer and thus is restricted to oscillatory motions of $\pm 90°$ or less. Such systems are capable of motion at hundreds of cycles per second.

---

[1] Nanomover, *Optics Guide 5,* pp. 27-1 to 27-25, Melles Griot, 1770 Kettering St., Irvine, CA 92714, phone (800) 835-2626.

[2] S. Jordan, "New Directions in Micropositioning," *Lasers and Optronics,* June 1991, pp. 37–42.

[3] General Scanning, Inc., 500 Arsenal st., Watertown, MA 02174, phone (617) 641–2702.

Piezoelectric actuation was relatively rare until recently, but now equipment is available from several manufacturers. The motion produced is the elastic deformation of a piezoelectric material when a voltage is applied to it, so the maximum motion possible is quite small—from a few micrometers to about 1 mm—with the longer-stroke actuators using an elastic-lever principle to magnify the smaller piezoelectric motion. An exception to this general limitation is the Inchworm Motor[1] which uses a clever incrementing principle to allow full-scale travel of up to 200. mm. As in other motion-control systems, piezoelectric actuation may use either open- or closed-loop principles. Some systems[2] use strain gages cemented to the piezoelectric element for position feedback. While motions are small, some piezoelectric actuators can provide forces as large as 30,000. N. They sometimes require voltages of several thousand volts; however, the amplifiers needed are available from the piezoelectric drive manufacturers. An application emphasizing the fine-motion capability of this type of actuator is the *scanning-probe microscope*,[3] which gave us our first "pictures" of individual atoms. Here an *extremely* fine tungsten tip (it terminates in a *single* atom!) is piezoelectrically positioned with a resolution of 0.1 Å [1 angstrom (Å) is 10^{-10} m, the approximate size of an atom; 10 Å is the distance between atoms in silicon]. Piezoelectric drives are typically capable of motion frequencies of hundreds or thousands of hertz, with specially designed systems approaching 100,000. Hz. A five-axis positioner[4] with three linear and two rotary axes, coarsely positionable by micrometer and finely positionable by piezoelectric drive, is shown in Fig. 6.5.

Hydraulic or pneumatic motion control is usually less convenient than the electrical types but may be indicated when large strokes must be accomplished rapidly and/or high force is needed. Sustained high loads under stalled conditions, which cause overheating in many electric drives, also may be best provided with a fluid system. The simplest systems use actuators positioned against hard mechanical stops and require only simple on/off valves. Continuous motion control and/or precise positioning to arbitrary locations requires feedback control using servovalves. If the experimenters have some feedback-system design experience, they

[1] *The Micropositioning Book, The Piezoelectric Book,* Burleigh Instruments, Inc., Burleigh Park, Fishers, New York 14453, phone (716) 924-9355.

[2] *Piezo Guide,* Parts 1 and 2, 1994, Polytec Optronics, Inc., 3001 Redhill Ave., Bldg. 5-104, Costa Mesa, CA 92626, phone (714) 850-1835.

[3] *The Scanning-Probe Microscope Book,* 1994, Burleigh Instruments, Inc., Burleigh Park, Fishers, New York.

[4] Polytec Optronics Inc.

FIGURE 6.5
Five-axis micropositioner with piezoelectric fine motion.

can construct their own apparatus; otherwise it can be purchased from various manufacturers. High-performance pneumatic servo systems have not been widely available off the shelf until recently, when German advancements in pressure-control servovalves and advanced control algorithms appeared.[1] Precision hydraulic positioning is a well-established technology available from a number of sources.[2] An example from this area is given in Fig. 6.20.

Our general topic of motion control includes also the function of

---

[1] Rexroth Corp., Pneumatic Division, PO Box 13497, 1953 Mercer Rd., Lexington, KY 40511, phone (606) 254-8031.

[2] MTS Systems Corp., Box 24012, Minneapolis, MN 55424, phone (612) 937-4000. Schenck Pegasus Corp., 2890 John R. Road, Troy, MI 48043, phone (313) 689-9000.

preventing unwanted motion, an important requirement in some experimental apparatus. We will be concerned here mainly with vibration-isolation systems to shield sensitive apparatus from floor vibrations. Optical tables[1] are perhaps the most widely available version of such equipment, and they combine the function of vibration isolation with that of providing a convenient mounting surface for components and a large selection of compatible mounting hardware to allow easy configuration of breadboard optical systems. Such commercially available tables may, of course, also serve for mounting systems other than optical types. Figure 6.6a shows a Melles Griot table together with some of the available mounting components. The tabletop itself is a "high-technology" component carefully designed to provide a rigid and flat mounting surface that is stable over time and temperature. It has internal damping built into it to suppress the transmission to other components of any vibrations that might arise from rotating equipment *on* the table. A sandwich construction of clad-metal honeycomb material between the top and bottom steel plates provides these desirable features. The mounting surface has a grid pattern of tapped holes for mounting components. Rather than tapping the threads directly into the top plate of the table, hardened, roll-formed threaded inserts are used. These provide increased strength and wear resistance, do not create machining debris, and are bottom-sealed to prevent entry of contaminants into the interior structure.

The tabletop itself is designed to be light and rigid so that it has no resonances below about 100. Hz and does not readily transmit vibrations from one part of the table to another. To isolate the entire tabletop from floor vibrations, it must be properly mounted on soft spring mounts, to create a natural frequency *lower* than the lowest important frequency present in the floor vibration. Most building vibrations are in the frequency range of 10 to 50 Hz for vertical motion and 1 to 20 Hz for horizontal. Passive air mounts are the simplest and cheapest, but they can only achieve a natural frequency of about 5 Hz. This provides adequate isolation (about a 10-to-1 amplitude reduction at 30 Hz) for many situations. Active (feedback-type) air mounts cost more but provide both a self-leveling feature, lower natural frequency (about 1.5 Hz), and less vibration transmission both at resonance and above (about a 50-to-1 amplitude reduction at 30 Hz). Although vertical vibration is most often encountered, building sway can cause significant horizontal vibration at

---

[1] *Optics Guide 5,* 1991, pp. 20-1 to 20-66, Melles Griot.

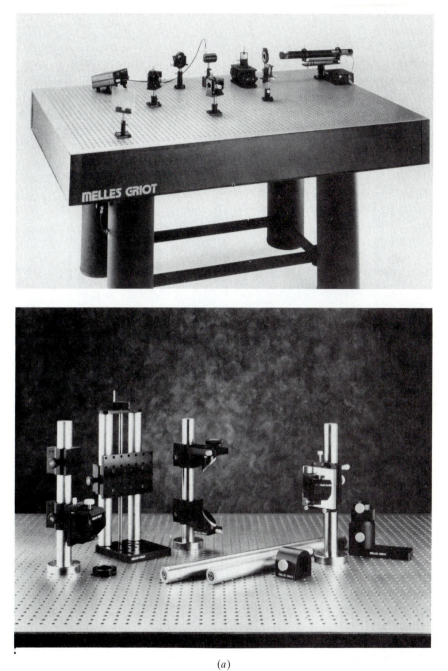

(a)

FIGURE 6.6
(a) Optical table and mounting hardware and (b) modular frame assembly system.

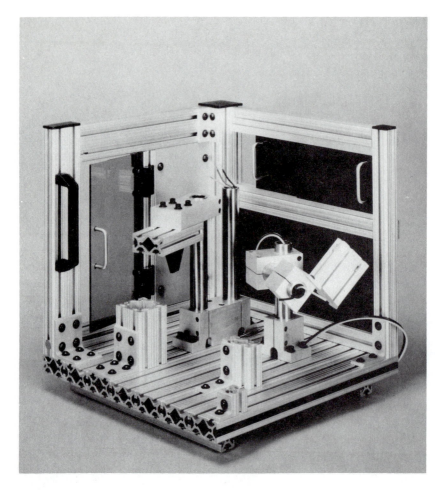

(b)

FIGURE 6.6
(Continued).

the upper floors. Special isolators designed to deal with horizontal motion
are available.

6.3.2 Mechanical Support Structure and Machine Parts

Most experimental apparatus requires some sort of supporting framework
and may also involve some kind of "machinery." While tables and
supports can often be quite simple and thus best implemented as

homemade, be aware that standardized commercial products of this kind *are* available when needed. Heavy-duty, welded-steel machine bases in several standard sizes (24×36 to 42×48 in top) can be delivered in 24 hours.[1] Variations on these products to include special nonstandard features needed in your apparatus can be provided, but require longer delivery times.

Much more versatile is a modular assembly system using standardized components to build up support structures, enclosures, and housings. Components are assembled with a screwdriver and hexagonal wrench and can later be disassembled and reused for a different purpose if desired. One of the standard components is a card cage to hold electronic circuit boards, useful for systems which are not entirely mechanical. Several vendors[2] provide such systems; Fig. 6.6b shows an example (80/20 Inc.). Aluminum-alloy extrusions are widely used, being lightweight and corrosion-resistant.

The prototyping and breadboarding of machines is facilitated by the availability of engineering kits of parts,[3] which can be assembled into operating equipment and include both the support structure and the moving parts, such as shafts, bearings, gears, couplings, cams, and small motors.

6.3.3 High-Pressure and Vacuum Systems

In some experiments, parts of the apparatus must be subjected to quite high or low absolute pressures. Pressure and vacuum vessels and associated equipment present specialized design and operational problems that may be unfamiliar to the experimenters and thus require assistance from specialty manufacturers in these fields. Vessel design and fabrication is often covered by applicable codes, but interpretation and implementation of these codes may require a specialist's expertise. Study of pertinent textbooks[4] is usually a worthwhile preliminary step before

---

[1] American Grinding and Machine Co., 2000 N. Mango Ave., Chicago, IL 60639, phone (312) 889-4343.

[2] Techno Isel, 2101 Jericho Tpke., Box 5416, New Hyde Park, NY 11042, phone (516) 328-3970. Item Products, 6703 Theall Rd., Houston, TX 77066, phone (800) 333-4932. 80/20 Inc., 1830 Wayne Trace, Fort Wayne, IN 46803, phone (219) 422-5860.

[3] Automat engineering kits, Stock Drive Products, 2101 Jericho Tpke., New Hyde Park, NY 11040, phone (516) 328-3300.

[4] W. R. D. Manning and S. Labrow, *High Pressure Engineering,* CRC Press, Cleveland, OH, 1971. I. L. Spain and J. Paauwe (eds.), *High Pressure Technology,* vol. 1: *Equipment Design, Materials and Properties,* Marcel Dekker, New York, 1977. J. F. O'Hanlon, *A User's Guide to Vacuum Technology,* 2d ed., Wiley, New York, 1989. A. Roth, *Vacuum Technology,* 3d ed., Elsevier, New York, 1990.

undertaking a homemade design or contacting commercial vendors to explore the purchase of design services and/or equipment. The field of vacuum has its own technical society [American Vacuum Society, 335 E. 45th St., New York, NY 10017, phone (212) 661-9404] and journal (*The Journal of Vacuum Science and Technology,* same address).

With regard to stress-induced failure and safety considerations, note that vacuum chambers generally have atmospheric pressure on the outside and can only approach zero absolute pressure inside, giving a modest 15-psi pressure differential tending to cause failure (if it did occur, it would be an implosion rather than an explosion). High-pressure vessels can, of course, be required to sustain thousands of pounds per square inch of pressure difference, with dangerous potential for leaks and explosions. Vacuum systems can, however, display a number of more subtle problems, which were conveniently summarized in a recent article[1] from which we now take some excerpts. Large gate valves, widely used to allow access to chambers, can be distorted and damaged if they are opened before the *pressure differential* has been reduced to recommended limits. *Virtual leaks* are caused by air trapped within the system when it was fabricated, often between two metal surfaces that are improperly welded or brazed. Full-penetration welds and furnace brazing techniques can help reduce this problem. Welding of vacuum equipment has its own set of rules which may be unknown to ordinary welders. *Cost cutting* by using vacuum-inexperienced fabricators who bid low can lead to poorly cleaned interior surfaces that contribute contaminant particles and/or take a long time to pump down to high vacuum because of outgassing. High-quality vacuum welding requires use of helium mass spectrometers and residual gas analyzers to check for cleanliness.

Generation of *contaminating particles* can be reduced by use of special valve designs which decrease wear. While undersizing does not seem to be a common problem with valves, it *is* with the plumbing connected to them, causing a loss of *conductance.* (In vacuum work, conductance refers to the flow capacity of a device.) *Sizing of vacuum components and design of complete systems* can be aided by software packages available to manufacturers[2] of vacuum components and complete systems. The referenced manufacturer has a system design program based on an Autocad platform which accepts as input the basic design requirements and produces three-dimensional views for checking of

---

[1] T. Studt, "Design away Those Tough Vacuum System Riddles," *R&D,* October 1991, pp. 104–108.

[2] Huntington Laboratories, 1040 L'Avenida, Mountain View, CA 94043, phone (800) 227-8059.

interferences, a complete set of drawings, and a price and delivery quotation. The *cyclic life* of valves can be a problem; a system cycling once a minute accumulates about 0.5 million cycles per year, a typical valve life. Some new valve designs have lifetimes exceeding 1 million cycles. *Misalignment* of adjacent chambers can cause distortion of a thin gate valve when it is mounted to connect the two chambers, resulting in valve leakage and/or jamming. *Materials* used in system design are determined by the pressure, temperature, and corrosion resistance; however, AISI-type 304 stainless steel is often most cost-effective. Its desirable properties are improved by *electropolishing*. This process attacks the microscopic peaks of the surface more rapidly than the valleys, thereby maintaining critical dimensions while improving surface finish and outgassing. If greater corrosion resistance is needed, 304L, 316, or 316L stainless steels offer special characteristics. Aluminum-alloy (6061) valves are less expensive but cannot reach the lowest pressures, are less rugged, and use only polymer seals, not the more desirable metal seals used in stainless-steel valves.

Laboratory-scale high-pressure vessels are available from a number of manufacturers, and we now discuss briefly the characteristics of a representative line of such products.[1] Standard vessels are available with internal volumes from 25 mL to 5 gal and with pressure ratings up to 8500 psi. Parr recommends that vessels not be filled to more than 75 percent of the available free space. The highest pressure ratings are restricted to the smaller volumes, and pressure ratings also must be decreased for high-temperature operation. For example, a 71-mL vessel rated at 8500 psi at 350°C can only stand 1850 psi at 540°C. Vessels used in experimental work must often be opened and closed repeatedly, with clear access to the interior, so designers have devised various schemes to provide this ready access and still maintain safe and leak-free sealing. The Parr approach, a patented split-ring closure, is shown in Fig. 6.7. The closing force is developed by tightening a set of cap screws located in the ring sections. Several different types of gaskets and gasket materials are needed, depending mainly on the operating temperature. Flat gaskets of Teflon are good to about 350°C with Grafoil (a special graphite material) being used for the higher temperatures. Stainless-steel ring gaskets with a diamond-shaped profile are used for some high-temperature applications. Some low-temperature vessels use self-sealing O-ring gaskets. These do not require a sealing force from external cap screws but rather develop the sealing force from the pressure in the vessel, with the sealing force

---

[1] Parr Instrument Company, 211 Fifty-third St., Moline, IL 61265, phone (800) 872-7720.

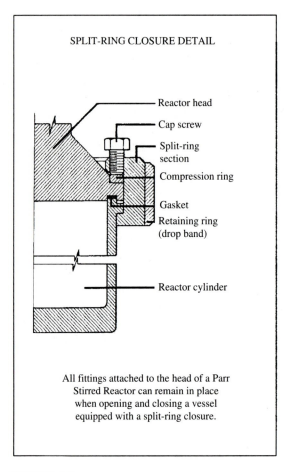

SPLIT-RING CLOSURE DETAIL

Reactor head

Cap screw

Split-ring section

Compression ring

Gasket

Retaining ring (drop band)

Reactor cylinder

All fittings attached to the head of a Parr Stirred Reactor can remain in place when opening and closing a vessel equipped with a split-ring closure.

FIGURE 6.7
Pressure-vessel sealing systems.

automatically increasing as the pressure increases. Synthetic rubber O-rings are good to 150°C, Viton to 225, and Teflon to 300.

Since many pressure vessels are used in studies of chemical reactions, built-in stirring devices are often needed to ensure proper mixing. This presents a dynamic sealing problem since a rotating stirrer shaft must somehow be provided inside the vessel. Because the penetration of a pressurized vessel wall with a moving shaft is a *general* problem not limited to the provision of a stirring function, we show two common methods of accomplishing this. Figure 6.8 shows some details of Parr

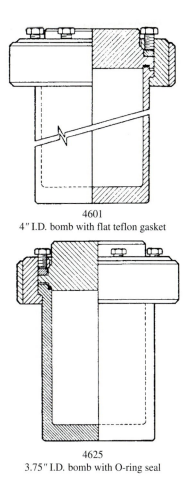

4601
4″ I.D. bomb with flat teflon gasket

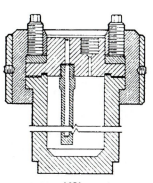

4651
2.5″ I.D. high-pressure bomb
with contained flat gasket

4625
3.75″ I.D. bomb with O-ring seal

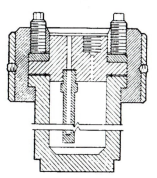

4651
2.5″ High-pressure bomb
with metal gasket

FIGURE 6.7
(Continued).

Instrument's version of these general solutions. For pressures up to about 2000 psi, or vacuum to 1 mmHg, the self-sealing packing gland provides a compact and inexpensive rotating seal for speeds up to about 2500 rpm. For higher pressures and/or greater reliability, the larger and more complex magnetic drive may be required. Here there are no rotating seals at all. The stirrer shaft is attached to an inner magnetic rotor inside a sealed housing made of a nonmagnetic material so that an external magnetic field can easily penetrate it. Torque to drive this rotor is provided through the housing wall by an outer motor-driven rotating

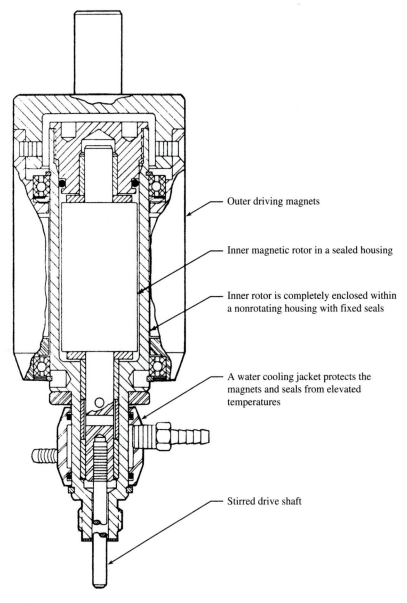

Outer driving magnets

Inner magnetic rotor in a sealed housing

Inner rotor is completely enclosed within a nonrotating housing with fixed seals

A water cooling jacket protects the magnets and seals from elevated temperatures

Stirred drive shaft

PARR A1120HC MAGNETIC DRIVE

FIGURE 6.8
Transfering rotary motion through pressurized walls.

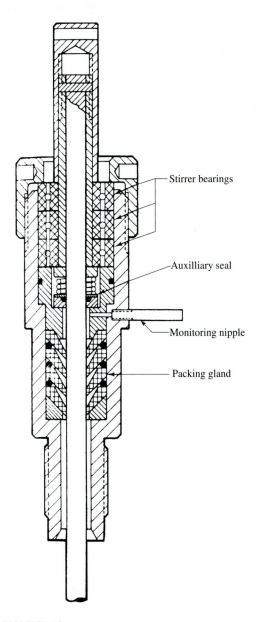

FIGURE 6.8
(Continued).

member containing permanent magnets. Up to 120 in·lb of magnetic torque can be provided to drive the stirrer shaft. As mentioned before, this magnetic drive principle is not limited to stirring functions; it has been used for various purposes wherever a leak-free rotating "seal" is needed.

Vessels can be fabricated from various materials which are selected mainly on the basis of temperature and corrosion resistance, but type-316 stainless steel meets many common requirements. The Parr catalog gives extensive guidance on this question. Internal heating or cooling systems and the associated control apparatus are also available. Some low-pressure vessels are made of plastic so that the contents can be heated by microwave methods. With regard to safety, the split-ring closure is designed to start leaking when the internal pressure is about 2 to 3 times the rating, well below the explosion point of the vessel itself. In addition, pressure relief valves and safety rupture disks can be set to relieve pressures as the pressures begin to exceed the rated value. Often rupture disks alone are used, but they tend to be unreliable below about 1000 psi, so relief valves are indicated for this range. Quartz or sapphire viewing windows can be provided up to about 3000 psi, and sealed electric leads can penetrate the vessel wall up to about 10,000 psi. Most chambers come equipped with a simple Bourdon-tube pressure gage.

6.3.4   Optical and Electro-optical Components and Systems

Some experiments require the use of optical and/or electro-optical devices. Information on the availability and application of such equipment can be found in several manufacturers' publications. One which we found particularly useful is the Melles Griot *Optics Guide*.[1] This is a combination of a catalog and a practical optics handbook of about 800 pages, and it was free last we knew. The handbook portion is very extensive and provides a wealth of useful formulas and application tips for the experimenter with a limited background in optics. This kind of information may be of more use to the experimenter with a problem to solve than the more conventional optics textbooks are, since only information of immediate practical utility is emphasized. Also each section of tutorial text is followed by catalog-type information showing available components of the kind just discussed.

The guide is well organized into four major sections and 28 chapters, as follows:

---

[1] *Optics Guide 5*, Melles Griot, 1770 Kettering St., Irvine, CA, 1991.

OPTICS—MATERIALS AND COATINGS

Fundamental optics
Optical specifications
Optical materials
Optical coatings

OPTICAL COMPONENTS

Singlets	Doublets and triplets
Lens kits	Cylindrical optics
Condensers	Prisms and retroreflectors
Filters and etalons	Mirrors
Polarization components	Windows, beam splitters, optical flats
High-energy laser optics	Microscopes, shutters, apertures

LASERS, ACCESSORIES, AND DETECTORS

Helium-neon lasers	Gaussian beam theory, accessories
CO_2 lasers and optics	Diode lasers, optics, accessories
Laser scan lenses	Detectors

OPTOMECHANICAL MOUNTING SYSTEMS

Component holders	Post-mounting system
Stable-rod system	Rails, bases, positioning stages
Nanomover system	Tables and breadboards

Figure 6.6, used earlier to illustrate vibration-isolated tables for mounting apparatus, also shows how optical components can be conveniently assembled into a working system, using this type of table. Alignment and focusing of optical components can be accomplished via the motion-control systems also discussed in that section. While we have emphasized the utility of manufacturers' technical notes, such as the Melles Griot *Optics Guide,* you certainly should not overlook the wealth of information available in standard texts.[1]

6.3.5 Heating and Cooling Equipment

Since temperature affects the behavior of so many materials and devices, often there needs to be a space in the apparatus where we can provide either a constant or time-varying known temperature. If the apparatus is located in an ordinary room with conventional temperature control for

---

[1] P. R. Yoder, Jr., *Opto-Mechanical System Design,* Marcel Dekker, New York, 1986. D. C. O'Shea, *Elements of Modern Optical Design,* Wiley, New York, 1985.

human occupancy, a temperature-controlled environment already exists. Building-temperature control systems usually hold the temperature constant within a few degrees of the set point and may allow setpoints within about ±5°C of the usual setting of 70°F (21°C), assuming that both heating and cooling are available at all times. If this range and accuracy are acceptable, no special heating or cooling provisions are necessary for the experiment. If our needs are outside these ranges, then we must design and construct, or purchase ready-made, a suitable temperature control system.

In most applications, the requirements for temperature control are sufficiently stringent that a feedback (closed-loop) type of control system, rather than the simpler open-loop system, will be needed. Design techniques for feedback systems in general[1] and temperature controls in particular[2] are available. Of course, we cannot duplicate this voluminous material here, so we concentrate on describing ready-made systems since they are often used. (It is also possible, if our apparatus already includes a digital computer for other purposes, to include temperature control as another of its tasks. This requires the acquisition of suitable software and the provision of interface devices for sensors and heating or cooling devices.)

The main components of ready-made systems are a source of heating and/or cooling, a temperature sensor, and a system controller. These days, system controllers are usually digital electronic devices and are quite general-purpose. That is, a single controller model could be used to implement *many* different control systems using a variety of sensors and many types of heating and/or cooling devices. In fact, a controller used for temperature control could also be used for pressure, flow, humidity, etc.; i.e., it is a "generic" device. Most controllers can be programmed to provide a variety of control laws or algorithms, the most versatile and accurate being some version of the so-called PID (proportional, integral, derivative) control. If one decides to use PID control, then it is necessary to "tune" the controller settings for optimum response of the particular system. That is, the *amount* of each of the three modes which produces the best response is different for each application and must be determined by a combination of theoretical prediction and trial and error. Many controllers now have a *self-tuning* feature which assists or completely takes over the tuning function. If your apparatus *changes* its thermal behavior (change in the mass to be heated, change in

[1] Doebelin, *Control System Principles and Design.*

[2] W. K. Roots, *Fundamentals of Temperature Control,* Academic Press, New York, 1969. J. R. Leigh, *Temperature Measurement and Control,* Peter Peregrinus, London, 1988.

heat-transfer coefficients due to different flow rates, etc.), the controller may have to be retuned.

A practical approach often taken is to purchase a general-purpose PID controller and then tune it by using the self-tuning feature (if available) plus whatever final manual tuning is needed. Such controllers are available from many manufacturers, most of whom will also recommend and supply a suitable temperature sensor. We thus concentrate in the remainder of the discussion on the various means of actually providing the desired heating or cooling effects needed for the experiment. We can first list the basic physical effects which produce heating or cooling and then describe some of the commercially available equipment for producing these effects:

HEATING METHODS

- Electric resistance
- Electric induction
- Microwave
- Thermoelectricity
- Combustion
- Radiation (lamps, lasers, solar)
- Chemical reaction (exothermic)
- Flow of hot fluids, plasmas
- Friction

COOLING METHODS

- Mechanical refrigeration
- Liquefied gases
- Vortex tube
- Thermoelectricity
- Water ice, dry ice
- Chemical reaction (endothermic)
- Flow of cool gases, liquids
- Adiabatic demagnetization

The above list is not exhaustive but includes most of the more common methods.

Electric heating in some form is probably most frequently used, with *resistance* heating the most common mode. Catalogs[1] for electric resistance heaters display a wide variety of forms and include technical information useful for selecting a unit for specific application characteristics. Heat is generated internally in the metallic elements (usually made of some nickel-chrome alloy) at a rate I^2R watts. This heat must then be somehow communicated to the specimen or space where we wish to establish a certain temperature. Sometimes we can place the heater and heated object in direct contact and can induce heat transfer mainly by conduction. Cartridge heaters are cylindrical and can be inserted into holes drilled into some portion of the article to be heated (if this is acceptable) and typically produce temperatures from 250 to 1000°F.

---

[1] *Chromalox Industrial Heating Products Stock Catalog,* Chromalox-E. L. Wiegand Division, Emerson Electric Co., 641 Alpha Dr., Pittsburgh, PA 15238, phone (412) 967-3800.

Their diameters go from 0.25 to 1 in, their lengths from 1.5 in to 3 ft, and their wattage ratings from 50 to 5000 W. Tubular heaters have diameters of 0.2 to 0.48 in, lengths up to 15 ft, wattage ratings up to 7500 W, and working temperatures of about 250 to 750°F. Some can be bent into shapes needed for a specific application. Strip heaters are rectangular, 0.75 to 2.5 in wide, 5 to 96 in long, and 0.38 in thick, with ratings from 150 to 4500 W. Ring elements are 0.38 in thick with outside diameters from 3 to 11 in and ratings from 125 to 1800 W. Disk elements are available in 2.3- and 3.3-in diameters with ratings from 150 to 600 W. Strip, ring, and disk heaters are used to produce temperatures from 250 to 950°F. Heating tape or cable can be wound around objects to be heated and can be any length. A typical rating is 8 W/ft of cable length. For winding around pipe or other cylindrical objects, changing the pitch of the winding allows one to adjust the wattage per foot of pipe over a wide range. For example, for 8 W/ft cable on a 24-in-diameter pipe, changing the winding pitch from 2 to 90 in gives heating rates from 260 to 11 W/ft of pipe. Band and nozzle heaters are designed to clamp around cylindrical objects with diameters from 1.3 to 20 in.

A special form of resistance heater called Thermofoil[1] uses techniques adapted from printed-circuit board manufacture to fabricate thin, flexible heaters in almost unlimited shapes and sizes. The thin heating element is sandwiched between layers of flexible insulation, the entire sandwich being about 0.01 to 0.02 in thick. Its flexibility allows it to be conformed to curved surfaces, and it can be attached with various adhesives, shrink bands, stretch tapes, or clamps. Power densities up to about 60 W/in^2 are possible. The shape of the film heating element can be tailored to the needs of the specific application, putting the local heating power exactly where it is needed, so as to give uniform or nonuniform temperature, as needed. Heaters can be supplied with temperature sensors built in, facilitating manual or automatic control.

The basic heating elements described above can be adapted to many specific applications. Sometimes we attach them directly to a solid object that we wish to heat. At other times the heater is used to heat a fluid which we pass over the object to be heated. In addition to the general-purpose elements above, more specialized resistance-heating apparatus is available. We will just list the names of some of these to indicate the type of application: immersion heaters, circulation heaters, process air heaters, flexible tank heaters, vaporizers, steam boilers, hot water boilers, and cast-in heaters.

---

[1] Minco Products Inc., 7300 Commerce Lane, Minneapolis, MN 55432, phone (612) 571-3121.

If you have studied thermodynamics and heat transfer, you might want to use this background when calculating the proper kilowatt rating of a resistance heater for a particular application. Unfortunately, conditions of the practical application often deviate significantly from the simple assumptions used in theoretical treatments, making such calculations quite uncertain. To allow more accurate sizing of heaters, the manufacturers combine theoretical principles with experimental correction factors in their calculations. The Chromalox catalog mentioned above gives detailed procedures for estimating heater sizes, and we now briefly review some of these. General factors which need to be considered include

1. Time allowed for heat-up
2. Heat losses to the surrounding medium
3. Thermal properties of material being heated and any insulation used
4. Makeup material supplied, per unit time
5. Surface heat loss of material and container
6. Dimensions and weight of container and material being heated
7. Heat carried away by products being processed through the heated area

A three-step calculation method is recommended:

1. Determine the kilowatt capacity required to bring the material temperature up to the desired value in the desired time.
2. Determine the kilowatt capacity required to maintain the operating temperature steady at the desired value.
3. Select the number and type of heaters required to supply the greater of the two capacities found in steps 1 and 2. Add 20 percent to this rating to allow for uncertainties.

Any actual operating experience with the apparatus or similar ones, if available, should be used to adjust the theoretical calculations.

The referenced catalog shows a number of sample calculations, and we now reproduce one to illustrate some typical numbers. An uninsulated covered steel tank weighing 50 lb is 2 ft in diameter and 4 ft high and contains 60 gal of water at 70°F. We want to heat this water to 180°F in 3 h and then heat 100 gal/h of water from 70 to 180°F thereafter. By using water and steel specific heats, respectively, of 1.0 and 0.12 Btu/(lb·°F) and using a catalog chart to estimate the tank surface heat loss as 66 W/ft² at the final temperature of 180°F, the step 1 calculation goes as follows:

$$\text{Total energy} = \frac{(60)(8.35)(1)(180 - 70) + (50)(0.12)(180 - 70)}{3412}$$

$$+ \frac{(66)[(3.14)(2)(4) + (3.14)(2^2/4)(2)]}{(1000)(2)} \quad (3)$$

$$= 16.3 + 3.1 = 19.4 \text{ kWh}$$

$$\text{Average heating rate} = \frac{19.4}{3} = 6.5 \text{ kW} \quad (6.1)$$

Note that in the heat loss calculation an *average* heat transfer rate equal to one-half the rate at the final (180°F) temperature is assumed. Mixed SI (W) and British (Btu, lb) units are typically used in this industry and require the conversion factor of 3412. Step 2 in the sizing calculation is as follows:

$$\text{Heating rate} = \frac{(100)(8.35)(180 - 70)(1)}{3412}$$

$$+ \frac{(66)[(3.14)(2)(4) + (3.14)(2^2/4)(2)]}{1000}$$

$$= 26.9 + 2.1 = 29.0 \text{ kW} \quad (6.2)$$

Since the steady-state heating requirement exceeds the initial heat-up requirement, and because the catalog recommends a 20 percent safety factor, the necessary heater rating is 34.8 kW. This heater will, of course, be capable of a much faster initial heat-up than originally specified; however, faster heat-up is usually acceptable, in fact, desirable. Note that an "oversized" heater may be undesirable from an initial-investment viewpoint, but does *not* represent an increased ongoing cost of operation because the temperature control system will set the *actual* power input so that it exactly matches the heat loss from the system. That is, a 34.8-kW heater is *capable* of producing this heating rate but actually produces only the rate that we set.

Resistance heating can be used to create a stream of hot fluid which in turn can be used to control the temperature of the apparatus or specimen. The Chromalox catalog referenced above lists some ready-made units usable with a variety of gases, liquids, steam, etc. Some units are heaters only and require in addition a pump or blower to move the fluid through the heater and to the apparatus. Others are sold complete with the necessary pump or blower. Typical process heaters using liquids have heat ratings from 20 to 150 kW, circulate the fluid at 35 to 150 gal/min, and create working temperatures up to 650°F. By using air, working temperatures up to 1200°F are possible. At 500 ft³/min of air, a 50°F temperature rise through the heater requires about 8.5 kW while a 600°F rise takes 102 kW. Other flow rates and/or temperature rises can be easily estimated from these values since the relations can be treated as linear in both these variables.

Heating by electric induction, dielectric processes, or microwaves is fairly common in industrial applications but is used less often in laboratory experiments since the commercially available heating equipment tends to be large, expensive, and specially designed for each application, rather than general-purpose, off-the-shelf items. These modes of heating would thus be considered for a lab application only if their characteristics were exceptionally well matched to apparatus needs and if simpler methods had been found not feasible. Induction heating[1] requires that the heated object be a good electric conductor, usually some metal, but ionized solutions, molten salts, and glass at high temperature have also been successfully heated. A suitably shaped coil is energized with an alternating voltage, produces a magnetic field, and induces into the workpiece eddy currents which cause heating of the workpiece due to its resistance. Coils may have one or more turns, may surround the work or fit inside hollow workpieces, and can be shaped to cause uniform or nonuniform heating, as desired. Since the work coil itself tends to heat up, it is usually made hollow for water cooling. Frequencies from 60 Hz up to about 3 MHz are used, with the higher frequencies often employed to concentrate the heat near the surface of the workpiece ("skin effect") when that is desired, as in surface hardening. Induction heating has no inherent temperature limitation since the heat is generated internal to the work, and thus the workpiece temperature rises so long as the heat supplied exceeds that lost from the work surface.

Dielectric heating requires nonconducting workpieces and places them between flat-plate (or other suitably shaped) electrodes which are energized with voltages of high amplitude (up to 20,000 V) and high frequency (5 to 50 MHz) to produce an electric field in the workpiece. Poor electric conductors are also usually poor thermal conductors, and thick sections are difficult to heat by applying *external* heat sources since the surface overheats before the interior reaches the desired temperature.

---

[1] S. Zinn and S. L. Semiatin, *Elements of Induction Heating: Design, Control, and Application,* American Society for Metals, Novelty, OH, 1988. *Induction Heating Guide,* Inductoheat, 32251 N. Avis Dr., Madison Heights, MI 48071, phone (313) 585-9393.

Dielectric heating, like induction heating, generates the heat internally in the workpiece and may be ideal for such applications. For a given voltage amplitude and frequency, the heating effect is proportional to the dielectric constant and power factor of the material being heated, the product of these two being called the *loss factor* of the material. The loss factor may depend on the frequency, so it may be necessary to explore a range of frequencies to find which gives the best performance. Since water has a much higher loss factor than most other materials, workpieces with high moisture content will heat rapidly; in fact, many industrial applications are really aimed at *drying* the workpiece. Microwave heating shares many of the characteristics of dielectric heating except that (1) the frequency range is somewhat higher (two main frequencies, 915 and 2450 MHz, are used) and (2) the power source is a magnetron rather than a high-power vacuum tube or solid-state oscillator.

Resistance-heated ovens and furnaces specially designed for incorporation into materials-testing machines are available from several manufacturers.[1] These come in several different configurations, such as box or split-tube types, and typically can reach maximum temperatures up to about 1700°C (3100°F). The test volume is enclosed by heavy insulation, and the resistance heating elements transfer heat to the specimen by free convection and radiation, there being no direct contact between the heaters and the specimen. Heat-up times from a cold start vary with the size and type of oven but typically range from 15 to 200 min. For specimens requiring controlled atmospheres or vacuum, special ovens called *retorts* are available. Figure 6.9 (reference 1) shows such a unit and makes clear the complexity of its design. Purchase, rather than in-house design and construction, of such specialized apparatus might often be wise unless the experimenter has previous experience.

Thermoelectric heating and cooling use the same phenomena employed for temperature measurement in the familiar thermocouple. Commercially available devices[2] are mainly used in cooling and are restricted to relatively low loads, such as cooling of radiation detectors in thermal imaging systems and "fogging mirrors" in humidity-measuring instruments. For loads within their capabilities they offer small size, no moving parts or noise, and long, reliable life.

Owing to the problems of possibly contaminating combustion products (soot, smoke, fumes, etc.) and generally poor controllability, combustion heaters are not much used in lab experiments (except

---

[1] Applied Test Systems, Inc., 348 New Castle Rd., PO Box 1529, Butler, PA 16003, phone (412) 283-1212.

[2] Fotodyne Inc., 16700 W. Victor Rd., New Berlin, WI 53151, phone (800) 362-3686.

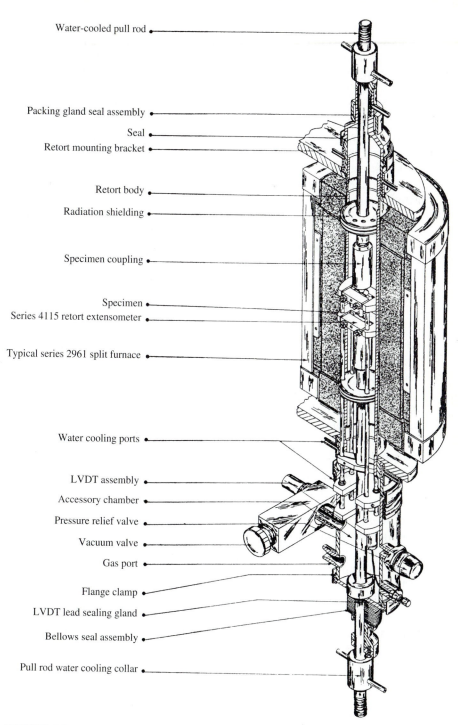

Water-cooled pull rod

Packing gland seal assembly

Seal

Retort mounting bracket

Retort body

Radiation shielding

Specimen coupling

Specimen

Series 4115 retort extensometer

Typical series 2961 split furnace

Water cooling ports

LVDT assembly

Accessory chamber

Pressure relief valve

Vacuum valve

Gas port

Flange clamp

LVDT lead sealing gland

Bellows seal assembly

Pull rod water cooling collar

FIGURE 6.9
Controlled-atmosphere oven (retort).

perhaps for Bunsen burners) unless they are uniquely suited to some specific experimental need.

While certain ovens or furnaces using resistance heating elements and described above transfer some of their heat by radiation, we next consider devices specifically designed and used to emphasize radiant heating. Perhaps most widely used as a basic radiant energy source are the quartz infrared lamps[1] available in lengths from about 2 to 50 in and with $\frac{3}{8}$-in diameter. By using tungsten filaments (resistance heating) inside a clear quartz envelope (translucent envelopes are used when diffused, rather than sharply focused, heating is wanted), radiation which peaks in the range of 890 to 1500 nm is produced and transmitted through the envelope to the specimen. Of the total heat produced, 72 to 82 percent is radiant and the rest is combined conduction and convection. Other types of "radiant" heaters use different filament materials, run at lower temperatures, and transmit as much as 80 percent of their total heat by conduction and convection.

Quartz lamps reach 80 percent of their maximum radiant output in less than 1 sec after being turned on, and they cool off in about 1 min after shutoff. This speed is much better than that of most other heat sources and is a main advantage, as is the cleanliness. With a radiant heat source, it is also possible to leave the lamp on continuously and use a mechanical shutter between the lamp and specimen to rapidly manipulate the applied heating rate. Lamps last about 5000 h at rated voltage, but much longer life can be obtained if lower voltages are acceptable since lamp life varies as the 13th power of voltage, giving an estimated 50 million h of life at 50 percent of the rated voltage. Lower voltages, however, reduce the power radiated. Voltages beyond 225 percent of rated result in almost instant failure.

In addition to the quartz lamps themselves, the referenced company also offers a wide variety of heaters, chambers, and controllers based on these lamps. Many of these take advantage of the focusability of radiant energy to provide highly concentrated spots or lines of heating. A *spot heater* uses an ellipsoid reflector with the lamp filament at one of the foci of the ellipse to produce a sharply focused heat spot at the other focus. A spot of $\frac{1}{4}$-in diameter has a heating density of 650 W/in^2. (Cartridge resistance heaters discussed earlier are limited to about 300 W/in^2.) Specimen temperatures up to about 2000°F are possible with small metal targets in less than 1 min. *Line heaters* use elliptical reflectors to produce a thin line of heat flux 5 to 25 in long. When a wider heated area is

---

[1] Research Inc., Box 24064, Minneapolis, MN 55424, phone (612) 941-3300.

desired, parabolic reflectors produce a parallel beam when the lamp filament is placed at the parabola's focus.

A variety of heating chamber designs are possible by using combinations of heating elements and suitable reflectors. Figure 6.10[1] shows a four-lamp unit with elliptical reflectors and some time-response curves for typical specimens. Heat-flux densities of about $100 \, kW/ft^2$, transient temperature of 2700°F, and steady-state temperatures of 2000°F are achievable. Another unit uses a bank of 24 lamps, individually controlled in pairs, to provide the capability of producing a temperature *profile* along a specimen's length. In addition to producing a desired *nonuniform* profile, such a device can be used to provide extra heating at the ends of the specimens (where attached load grips cause cooling) and thus give a desired *uniform* temperature profile. In all these radiant-heating applications, the highest heating rates and specimen temperatures will be achieved when the specimen has a high absorptivity (emittance). This is encouraged by rough, blackened specimen surfaces, if that is allowed by experiment requirements.

Laser beams of various types can concentrate very large heating rates into very small areas, can be modulated rapidly in time, and can be "steered" with galvanometer-driven mirrors or other optical means to precise positions on a specimen. Originally limited to rather low total power, the largest commercial lasers are now capable of kilowatts of power, with the CO_2 laser being perhaps the most common high-power type. While a laser beam can be defocused to spread the energy over a wider area and then moved rapidly to cover a large surface, lasers in general are not very efficient bulk heating devices and have been used as heaters (rather than cutters or welders) mainly for surface hardening of metals. Also, most metals, unless treated with absorbent coatings, do not readily absorb the wavelengths of laser radiation ($10.6 \, \mu m$ for the CO_2 laser) until the metal starts to melt or vaporize. Nonmetals are usually much better absorbers and thus are easier to heat. Power densities for heating applications are about 1000 to 10,000 W/cm^2—much less than the values (about $10^7 \, W/cm^2$) used for cutting, welding, and drilling. We can see that lasers are not well suited to general-purpose heating tasks and will be selected for use in an experiment only when their unique characteristics meet certain specific needs. One such application is the measurement of the thermal transport properties (thermal diffusivity, specific heat, etc.) of materials. The *laser pulse* method of measurement[2]

---

[1] Research Inc.

[2] H. Koebner, *Industrial Applications of Lasers,* Wiley, New York, 1984, pp. 199–205.

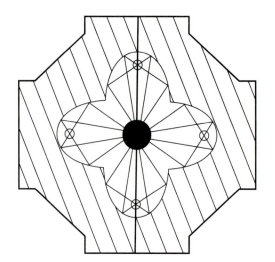

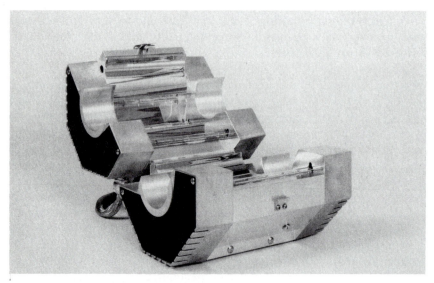

FIGURE 6.10
Quartz-lamp radiant heater and performance curves.

relies on the short duration of a laser heating pulse, and the thinness of the heated surface layer, to satisfy the simplifying assumptions of a theoretical analysis, which leads to a simple measurement of thermal diffusivity. *Electron beams* share many of the characteristics of laser heat sources but require operation in a vacuum, further complicating their use.

We turn now to *cooling methods.* Commercially available chambers

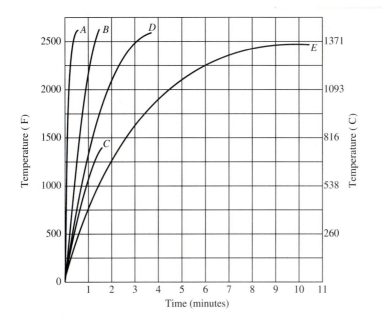

Curve	Unit used	Lamps	Voltage	Specimen material	Specimen size	Specimen condition
A	E4-10	2000T3	240	S.S. rod	1/4″ (0.63 cm) dia.	Oxidized
B	E4-10	2000T3	240	S.S. rod	1/2″ (1.27 cm) dia.	Oxidized
C	E4-10	2000T3	240	S.S. rod	7/8″ (2.22 cm) dia.	Unoxidized
D	E4-10	2000T3	240	S.S. rod	7/8″ (2.22 cm) dia.	Oxidized
E	E4-10	2000T3	240	S.S. rod	1-3/8″ (3.45 cm) dia.	Oxidized

FIGURE 6.10
(Continued).

for producing low temperatures either will utilize some type of refrigeration "machine," which achieves its cooling effect by a continuous energy conversion process employing electric power as its input, or will instead consume cold liquefied (or solidified) gases, purchased as needed, as the cooling source. (The purchased liquid nitrogen, solid carbon dioxide, etc., were of course themselves produced by refrigeration machines.) The essential tradeoff between these two cooling techniques is that of equipment cost, complexity, and size of the machine approach versus the higher operating cost of purchasing liquefied gases. The major application criteria of a cooling system are the lowest temperature attainable and the heat load. While aimed at the specific area of electronic equipment cooling, Figure 6.11[1] gives a useful overview of cooling methods from this viewpoint.

Unless our experiment requires a controlled atmosphere for some reason, in most applications the specimen or experimental space is cooled by using a flow of cold air. One family of commercial units[2] uses bottled liquid carbon dioxide or nitrogen, which is atomized into the airstream to cool it by absorbing energy according to the heat of vaporization of the liquefied gas. Temperatures of $-73°C$ can be reached with CO_2 and $-184°C$ with N_2. Chamber sizes from about 0.4 to 6.2 ft^3 are available, and typical cooling rates are 0.25 to 1.0°C/s. Airflow rates range from 60 to 600 ft^3/min, and consumption of liquefied gas is from 0.7 to 35. lb/h. These units also provide electric resistance heating since they are intended mainly for temperature-cycling tests on electronic components. For applications requiring extended periods of cooling, the manufacturer recommends use of refrigeration machines, rather than bottled gas, as economically superior.

A simple refrigeration "machine" usable for temperatures down to about $-40°F$ and producing up to 6000 Btu/h is the *vortex tube* (or *Hilsch tube*).[3] A small tubular device with no moving parts, the vortex tube runs on 100-psi shop air at 70°F and produces a flow of cold air at one end and hot air at the other. The vortex tube produces a significant level of acoustic noise, but usually this can be made acceptable with suitable mufflers.

While refrigeration machines used to produce liquid helium (4 K) on an industrial scale are in operation, most laboratory-scale apparatus requiring very low temperatures will use purchased liquefied gas and evaporate it to achieve the desired temperature. If this is uneconomical due to long test times, small-scale refrigeration machines of various types[4] are available. Experimentation at very low temperatures requires many specialized techniques in addition to the attainment of the desired temperature. These are discussed at length in several texts.[5] Attainment

[1] C. J. Morrissey, *Survey of Cooling Techniques,* JPL Invention Report NPO-17457/6964, Jet Propulsion Lab, Pasadena, CA, July 1989.

[2] Sun Electronic Systems, Inc., 5307 N.W. 35th Terrace, Fort Lauderdale, FL 33309, phone (305) 739-7004.

[3] Vortec Corp., 10125 Carver Rd., Cincinnati, OH 45242, phone (513) 891-7475.

[4] E. A. Avallone and T. Baumeister III, *Marks' Standard Handbook for Mechanical Engineers,* 9th ed., sec. 19.2, cryogenics, McGraw-Hill, New York, 1987.

[5] A. C. Rose-Innes, *Low Temperature Laboratory Techniques,* 2d ed., Crane, Russak & Co., New York, 1973. G. K. White, *Experimental Techniques in Low-Temperature Physics,* 2d ed., Oxford, London, 1968.

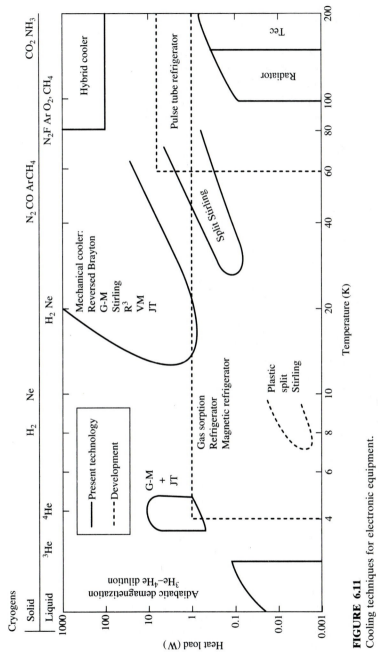

FIGURE 6.11

Cooling techniques for electronic equipment.

401

of the very lowest temperatures (within a few microkelvins of absolute zero) involves the use of adiabatic demagnetization, which is explained in chapter 9 of White.

The subject of this section—ready-made apparatus, subsystems, and components—is obviously vast and could fill entire volumes. However, at this point we have included enough material for the purposes of this book. We have given considerable detail on several major areas and have familiarized readers with the general approach to the subject so that they can efficiently pursue, on their own, topics not covered here that they might need for a particular application.

6.4 USING SPECIALTY MATERIALS TO ADVANTAGE

Most design engineers or students have had some exposure to the general subject of materials selection in courses on materials and/or design. We do not review or duplicate this general background here. There are, however, some helpful hints relevant specifically to apparatus and instrument design that are not usually brought out in general materials and design courses that we feel should be part of this text.

While materials designed for *strength* at extreme temperatures *are* usually familiar, those intended to minimize thermal expansion, or changes in modulus of elasticity, over ranges of temperature are less so. Metals which exhibit minimal thermal expansion[1] include Invar and Super Invar 32-5. Super Invar has an average expansion coefficient of 0.35×10^{-6}/°F from -60 to 200°F and is used in many optical instruments where dimensional stability is vital.[2] Zerodur[3] and ULE[4]—glass ceramics—have expansion coefficients of about 5×10^{-8}/°C for 0 to 50°C. Zerodur also exhibits a low aging coefficient, $<1 \times 10^{-7}$ per year. Devices (load cells, springs, diaphragms, tuning forks, etc.) which require stable elastic deflection coefficients may be made from materials whose modulus of elasticity varies little with temperature. Two such materials are

---

[1] Carpenter Technology Corp., 101 W. Bern St., Box 14662, Reading, PA 19612, phone (215) 371-2000.

[2] J. W. Berthold, S. F. Jacobs, and M. A. Norton, "Dimensional Stability of Fused Silica, Invar, and Several Ultralow Thermal Expansion Materials," *Applied Optics,* 15(8); 1898–1899, August 1976. D. G. Chetwynd, "Selection of Structural Materials for Precision Devices," *Precision Engineering,* 9(1) January 1987.

[3] Schott Glass Technologies Inc., 400 York Ave., Duryea, PA 18642, phone (717) 457-7485.

[4] Corning Glass Works, MP-21-4, Corning, NY 14031, phone (607) 974-7709.

NI-SPAN-C alloy 902[1] and Iso-Elastic.[2] NI-SPAN-C can be heat-treated to give a positive, negative, or near-zero temperature coefficient for the elastic modulus, over the range of 0 to 250°F. Iso-Elastic has similar properties and also exhibits very low creep and hysteresis. Bases, frames, and structural elements of precision apparatus and machinery are sometimes fabricated from natural granite.[3] This material exhibits long-term dimensional stability, good vibration-damping properties, freedom from corrosion, temperature stability, and wear resistance. Synthetic granite,[4] a composite material made from granite chips embedded in a polymer matrix, offers many of these same properties plus the ability to be cast in precision molds to form parts of complex shape.

The temperature sensitivity of mechanical damping devices can be reduced by using special silicone damping fluids[5] whose viscosity change with temperature is low. Fluids to enhance free-convection heat transfer (which is driven by temperature-induced density differences) have been designed to have thermal expansion coefficients 8 times that of water and densities 75 percent greater than that of water.[6]

Natural and synthetic diamonds[7] and the recently developed deposited diamond films[8] have several unique and useful properties. These include extreme hardness, chemical inertness, optical transparency over wide spectral ranges (pure type 2A diamonds), and the highest thermal conductivity of all known materials. A listing of some current applications may suggest potential uses in experimental apparatus:

Heat sinks for microcircuits	Infrared optical components
Fiber-optic cleaving knives	Lathe cutting tools
Cutterheads for video disks	Chemically inert coatings
Wear-resistant parts	Light- and heat-conducting surgical blades

---

[1] Inco Alloys International, Inc., Huntington, WV 25720.

[2] J. Chatillon & Sons, 7609 Business Park Dr., Greensboro, NC 27409.

[3] Tru-Stone Corp., Box 430, 1101 Prosper Dr., Sundial Industrial Park, Waite Park, MN 56387, phone (612) 251-7171.

[4] *Anonite Polymer Composite,* Anorad Corp., 110 Oser Ave., Hauppauge, NY 11788, phone (516) 231-1995.

[5] Dow-Corning Corp., Box 0994, Midland, MI 48686, phone (517) 496-4000.

[6] Fluoinert, 3M Industrial Chemical Products Division, Bldg. 223-6S-04, 3M Center, St. Paul, MN 55144.

[7] Dubbeldee Diamond Corp., 100 Stierli Ct., Mt. Arlington, NJ 07856, phone (201) 770–1420.

[8] Genasystems, Inc., 6325 Huntley Rd., PO Box 568, Worthington, OH 43085, phone (614) 438-5600. D. S. Hoover and S. Y. Lynn, "Diamond Thin Film: Applications in Electronics Packaging," *Solid State Technology,* February 1991, pp. 89–92.

The thermal conductivity of type 2A diamonds is about 5 times better than that of copper, the next best conductor.

Shape-memory alloys[1] have a unique behavior related to temperature. In the one-way shape memory effect, a wire bent into some shape at room temperature returns to its unbent shape when heated. Cooling back to room temperature causes no further shape change. In the two-way effect, a wire bent into some shape at room temperature returns partway to its unbent shape when heated. If it is cooled to room temperature, it returns to its bent shape. Subsequent temperature cycles produce a repeated shape change between these two partially deformed shapes. These materials are *not* the familiar bimetals but are a single material, usually special brass or nickel-titanium alloy. Applications include pipe and tube couplings, thermal actuators, circuit breakers, and slack adjusters for automotive gearboxes.

Structural ceramics are finding increasing application in precision measuring equipment and other experimental apparatus because of their favorable characteristics with respect to hardness, wear and corrosion resistance, electrical insulating properties, stiffness, and thermal stability. Figure 6.12 compares several competitive materials.[2] In addition to ready-made slides, *x-y* tables, etc., one can purchase component parts such as surface plates, squares, straightedges, gage stands, and structural kinematic pairs. A recent coordinate-measuring machine uses ceramics for much of its mechanical structure.[3]

Occasionally we require materials with the highest possible density to fabricate masses of the smallest possible size. Typical applications include counterweights for flight control systems, fixed-wing and helicopter balance weights, engine balance weights, and gyro rotors. Tungsten alloys[4] can provide a density of about 18. g/cm^3, about 65 percent more than lead, and have good strength (about 95,000-psi yield strength).

---

[1] C. Robinson, "Shape Memory Alloys Find Expanding Applications," *InTech*, March 1987, pp. 55–58. T. Borden, "Shape-Memory Alloys: Forming a Tight Fit," *Mechanical Engineering*, October 1991, pp. 67–72. J. C. Falcioni, "Shape Memory Alloys," *Mechanical Engineering*, April 1992, p. 114. Sources for materials: Raychem Corp., 300 Constitution Dr., Menlo Park, CA 94025, phone (415) 361-5887; Memory Metals Inc., 84 West Park Pl., Stamford, CT 06901, phone (203) 357-9777. Information source: Battelle Columbus Labs, 505 King Ave., Columbus, OH 43201, phone (614) 424-4369.

[2] Toto National Machine Systems Inc., 137 Bristol Lane, Orange, CA 92665, phone (714) 921-0630.

[3] CS series, Numerex Corp., 7008 Northland Dr., Minneapolis, MN 55428, phone (612) 533-9990.

[4] Powder Tech, 31 Daniel Rd., Fairfield, NJ 07006, phone (201) 575-8661.

Comparison table of physical properties

Property	Alumina ceramics		Cast iron (FC25)	Carbon steel	Granite
	AC270	AL170			
Specific gravity	3.4	3.8	7.8	8.0	3.0
Water absorption (%)	0	0	0	0	0.03–0.3
Flexual strength ($\times$ kg/cm^2)	3200	3400	4000	3500	300–500
Young's modulus ($\times 10^6$ kg/cm^2)	2.4	2.5	1.1	2.0	0.3–0.9
Thermal conductivity [cal/(cm $\cdot$ s $\cdot$°C)]	0.04	0.05	0.11	0.06	0.003
Coefficient of thermal expansion* ($\times 10^{-6}/$°C)	4.3	4.3	11.0	12.0	5.0

*Temperature range of tests: 15° to 25°C.

FIGURE 6.12
Ceramics compared with conventional materials.

Depleted uranium,[1] a by-product of the nuclear fuel enrichment process, has a similar high density and is used for similar applications. In addition, it is used for radiation shielding and kinetic energy armor-piercing projectiles. In shielding applications it is 5 times more effective than lead, and due to its better strength and hardness, depleted uranium can serve as both the structural material and the shield.

Aerospace optical systems often utilize mirrors as critical elements. The "perfect" mirror material has been described[2] as having the structural properties of beryllium, the conductivity of copper, the thermal stability and polishability of Zerodur, and the price and availability of aluminum. Unfortunately, no single material has all these attributes so

[1] Nuclear Metals Inc., 2229 Main St., Concord, MA 01742, phone (508) 369-5410.

[2] J. R. Bolch and R. J. Drake, "Silicon Carbide Makes Superior Mirrors," *Laser Focus World,* August 1989. United Technologies Optical Systems, PO Box 109660, W. Palm Beach, FL 33410.

the usual compromises must be struck. Stiffness and low weight are two major considerations, and a useful "figure of merit" is given by the ratio of elastic modulus to density. A good indicator of thermal stability is the ratio of the thermal conductivity to the coefficient of thermal expansion. Beryllium has an elastic modulus higher than that of steel (44 versus 30 million psi) and a density less than that of aluminum (0.067 versus 0.1 lb/in^3), desirable for stiff, lightweight structures which also have high frequencies of natural vibration and thus are less easily deformed by vibration-exciting forces. While the referenced article emphasizes optical equipment applications, it gives a good overall discussion of many important factors to be considered in the design of any stiff, lightweight component. Texts on precision machine design often have useful sections[1] on thermal errors and means to control them.

A wide variety of gases and liquids are available for use in experimentation. Perusal of a commercial catalog[2] showed the products classified into pure gases or liquids, mixtures, and special application materials. Under pure gases, an alphabetical listing included acetylene, air, ammonia, argon, arsenic pentafluoride, arsine, ..., hydrogen, ..., isobutane, ..., krypton, methane, methyl bromide, ..., tungsten hexafluoride, and xenon. The more widely used gases are often available in several purity levels. For example, argon is listed in four grades ranging from 99.999 to 99.996 percent purity. Carbon dioxide comes in grades called *research, supercritical, anaerobic, instrument,* and *bone dry.* Hydrogen can be supplied in six grades, with the purest being 99.9999 percent pure. Off-the-shelf gas mixtures are mainly binary (two component gases) and are typically used for instrument calibration purposes. Multicomponent mixtures (up to 20 constituents) can be prepared upon special order. *Primary standard mixtures* are the most accurate made, using carefully weighed proportions of the constituents. *Certified standard mixtures* are the most widely used (as working standards) and are mixed by partial-pressure or volumetric means. A partial listing of off-the-shelf binary mixtures includes ammonia in argon (50 ppm to 4.5 percent), butane in helium (5 ppm to 1 percent), carbon monoxide in hydrogen (5 ppm to 50 percent), hexane in nitrogen (5 to 700 ppm), methane in ultra zero air (1 to 49 ppm), and sulfur dioxide in nitrogen (10 ppm to 0.5 percent).

---

[1] A. H. Slocum, *Precision Machine Design,* Prentice-Hall, Englewood Cliffs, NJ, 1992, pp. 96–103, 287–295.

[2] *Liquid Carbonic Specialty Gases,* 135 S. LaSalle St., Chicago, IL 60603, phone (312) 855-2500.

Special applications gases and mixtures are used in instrument calibration, emissions monitoring, lasers, supercritical fluid chromatography, and clinical and diagnostic uses. Emission monitoring gases include EPA protocol gases, stationary-source gases, mobile-source gases, and auto emission gases. Some other specialty gases include flame ionization fuel mixtures, furnace atmosphere gas, leak detection mixtures, proportional counting mixtures, quench gas, spark chamber mixtures, and x-ray fluorescence spectroscopy gas.

We conclude this section on materials with a final note on sources, mainly for solid materials. While we have cited specific sources for some materials, you also should be aware of distributors who specialize in supplying a wide variety of materials to the research and development community. We simply describe one such source[1] familar to us which has a very broad line of products. A brief overview of their 500-page catalog shows that materials are grouped into pure metals (68), alloys (230), ceramics (machinable glass ceramic, beryllia, zirconia, quartz), polymers, single crystals (30 pure metals, 50 alloys), composites, and honeycombs. Many of these materials are available in different forms, including bar, chopped fiber, fabric, fiber, filament, film, flake, floccullent, foil, granule, honeycomb, insulated wire, L-B film, laminate, liquid, lump, metallized film, microfoil (10^{-9} m thick!), nonwoven fabric, powder, rod, sheet, single crystal, sputtering target, staple fiber wool, tube, and wire. Since vendors of this class specialize in the small quantities usually needed for experiments, they can usually provide very quick service, often just a day or two.

6.5 DESIGN FOR SAFETY

Since safety of equipment and human operators is so important, perhaps we should have addressed it earlier in this chapter. We chose this rather late position so that we could discuss safety as it interacts with the functional aspects of design, which we have now mostly covered. *Safety engineering* is today recognized as an intrinsic part of the design procedure for products and processes,[2] including the process of ex-

---

[1] Goodfellow Corp., PO Box 937, Malvern, PA 19355, phone (800) 821-2870.

[2] *Public Safety: A Growing Factor in Modern Design,* National Academy of Engineering, Washington, 1970. W. P. Rodgers, *Introduction to System Safety Engineering,* Wiley, New York, 1970. R. A. Wadden, P. A. Scheff et al., *Engineering Design in the Control of Workplace Hazards,* McGraw-Hill, New York, 1987. G. Marshall, *Safety Engineering,* Brooks/Cole, Monterey, CA, 1982.

perimentation which is the subject of this book. As usual, while safety considerations in the design and use of experimental apparatus share many common features with those associated with products mass-produced for the consumer, there will also be significant differences in emphasis. These are mainly due to the one-of-a-kind nature of the equipment and the level of training and sophistication of the apparatus user.

As the design of our experimental apparatus and the procedures for using it develop, at all stages we should periodically stop and consider safety aspects. By including safety considerations right from the earliest system concepts, we make sure that we don't finalize a design that has serious safety flaws. We can place safety concerns into two major classes: hazards to equipment and hazards to human operators. Since no machine or process can ever be made *perfectly* safe, we need to recognize that judgment will be necessary in balancing safety with other important factors such as cost (first cost and operating cost), convenience of use, and ease of maintenance and repair. Training and supervision of operators and the quality of instruction manuals are clearly important as well.

Safety hazards associated with any apparatus must be first identified and then either eliminated by redesign or controlled in some acceptable manner. Identification of hazards requires a careful consideration of the proposed apparatus design and its method of use. The level of detail of such consideration clearly depends on the stage of design that has been reached. As we move toward the final detail design where the hardware description has become very precise, we are able to consider increasingly more complete versions of the system. At earlier stages, we are only able to evaluate rather general features for their safety implications. In identifying and evaluating hazards, it may be helpful to classify them into standard severity levels. A document which provides such a classification and which is also a good source for general safety information is *Military Standard for System Safety* (Mil-Std-882), published by the U.S. Department of Defense. Here, four levels of severity are defined:

1. *Negligible*—will not result in personnel injury or system damage

2. *Marginal*—can be counteracted or controlled without injury to personnel or major system damage

3. *Critical*—will cause personnel injury or major system damage or will require immediate corrective action for personnel or system survival

4. *Catastrophic*—will cause death or severe injury to personnel, or system loss

In addition to defining hazard severity levels, this standard gives a useful list of safety criteria and considerations. "System designs and operational procedures...should consider, but not be limited to, the following."

1. Avoiding, eliminating, or reducing significant hazards identified by analysis, design selection, material selection, or substitution. Composition of a propellant, explosive, hydraulic fluid, solvent, lubricant, or other hazardous material shall provide optimum safety characteristics.

2. Controlling and minimizing hazards to personnel, equipment, and material which cannot be avoided or eliminated.

3. Isolating hazardous substances, components, and operations from other activities, areas, personnel, and incompatible materials.

4. Incorporating fail-safe principles where failures would disable the system or cause a catastrophe through injury to personnel, damage to equipment, or inadvertent operation of critical equipment.

5. Locating equipment components so that access to them by personnel during operation, maintenance, repair, or adjustment shall not require exposure to hazards such as chemical burns, electric shock, electromagnetic radiation, cutting edges, sharp points, or toxic atmospheres.

6. Avoiding undue exposure of personnel to physiological and psychological stresses which might cause errors leading to mishaps.

7. Providing suitable warning and caution notes in operations, assembly, maintenance, and repair instructions; and distinctive markings on hazardous components, equipment, or facilities for personnel protection. These shall be standardized in accordance with the requirements of the procuring activity.

8. Minimizing severe damage or injury to personnel and equipment in the event of an accident.

In satisfying the safety requirements, we should use the following approaches, in order of preference:

1. *Design for minimum hazard.* The major effort throughout the design stages shall be to select appropriate safety design features, e.g, fail-safe, redundancy.

2. *Safety devices.* Known hazards which cannot be eliminated through design selection shall be reduced to an acceptable level through the use of appropriate safety devices.

3. *Warning devices.* Where it is not possible to preclude the existence or occurrence of an identified hazard, devices shall be employed for the timely detection of the condition and the generation of an adequate warning signal. Warning signals and their application shall be designed to minimize the probability of incorrect personnel reaction to the signals and shall be standardized within like types of systems, in accordance with the directives of the procuring activity.

4. *Special procedures.* Where it is not possible to reduce the magnitude of an existing or potential hazard through design or through the use of safety and warning devices, we should develop special procedures. These should be carefully documented and explained to the personnel involved, and suitable training measures instituted.

Analysis of the hazards present in a specific apparatus goes through several phases as the system design develops. These phases have been called *preliminary hazard analysis, subsystem hazard analysis, system hazard analysis,* and *operating hazard analysis.* Areas to be considered in *preliminary hazard analysis* include, but are not limited to, the following.

1. Isolation of energy sources

2. Fuels and propellants: their characteristics, hazard levels, and quantity-distance constraints; handling, storage, transportation safety features, and compatibility factors

3. System environmental constraints

4. Use of explosive devices and their hazard constraints

5. Compatibility of materials

6. Effect of transient current, electrostatic discharges, electromagnetic radiation, and ionizing radiation to or by the system; design of critical controls to prevent inadvertant activation and employment of electrical interlocks

7. Use of pressure vessels and associated plumbing, fittings, mountings, and hold-down devices

8. Crash safety

9. Safe operation and maintenance of the system

10. Training and certification pertaining to safe operation and maintenance of the system

11. Egress, rescue, survival, and salvage

12. Life support requirements and their safety implications in staffed systems

13. Fire ignition and propagation sources and protection

14. Resistance to shock damage

15. Environmental factors such as equipment layout and lighting requirements and their safety implications in manual systems

16. Fail-safe design considerations

17. Safety from a vulnerability and survivability standpoint; e.g., application of various types of personnel armor (metal, ceramics, glass), fire suppression systems, subsystems protection, and system redundancy

18. Protective clothing, equipment, or devices

19. Lightning and electrostatic protection

20. Human error analysis of operator functions, tasks, and requirements

Subsystem hazard analysis is an expansion of the preliminary hazard analysis. It is performed to determine, from a safety standpoint, the functional relationships of components and equipment comprising each subsystem. Such analysis shall identify all components and pieces of equipment whose performance degradation or functional failure could result in hazardous conditions. The analysis should include a determination of the modes of failure and the effects on safety when failures occur in subsystem components.

System hazard analysis is a review of the safety integration and interface requirements of the total system. Analyses are performed of subsystem interfaces to determine safety problem areas of the total system. Such analyses include, but are not limited to, review of subsystem interrelations for

1. Compliance with safety criteria.

2. Possible independent, dependent, and simultaneous failures that could present hazardous conditions.

3. Ensuring that the normal operation of a subsystem cannot degrade the safety of another subsystem or the total system. When changes occur within subsystems, the system safety hazard analysis shall be updated accordingly. In staffed systems, consideration shall be given to crash safety, escape, egress, rescue, and survival.

Operating hazard analysis determines safety requirements for personnel, procedures, and equipment used in installation, maintenance, support, testing, transportation, storage, operations, emergency escape, egress, rescue, and training during all phases of intended use as specified in the system requirements. Engineering data, procedures, and instructions developed from the engineering design and initial test programs are used in support of this effort. Results of these analyses provide the basis for

1. Design changes where feasible to eliminate hazards or provide safety devices and safeguards
2. The warning, caution, special inspections, and emergency procedures for operating and maintenance instructions including emergency action to minimize personnel injury
3. Identification of a hazardous period time span and actions required to preclude such hazards from occurring
4. Special procedures for servicing, handling, storage, and transportation

While each experimental apparatus will have its own peculiarities, there are some general safety concerns that can be included in *checklists*[1] to form the foundation for a systematic search for hazards. Some lists based on the cited reference and organized according to the type of system or component involved are given below:

MECHANICAL SYSTEMS

1. Use protective guards for parts that move, are hot, etc.

2. Design to minimize wear, chafing, and abrasion of moving parts.

3. Provide adequate lighting.

4. Be sure that acoustic noise levels are within safe limits.

5. Comply with federal and state inspection requirements and codes for boilers, elevators, pressure vessels, and fire equipment.

6. Use reliable locking techniques for fasteners which could vibrate loose.

7. Test high-pressure systems for leaks and overstress before starting routine operations.

---

[1] W. P. Rodgers, *Introduction to System Safety Engineering*, Wiley, New York, 1971, pp. 91–99.

8. Test high-speed systems at lower speeds to detect and correct any malfunctions before going to maximum speed.

9. Provide adequate clearances for hands and tools needed to make repairs and adjustments. Avoid sharp edges and points.

ELECTRICAL/ELECTRONIC SYSTEMS

1. Route wires and use strain reliefs to prevent excessive strain on wiring and termination points.

2. All electric connectors should be coded, keyed, or otherwise designed to prevent mismatching.

3. Use moisture- and corrosion-protected connectors if necessary.

4. Components used in areas with flammable fluids should be incapable of causing ignition.

5. Interlocks, shielding, safety guards, barriers, and warning markings should be specified where personnel hazards exist.

6. Protective devices or circuits should be used to provide circuit and equipment overload protection in power circuits.

7. Means should be provided to isolate all power from specific pieces of equipment to allow maintenance or removal operations.

8. Supports should be provided to prevent abrasion or chafing of wires and cables.

9. Dissimilar-metals contact should be avoided unless corrosion protection is provided.

10. Do not route power and signal leads through adjacent pins of connectors. Provide proper shielding and grounding.

11. Allow sufficient working space for engaging and disengaging connectors. Provide pressure and moisture seals if needed.

12. Provide personnel protection from accidental contact with voltages exceeding 30 V. Use interlocks where possible.

13. Design switches and controls to prevent accidental activation.

CHEMICAL SYSTEMS

1. Provide protective clothing for chemical handling operations.

2. Piping and transfer systems should be designed to minimize the effects of leakage.

3. Filters, lines, and connectors should be designed to preclude blockage by particulates and/or contamination.

4. All cleaning agents should be removed from the system after cleaning. Materials must be compatible with chemicals handled.

5. Only trained personnel may handle dangerous chemicals.

6. Emergency equipment and procedures must be available and clearly understood.

7. For toxic materials, a leak detection system should be installed where materials are handled and stored.

8. Aging and deterioration characteristics should be evaluated for all chemical storage requirements.

9. Proper fire protection should be provided for all combustible chemicals. Vapors or wastes must not jeopordize the surrounding community.

The above lists are not exhaustive but provide a good starting point in evaluating safety of experimental apparatus.

Color coding of apparatus parts can be helpful in improving safety. A standard scheme of colors exists and should be used unless there are good reasons for deviating from it.

Red	*Danger* (high-voltage signs, physical hazards, emergency-stop devices, fire control equipment, safety cans, etc.)
Orange	*Dangerous parts of equipment* (exposed edges and dangerous surfaces, guards for moving parts, inside of switch box doors to indicate that they are open, etc.)
Yellow	*Caution* (materials handling equipment, warnings of unsafe practices, heavy construction equipment, guardrails, etc.)
Green	*Safety* (signs describing safe practices, first aid equipment, start buttons, deluge showers, etc.)
Blue	*Do not touch* (to prevent accidental starting or moving of equipment being repaired or adjusted—markings located at starting points or power sources)
Purple	*Radiation hazard* (for rooms and areas where radioactive materials or items which have been dangerously contaminated with radioactive materials are stored or handled: disposal cans for radioactive materials and wastes)
Black/ White	*General information* (markings and signs to point the way to fire escapes, stairways, refuse cans, etc.; display facts on established procedures)

This section on safety is of necessity rather brief, but at least it calls attention to the need for careful consideration of these questions. It also provides an organized approach to the subject and gives a few specific guidelines to get you started. Details not covered here can be found in the references cited at the beginning of the section.

6.6 SPECIAL DESIGN CONCEPTS

In addition to the general design procedure explained earlier in this chapter, we discuss a few special concepts that have been found to be useful in solving specific problems in apparatus.

6.6.1 Kinematic Mountings and Motions

When the operation of an apparatus requires frequent mounting and demounting of parts and/or specimens while maintaining accurate positioning and alignment, the classical technique of *kinematic design*[1] should be considered. These methods may also be applied to the provision of motions which allow the moving part certain degrees of freedom while restricting others. The general concept can be thought of in terms of the degrees of freedom of a rigid body. To locate such a body precisely, we must specify three translations (x, y, z) and three rotations, giving a total of 6 degrees of freedom. If all 6 degrees of freedom are restrained, the object is fixed in a specific location. If we restrain less than all 6 degrees of freedom, then some type of motion is possible.

For example, a ball resting in a triangular pyramidal hole has three point contacts with the three planes that form the sides of the pyramid; thus its 3 degrees of translational freedom are restrained (see Fig. 6.13*a*). It can, however, rotate freely in all directions. This would be a kinematic version of a ball-and-socket joint and could be used in its place if we provided some kind of preloading scheme to keep the ball always in contact with the pyramid faces as the desired rotations occurred. When we want to restrain all 6 degrees of freedom, as in mounting a part to a frame, a number of physical realizations are possible, with the three-ball, three-groove version of Fig. 6.13*b* being perhaps the most common and commercially available.[2] Melles Griot also markets several kinematic mirror mounts which allow 2 degrees of angular freedom for positioning mirrors about two axes of rotation.

Kinematic designs, when taken part and reassembled, will reproduce their location accurately. While wear will cause dimensional changes, it does not produce lost motion (play) in the joint. A "locator force" (gravity, magnetic force, springs) is necessary, and the point

[1] A. H. Slocum, *Precision Machine Design,* Prentice-Hall, Englewood Cliffs, NJ, 1992, pp. 401–412. P. H. Sydenham, *Mechanical Design of Instruments,* Instrument Society of America, Research Triangle Park, NC, 1986, pp. 43–47.

[2] Kinematic Base, Melles Griot, Irvine, CA.

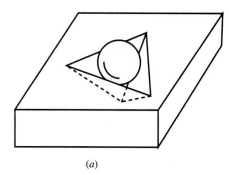

(*a*)

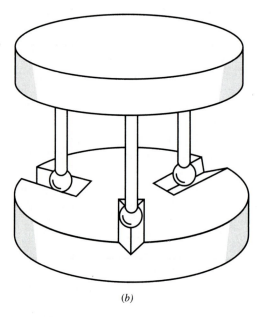

(*b*)

FIGURE 6.13
Kinematic principles and application.

contacts used are highly stressed, but usually the forces can be accurately calculated (statically determinate) and the contact stresses estimated. Theoretical and experimental studies[1] of kinematic designs provide information useful for their application.

---

[1] A. Slocum, "Kinematic Couplings for Precision Fixturing—Part I—Formulation of Design Parameters," *Precision Engineering*, 10(2): 85–91, April 1988; "Part II—Experimental Determination of Repeatability and Stiffness," *Precision Engineering*, 10(3): 115–121, July 1988.

6.6.2   Bearings for Accurate Motions

When mechanisms for precise motion control are designed or selected, a number of different bearing schemes can be used. Figure 6.14[1] shows several of the most popular, some of which use kinematic principles discussed in the previous section. Slides using the magnetic kinematic design suppress backlash without causing large and uneven forces on the rolling elements, thus giving precise and smooth motion. Note that the sidewise location of the upper slide member is determined entirely by the left ball and V grooves since the right ball's upper contact is with a *plane,* not a V groove. This is the kinematic feature. Magnetic forces keep the two slide members preloaded and free of backlash. Since the magnetic force changes only slightly for small changes in separation, the preload force is nearly constant over the travel of the slide, giving a smooth motion without binding. Actual measurements show typically less than 0.2 μm of vertical motion over 12. mm of slide travel.

Instead of V grooves, one can use guide rods in several ways in slide design. Guide rods with radial ball bearings are useful for larger motions in the horizontal direction. A kinematic design in which the left rod and its ball bearings do all the locating and the right rod allows sidewise slip is again used. Springs are used to maintain a preload and to suppress backlash. If a positioner is used in the vertical direction or if it will experience large torques, then linear ball bearings can be used. The guiding rod of each linear ball bearing is surrounded by several rows of rotating balls so that the bearing can support radial forces from any direction. To get uniform motion, the linear ball bearing must allow some backlash. In vertical use or with constant forces, this will not influence the positioning precision excessively. For the largest loads and a compact construction, crossed-roller bearings are available. Here the positioner moves on cylindrical rolling elements along the guiding rods. The axes of successive rollers are each rotated by 90° so that the positioner can accept tractive and compressive stress. An almost complete absence of backlash is attained by parallel adjustment of the guiding tracks and pressing the rims against the bearing rollers.

When only small translational or rotary motions must be accommodated, *flexural bearings*[2] should be considered. Here the relative motion of the two members is provided by elastic deflection of suitably shaped connecting parts. Since there is (at least at the macroscopic scale)

---

[1] Information MP 12 E, Physik Instrumente, Polytec Optronics, 3001 Redhill Ave., Bldg. 5, Suite 104, Costa Mesa, CA 92626, phone (714) 850-1835.

[2] Slocum, *Precision Machine Design,* pp. 521–538. Sydenham, op. cit., chap. 5.

THE VARIOUS GUIDING PRINCIPLES

1. The magnetic-kinematic principle
   (*Patented in most major countries*)

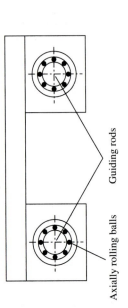

Ferromagnetic
steel inserts

Magnets

2. Guiding rods and radial ball bearings

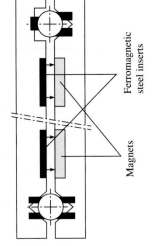

Guiding rods

Ball bearings

3. Guiding rods with linear ball bearings

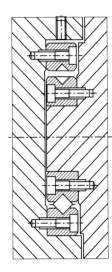

Guiding rods

Axially rolling balls

4. Crossed roller bearings

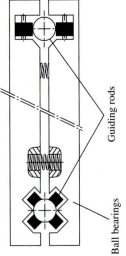

FIGURE 6.14
Bearing and preloading schemes for precise translatory motion.

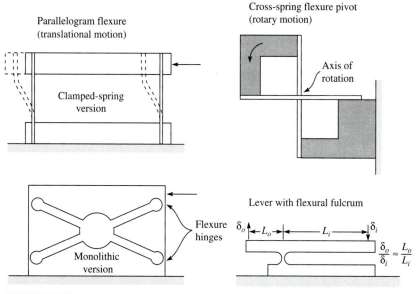

FIGURE 6.15
Flexural "bearings" for rotary and translatory motion.

no sliding or rolling contact, friction is extremely small and is mainly due to internal hysteresis in the stressed members. There is no wear and no free play or lost motion. Many different configurations are possible and are described in the references. Figure 6.15 shows a few of the most common. Two basic construction techniques, clamped flat spring and monolithic, are used. The clamped design is usually more economical and may be suitable for larger motions where its microslippage can be accepted. Monolithic designs start with a single block of material, and we cut away the parts not wanted. Intricate shapes may require electric-discharge machining techniques. Flexure bearings are often "home-made." However, a commercial version of the rotary-motion cross-strip flexure is available.[1]

Other bearing types which may be useful in apparatus design are hydro- and aerostatic and magnetic.[2] Hydrostatic (oil) and aerostatic (air) bearings use external pressurizing pumps to maintain a separating fluid

[1] Lucas Aerospace, 211 Seward Ave., PO Box 457, Utica, NY 13503, phone (315) 793-1200.

[2] Slocum, *Precision Machine Design,* chap. 9.

film between the bearing moving parts under all conditions, even when at rest, when ordinary hydrodynamic bearings have metal-to-metal contact. With no mechanical contact, these bearings exhibit only viscous friction, which gives a zero breakaway force and no wear. Magnetic bearings are relatively rare in commercial applications but have been used in experimental apparatus and instruments[1] for some time. They are generally feedback control systems which use displacement sensors to determine whether the "shaft" is centered in the "hole" and electromagnets to apply magnetic forces which return the shaft to the desired centered position when deviations are detected. Magnetic bearings can levitate objects and provide guidance for translational and rotary motions without physical contact.

6.6.3 Use of Feedback for Variable Manipulation

Many experimental setups require that we hold certain variables at values of our choice or, alternatively, to force them to change in a way that we prescribe. When simple open-loop control schemes are unable to meet our accuracy requirements, use of closed-loop (feedback) methods may give the performance needed. We have already given some examples (temperature and motion control systems) earlier in this chapter. For those unfamiliar with the basic concept, Fig. 6.16 shows a functional

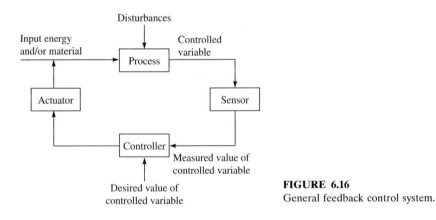

FIGURE 6.16
General feedback control system.

---

[1] R. H. Frasier, P. J. Gilinson, and G. A. Oberbeck, *Magnetic and Electric Suspensions,* MIT Press, Cambridge, MA, 1974.

block diagram representing any feedback control system. The *controlled variable* is the quantity that we wish to vary or hold fixed in our apparatus, e.g., the temperature of a test specimen.

The *process* is the means, perhaps an electric oven, by which we affect the controlled variable. The *sensor* (perhaps a thermocouple) is the measuring device which determines the actual value of the controlled variable and reports this information to the *controller*. The controller also must be told the *desired value* of the controlled variable. In our specimen temperature control example, this might be a dial calibrated in degrees Celsius, or maybe a digital keyboard might be used to enter the desired temperature into the controller. The controller first compares the actual and desired values of the controlled variable and then implements a control law or algorithm to calculate a controller output signal which is sent to the *actuator* (also called the *final control element*). Control laws range from simple schemes, such as the on/off type used in most home heating systems, to sophisticated adaptive and self-tuning controllers. In many cases, the PID (proportional, integral, derivative) control law can be "tuned" to meet the needs of the application. The actuator is the device which manipulates the energy and/or material input to the process so as to change the controlled variable when that is needed. In an electric heater, this might be a simple on/off power switch or a more complex silicon controlled rectifier (SCR) for smoothly modulating ac heating power. *Disturbances* are any inputs to the process that are not under our control but are capable of affecting the controlled variable. If our oven were located outdoors, the outdoor temperature (ambient temperature for the oven) would act as a disturbance.

The analysis and design of feedback control systems[1] is a well-developed area of technology which we do not want to duplicate here, and many practical systems for routine applications are satisfactorily selected and operated without extensive analysis. (This is largely due to the adaptability of the widely used PID control law. It can be experimentally adjusted, or tuned, to meet many needs.) For those with little or no experience, we mention a few guidelines and warnings. A general price paid for the speed and accuracy advantages of *any* feedback system is the potential for dynamic instability. *All* closed-loop systems can go into (sometimes violent) oscillation if the numerical values of all parameters in the system (not just the controller) are not proper. In fact, when we try to adjust the controller to get the maximum accuracy and/or speed of response, instability is almost sure to occur if we press these adjustments

---

[1] Doebelin, *Control System Principles and Design.*

too far. Furthermore, if we have carefully adjusted the controller to get good response for one set of apparatus conditions, then when we *change* those conditions, the controller settings may require readjustment. In fact, the new conditions may have caused instability! Nowadays, feedback systems are vital to the effective operation of many experimental studies, but we must be careful in their operation. For experiments sufficiently complex to warrant computer-aided operation, we may want to use the system computer for some of or all the feedback control tasks that we need. Commercial software is readily available for simple or complicated control schemes, and multitasking allows the computer to timeshare these control tasks with its other duties in managing the experiment.

6.6.4   Use of Symmetry, Guarding, and Compensation

Principles of symmetry are sometimes useful in deciding on the best configuration for an apparatus. Experiments for finding the thermal resistance R_t of insulating materials are often based on the equation

$$Q = \frac{\Delta T}{R_t} \tag{6.3}$$

where Q is the heat flow rate through the flat slab specimen and ΔT is the temperature difference measured across it. The heat flow rate is usually provided by an electric heater and is measured as the electric power input to the heater, which is much easier to measure accurately than the heat flow rate itself. This scheme requires, however, that *all* the heat from the heater go through the test specimen and none be lost as "leakage" to other parts of the apparatus. The simplest and most obvious implementation of Eq. (6.3) uses the arrangement of Fig. 6.17a, where thermal insulation is used to prevent the leakage heat loss. Since no thermal insulation is perfect, some leakage still occurs. The magnitude of the error caused by such leakage will depend on how much greater the thermal resistance of the insulation is than that of the specimen. If the thermal resistance of the insulation can be estimated, we can also estimate its heat loss and thereby correct the value of Q used in Eq. (6.3).

An alternative arrangement which makes use of symmetry principles is shown in Fig. 6.17b. This method requires the use of two identical specimens, and the R_t value obtained is the average for the two specimens. If this constraint is acceptable, we gain the advantage that now all the heater output goes through the two specimens, without the uncertainty of insulation heat leaks.

A little more thought about Fig. 6.17 reveals some additional problems, however. In both of the schemes shown, there will be error-causing heat losses at the "edges" of the specimens. One could

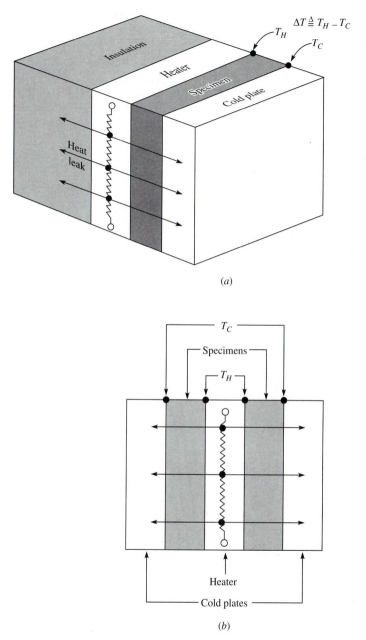

(a)

(b)

FIGURE 6.17
Use of symmetry principle in thermal resistance apparatus.

again try to insulate against these, but another concept called *guarding* has been found to be more effective and is in wide use.[1] The guarding principle is also used in certain fluid flow and electricity applications. In all cases, the idea is to design the apparatus such that the "flow lines" (heat flow in our present example) in the specimen are all parallel to its thickness, so that no flow out the edge occurs. Since the flows (thermal, fluid, or electrical) are *caused* by some potential difference, if we can ensure that no such difference exists in the edgewise direction, then there will be no flows in this direction. For our thermal apparatus, heat flows are caused by temperature differences, so we must minimize temperature gradients in the edgewise direction.

This is accomplished in the apparatus of Fig. 6.18 by using a *guard heater* in addition to the main heater. This guard heater, which surrounds the edges of the main heater, supplies the heat needed to balance the edge heat losses, so that the main heater (whose electrical input is measured and used as the Q value) need only supply the heat going through the central portion of the specimens. The power to the guard heater is adjusted such that the temperature difference (measured with differential temperature sensors) across the gap between the main and guard heaters is kept near zero. If this temperature difference were exactly zero, then all the heat from the main heater would be going through the specimen and none would be lost at the edges. While our explanation here describes the basic ideas, you should consult the reference for additional important details if you actually need to use such an apparatus. An application of the guarding principle to a fluid flow situation is shown in Fig. 6.3. There the purpose is to ensure that flow through the filter has no edgewise component, so that calculation of the filter fluid resistance, using a measured flow rate and pressure drop, is more accurate. In capacitance-type displacement transducers,[2] a guard ring around the probe disk reduces "fringing" (edgewise components) in the electric field, improving transducer characteristics.

A good example of the principle of *compensation* is found in a transient hot-wire apparatus for measuring the thermal conductivity of fluids.[3] A very fine platinum wire, centrally located within a cylindrical

[1] ASTM C 177-85, Standard Test Method for Steady-State Heat Flux Measurements and Thermal Transmission Properties by Means of the Guarded-Hot-Plate Apparatus, *1987 Annual Book of Standards,* ASTM, Philadelphia, pp. 21–36.

[2] Doebelin, *Measurement Systems,* p. 257.

[3] R. A. Perkins, H. M. Roder, and C. A. Nieto de Castro, "A High-Temperature Transient Hot-Wire Conductivity Apparatus for Fluids," *Journal of Research of National Institute of Standards and Technology,* 96(3); May-June 1991.

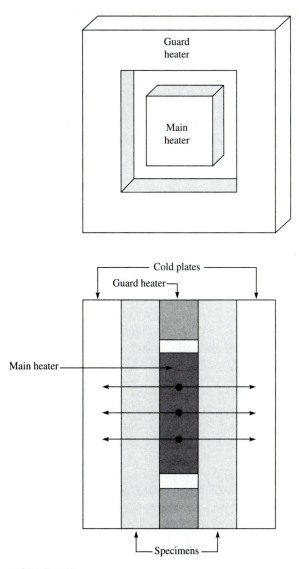

FIGURE 6.18
Use of guarding principle in thermal resistance apparatus.

sample cell containing the fluid, is subjected to a step change in applied voltage, causing a step change in electric heating to the fluid. In addition to serving as a heater, the wire is a sensor for measuring the fluid temperature. The conductivity measurement principle is based on a theoretical model of the situation as an ideal line heat source within an infinite body of convection-free fluid. Solution of the applicable differential equations shows that the fluid temperature at the wire surface experiences an instantaneous jump as the wire heating rate jumps, followed by a gradual rise proportional to the natural logarithm of elapsed time. [The actual apparatus cannot have an instantaneous temperature rise because the wire has some mass, but this rise time is very small with the fine wires (about 10. μm) used.] The fine wire also means that the wire temperature will closely follow the fluid temperature, allowing use of the wire as a resistance thermometer for measuring the fluid temperature.

After the initial sudden rise, the temperature T_2 at time t_2 is related to the temperature T_1 at time t_1 by the equation

$$T_2 - T_1 = \frac{q}{4\pi k} \ln \frac{t_2}{t_1} \tag{6.4}$$

where k is the fluid thermal conductivity and q is the constant heating rate of the wire per unit length. By plotting $T_2 - T_1$ versus $\ln(t_2/t_1)$ we should get a straight line whose slope gives us the numerical value of k. As time passes, pure-conduction heat transfer (which was *assumed* in the mathematical model) is "contaminated" by gradually increasing convection effects, causing errors. Fortunately, the onset of significant convection is signaled by curvature of the formerly straight line, allowing us to restrict our data taking to the early pure-conduction phase of the experiment.

Another potential error source in this method lies in the assumption of infinite length of the wire and fluid. This has to do mainly with conduction heat losses at the ends of the wire. These losses mean that not all the wire-heating rate goes into the fluid, causing a deviation from the assumed mathematical model. These conduction losses can be made smaller by increasing the length/diameter ratio of the wire (another reason for using fine wire). If we wish to reduce this error without going to large apparatus size, an elegant compensation scheme is available. It involves the use of *two* hot wires in the fluid, these two wires being located electrically in adjacent legs of a Wheatstone bridge used to measure wire resistance and thus temperature.

The two wires are identical except for their lengths; in the reference cited, the lengths were about 5 and 19 cm. Recall at this point a basic characteristic of resistance bridges. With the bridge initially balanced, equal changes in the resistance of adjacent legs will cause *no* unbalance. Thus if our two wires were the same length, application of the heating voltage to both would result in no output signal from the bridge. When the wires are unequal lengths, an output will occur. This output will be *compensated* for the wire-end heat losses because these losses will be essentially *the same* for the two wires and thus will cancel, since they appear in adjacent legs of the bridge. Since our data reduction formula [Eq. (6.4)] requires the heating rate per unit length, a wire length, resistance, and applied voltage are needed. In these calculations, we use for wire length the *difference* in the physical lengths of the two wires, in the above example, $19 - 5 = 14$ cm. Figure 6.19 shows some of the experimental arrangement. In designing any apparatus, we should make a list of all extraneous effects which may detract from the accuracy of our results and try to devise schemes such as guarding and compensation to mitigate their influence.

6.7 SIMULATION, COMPUTER AND PHYSICAL

By *simulation* here we mean the representation of an actual situation by a mathematical model (computer simulation) or laboratory apparatus (physical simulation). Pure computer simulation is a most powerful tool of modern engineering, but because it does not involve physical apparatus, we choose not to develop its methods here, even though simulation studies are sometimes called *computer experiments.* If we write differential equations of motion for an automobile passing over a terrain of specified contour and then study the system behavior by solving these equations numerically, using a general-purpose simulation language such as ACSL, then no "automotive apparatus" is used and we are doing pure computer simulation. On the other hand, if we build electrohydraulic servo systems to move the wheels of an actual automobile in accordance with electric motion commands derived from tape recordings of cars being driven over real roads, then we are doing physical simulation. Figure 6.20[1] shows an application of this type.

A *combination* of computer and physical simulation can also be a useful tool in product and process design and development. This approach was first used in the aircraft industry many years ago and is today quite common in many fields, so we describe it briefly. When a new design is first conceived, there are only drawings and equations; no hardware has been built yet. At this stage, we can do only pure computer simulation. As the first components of the overall system are built or purchased, it may be possible to remove their representation from the system simulation and replace it by the actual hardware. We now have a simulation which is partly equations and partly operating hardware, and

---

[1] MTS Systems Corp., 14000 Technology Dr., Eden Prairie, MN 55344, phone (612) 937-4000.

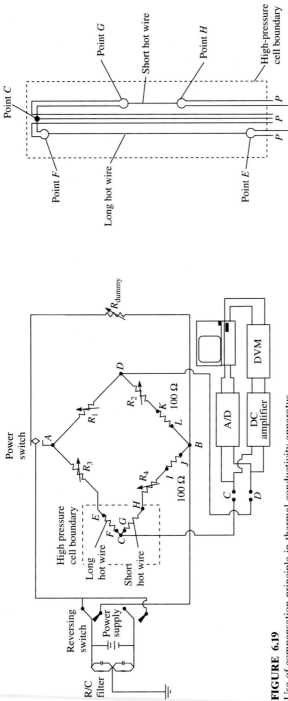

FIGURE 6.19
Use of compensation principle in thermal conductivity apparatus.

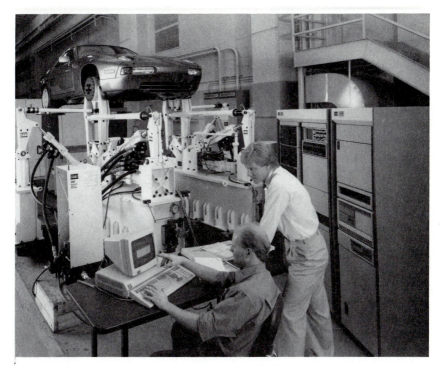

FIGURE 6.20
Electrohydraulic servo system use in automotive road simulator.

these two portions of the simulation must be interfaced to work together. Analog computers were originally used in both pure computer and combined computer and physical simulation, but have now been largely replaced by digital computers. When digital computers are used in combined simulation, we also need sensors, actuators, analog-to-digital (A/D) converters, and digital-to-analog (D/A) converters. The real hardware part of the simulation will have physical input and output variables which appear in the equations of the computer part of the simulation. The physical output variables must be sensed (transduced into proportional voltages) and sent through A/D converters to the digital computer. The computer computes the quantities that are inputs to the hardware. These first must be converted from digital to analog to produce proportional voltages. Then these voltages must be sent to actuators which transduce the voltages into the actual physical input variables of the hardware portion of the simulation.

The advantage of combined simulation over pure computer simulation is that equations are always approximations to real behavior, whereas no assumptions are involved in the operation of the actual hardware, so the combined simulation is more realistic. As development continues, more and more of the computer portion of the simulation is replaced by actual hardware, and realism increases. Finally, we have built the complete system and go into a phase of pure physical experimentation with no computer simulation remaining.

An excellent example of combined computer and physical simulation is shown in Fig. 6.21.[1] Aircraft Braking Systems Corporation of Akron, Ohio, uses this simulation facility in the design and development of brake control systems for aircraft. Starting with analog computers in 1968, the company now uses a combination of digital, hybrid, and analog computers. Note that when digital computers are used for *pure* computer simulation (no interfacing with actual hardware), the speed of solving the simulation equations is not critical since no matter how slow the solution proceeds, we *do* eventually "get the right answer." A fast solution is desirable but *not* necessary. For combined computer and physical simulation, the computer solution *must* be accomplished in "real time" since the computer is interfaced with real hardware which has no choice but to run in real time. That is, if part of our computer model represents, say, a structural member vibrating at 20. Hz, the computer must be capable of solving the vibration equations about 10 times in $\frac{1}{20}$ sec because it takes about 10 points per vibration cycle to clearly represent a vibratory motion. Whereas analog computers have been largely replaced by digital computers in *pure* computer simulation (where speed is not critical), analog computers persist to some degree in combined simulation because they are inherently faster than most digital machines.

Returning to the system of Fig. 6.21, we see that computer simulation is used to represent the braking torque of the brakes and all the motions of the aircraft on which the brakes are installed. While Aircraft Braking Systems Corporation manufactures braking systems and *not* aircraft, the design of the braking system is intimately linked to the behavior of the combined brake and aircraft system, so this entire assemblage must be simulated, not just the braking portion. If computer simulation can give a good modeling of the aircraft behavior, clearly it will be advantageous to use this approach rather than a real aircraft for our development testing. *Some* such complete-system testing will still be necessary toward the end of the development process, but it can usually be deferred until most of the design decisions have been made while using the combined computer and physical approach. The simplification,

---

[1] *Newsletter,* Spring 1992, Applied Dynamics, 3800 Stone School Rd., Ann Arbor, MI 48108, phone (313) 973-1300.

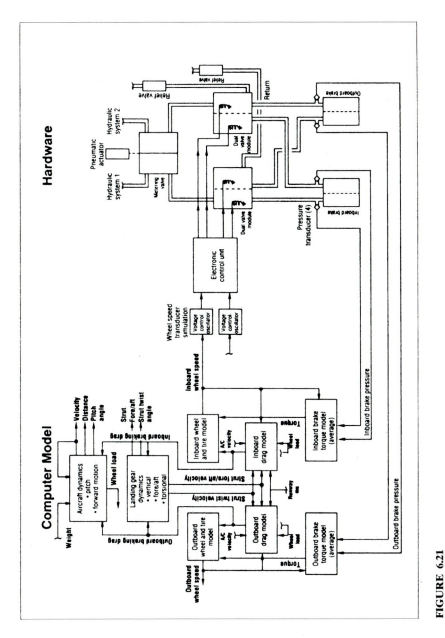

FIGURE 6.21
Combined computer and physical simulation for aircraft braking system development.

cost reduction, and speeding-up of development are the main benefits of the technique.

With regard to details, note that in Fig. 6.21 the only inputs which the computer simulation needs from the braking hardware are the hydraulic pressure signals from the brake cylinders. It is thus necessary to measure these pressures with transducers which convert them to proportional voltages and then pass these voltages through A/D converters (not shown in the diagram) into the digital computer, where the numerical values are needed in some of the equations. The braking hardware needs from the computer simulation the aircraft wheel speeds because these speeds are used in the actual braking system control algorithms. Here we need D/A converters (again not shown in the diagram) to convert the digital version of the wheel speed to a proportional voltage. Since the actual wheel-speed transducers produce a pulse rate proportional to the wheel speed, the wheel-speed voltage from the D/A converter is processed through a voltage-controlled oscillator (voltage-to-frequency converter) to get a signal compatible with the electronic control unit which is part of the braking system. When combined-simulation systems must be fast, as in this braking system, the digital-computer portion (hardware and software) may need to be specially designed for such applications. One may *not* be able to use an "ordinary" digital computer. Some manufacturers, such as Applied Dynamics in the reference, specialize in providing such computer systems.

Before leaving the general subject of simulation, we mention the possibility of using it to calculate, from those variables which *are* measurable in an apparatus, those variables which may be difficult or impossible to measure. This of course requires that we have a valid mathematical model (equations) relating the measurable and unmeasurable variables. If this is the case, we can interface the measurable variables (transduced to voltages) through A/D converters to a digitally simulated mathematical model which can compute the unmeasurable variables and even output them through D/A converters to recorders, if we wish. If real-time processing is not possible or desirable, the measurable variables can be recorded (analog or digital) and later "played" into the simulation model at the speed desired.

6.8   LABORATORY AUTOMATION

The design of any laboratory apparatus must reflect the degree of automation appropriate for the experiment. This can range from completely manual operation (no automation) to extremely sophisticated systems involving robots and little or no human activity. Automation is considered to involve not only sensor reading and data processing but

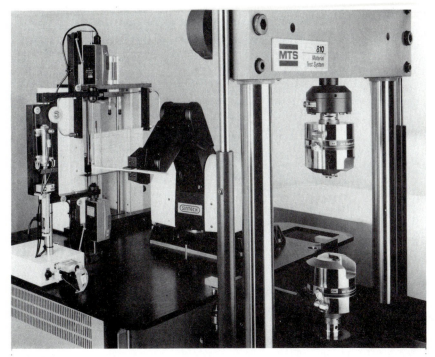

FIGURE 6.22
Automated tensile test system and performance curves.

also specimen loading and manipulation, parameter adjustment, control of operating conditions, alarming of critical or dangerous situations, operator instruction, time cycling and logging, etc. First we show some examples of off-the-shelf systems which can be purchased ready to run. These examples are of interest both for their own sake and as a source of ideas for designing your own systems to meet special needs.

Figure 6.22[1] shows an automated system for tensile testing of materials. It includes a sample feeder, a robot, a specimen gaging station, and a tensile test machine. When a large amount of relatively routine testing is necessary, such systems can cut costs and speed up data collection. By removing the uncertainties of human operator performance, the systems can also reduce variability and error in the data gathered. Figure 6.22*b* compares manual and automated operation,

[1] Sintech, 378 Page St., Stoughton, MA 02072, phone (617) 344-0941.

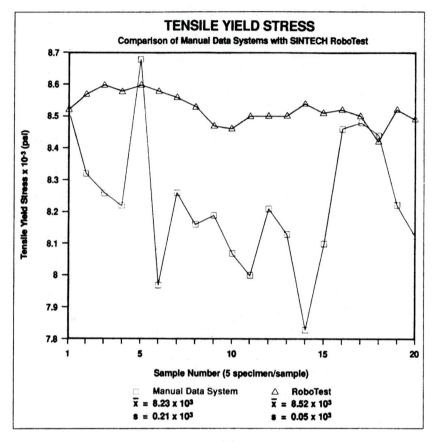

(b)

FIGURE 6.22
(Continued).

showing significant improvement in data uncertainty with the automated operation.

The sample feeder holds up to 125 eighth-inch-thick specimens. These are picked up by a pneumatic cylinder and suction cups and oriented for easy grasping by the robot. Specimens of different dimensions can be accommodated, up to 1 inch in width and 9 inches in length. A dual-contact gage measures specimen dimensions to 0.0002 in. The robot can be programmed to release the specimen during measurement, if there is concern that the robot gripper forces are distorting the specimen. The five-axis robot is sufficiently dextrous to manipulate specimens as a human would and is repeatable to 0.005 in. Its top speed is 720 in/min, allowing fast cycling. The payload is 3 lb, and the reach is 22 in.

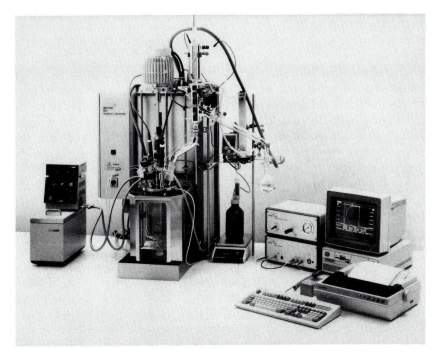

FIGURE 6.23
Automated reaction calorimeter system.

Our next example, shown in Fig. 6.23,[1] is a reaction calorimeter system using a personal computer for overall system control. This highly automated apparatus allows one to run and develop chemical processes under simulated plant conditions, with automatic determination of reaction data. Both glass and metal reactor vessels can be used, depending on the nature of the reaction studied. Temperature sensors monitor temperatures, and heat flows are determined from the temperature changes and flow rates of coolant flowing in vessel jackets. Addition of reactants is monitored by the weight change of supply vessels, by using electronic scales or volumetrically by using calibrated pumps. While a process is running in the reactor, you can use the computer to control parameters such as temperature, pressure, pH value, and other process variables. At the same time, you can dose reaction components automatically. The heat flow measurement is continuous and automatic, which

---

[1] RC1 reaction calorimeter system, Mettler Instrument Corp., Box 71, Hightstown, NJ 08520, phone (800) 638-8537.

enables you to determine quantitatively the course of chemical reactions with high accuracy. A sophisticated *safety system* with two main parts is included. One part monitors the electronic and mechanical instrument functions and ensures trouble-free operation of the instrument. The other part monitors process-related parameters and ensures adherence to limit values that have been defined specifically for the particular reaction.

The above two examples give a general overview of the type of ready-made automated apparatus that is available. We now discuss the design of laboratory automation for those applications where ready-made systems are not appropriate for technical and/or economic reasons. This general area is usually described as *computer-aided experimentation* because a digital computer (often a personal computer) is used to integrate all the needed operations. When a personal computer is used, experimenters themselves will often assemble the needed hardware and software rather than having this done by outside specialist companies. Our discussion will emphasize such situations because the low cost and high performance of personal computers, together with the wide availability of lab automation software and its increasing user-friendliness, make this a common choice. Because the personal computer and allied instrumentation fields change so rapidly, textbooks devoted to the design of computer-aided experimental apparatus may not be very helpful when you need advice on how to put together a system for a particular project.

Fortunately, vendors of the hardware and software often include in their catalogs extensive tutorials on how to choose and use their products. Also, the larger firms will have application engineers available to talk to you as you make decisions and define your system. In this brief section we use one of these catalog tutorials as the basis for the discussion of some of the major considerations. When you get ready to select an actual system, contact several vendors to find out which can meet your needs. After going through detailed studies of each of the competitive systems, you must decide which is best, based on the usual technical and economic tradeoffs. We used a particular vendor's catalog[1] as background for the following condensed treatment; more details can be found in the referenced document. As usual, this reference to a specific commercial firm does not constitute an endorsement or preference; we just wanted to show you what *type* of information is available from this kind of source.

Personal-computer systems for lab automation can be configured in

---

[1] *1992 Catalog,* National Instruments, 6504 Bridge Point Parkway, Austin, TX 78730, phone (800) 433-3488.

several ways; the reference cited lists four. Many freestanding instruments such as digital voltmeters, frequency counters, and digital thermometers support the IEEE-488 (also known as GPIB or HP-IB) interconnection scheme. We can control up to 14 such instruments from our computer, using cable lengths up to 20 m and maximum data transfer rates of 1 Mbyte/sec. For simpler instruments and lower performance (data rate less than 1 kbyte/sec), the RS-232 serial remote-control protocol is available. This can be used for longer distances and control of remote data acquisition subsystems, and the interface is standard on many personal computers. For sophisticated, high-performance instruments, the VXI bus combines the VME backplane with the IEEE-488 protocols. Finally, the "plug-in data acquisition" approach is very popular for the many applications which require moderate performance and low cost. Further discussion will emphasize this fourth type of system.

6.8.1 The Personal Computer

If you already have a personal computer available, then you must choose data acquisition hardware and software compatible with this machine. If not, definition of a complete system requires choice of three basic items; a personal computer (PC), data acquisition hardware, and software for data acquisition, analysis, and presentation. The PC used will determine the overall processing speed. Real-time processing of high-frequency signals (such as acoustics and vibration, which may go to several thousand hertz) will need a high-speed, 32-bit processor and accompanying mathematical coprocessor, or else a dedicated plug-in processor such as a digital signal processing (DSP) board. If only a few signals need to be sampled several times a second, a "low-end" PC may be adequate. You should also consider the level of technical expertise and experience of the system users; this will influence the user-friendliness required of the operator interface. Graphical user interfaces (GUIs) such as Windows and Macintosh can greatly simplify the user's tasks, but may require a more sophisticated PC.

6.8.2 Signal-Conditioning Hardware

The electric signals generated by your measurement transducers must be converted to a form that the data acquisition board can accept. Some transducers (strain-gage bridges, LVDTs, etc.) require electrical excitation. Signal-conditioning hardware provides these kinds of functions. Amplification to raise the transducer signal level (often millivolts) to the

A/D converter level (say, 10 V) is often needed. Simpler amplifiers can change gain only "manually" and may require *all* the channels to share the same gain. Programmable-gain or auto-ranging amplifiers can select gain to suit the signal level and assign optimum gains to each channel, under software control. Some signals may require differential input, and others single-ended input. Thermocouple signals usually require both cold-junction compensation and linearization. Linearization can be done in either hardware or software. Isolation amplifiers may be needed to suppress noise and/or for safety purposes. Multichannel systems require multiplexers. If you need to avoid time skew among several signals, you must use *simultaneous* sampling rather than sequential. This is expensive since it requires one A/D converter per channel. If desired signals are contaminated with noise, appropriate frequency-sensitive filtering (low-pass, high-pass, bandpass, band-reject) may be desirable. Some filters provide software-selectable bandwidths.

6.8.3 Data Acquisition Hardware

Many input signals are analog and thus require analog-to-digital (A/D) conversion. In choosing a particular A/D board, you should consider the number of channels, sampling rate, resolution, range, accuracy, noise, and nonlinearity. Boards may provide both single-ended and differential inputs. Single-ended inputs are all referenced to a common ground point. They are usually used when the input signals are large (about 1 V or greater), lead wires are short (less than about 15 ft), and all input signals share a common ground reference. If these criteria are not met, differential input may need to be used. For boards which offer a choice of single-ended or differential input, the number of available differential-input channels will be one-half of the number of single-ended channels.

   Often one A/D converter is time-shared (multiplexed or switched) among many channels. This means that if the basic sampling rate is, say, 100,000 samples per second, then if we have 10 channels, the sampling rate for each channel will be only 10,000 samples per second. Remember that to preserve information content up to a certain desired frequency, we must sample at a rate *at least* twice that frequency, usually somewhat more. Also, *before* doing the sampling, we must *antialias filter* (low-pass filter) the signal to remove all frequencies above the highest that we wish to measure. That is, if we are measuring an acoustic signal containing frequencies up to 15,000 Hz, and if we want to measure the frequency content up to 8000 Hz, we must low-pass filter the data to remove frequencies above 8000 Hz before we do the sampling. Then we must sample at 16,000 samples per second or higher.

   Resolution is the number of bits that the A/D converter uses to represent the analog signal. For example, a 16-bit unit with a range of 0 to 5 V and an amplifier gain of 500 V/V has a resolution of $5/(500 \times 2^{16}) = 153$ nV. That is, a signal change of 153 nV will cause a one-digit change in the least significant digit. Whether such a small change is actually meaningful depends on factors such as the noise level, non-linearity, and settling time. For high gains and sampling rates, the amplifier may not settle to within the resolution level in the time available between samplings. Some boards are capable of sampling the channels in a selected sequence and/or at different rates for each channel. Simpler boards have a fixed rate and sequence. Interval scanning [scan all channels at one interval and use a second (different) interval before repeating the scan] can also be obtained with some boards.

6.8.4 Analog Outputs

When the computer must send analog signals to hardware that is part of the experimental apparatus, we need digital-to-analog converters (D/A converters) in our system. For example, if our apparatus uses an electropneumatic transducer to convert voltage signals to proportional air pressures, and if we want to command this device from our computer, we need D/A conversion capability. As in the A/D converter, we again need to choose a suitable resolution. The settling time and slew rate may also need to be considered. The slew rate is the maximum rate of change (volts per second) that the unit can produce in the output signal. Most D/A converters cannot provide much current or power in their output signals; a ±10-V unit might provide about 3 mA of current at full-scale voltage. If the experimental apparatus requires high power, a power amplifier between the D/A converter and the apparatus will be needed.

6.8.5 Digital I/O Signals

We may want to send digital (logic) signals into and out of our computer. For example, if our apparatus includes some microswitches whose open or closed state is used to tell the computer whether some hydraulic cylinders are in the extended or retracted position, our system needs some digital input capability. If the computer needs to tell some apparatus, such as a heater, to turn on or off, we need digital output. A single digital I/O board can provide several channels each of such functions. Such units usually themselves use small plug-in components which can be selected to suit the types (say, ac or dc voltages) and ranges of signals needed. Digital I/O signals can also be used for *data* transfer between the computer and external data loggers, printers, etc. There may also be "hand-shaking" circuitry for communication synchronization purposes.

6.8.6 Timing I/O Signals

Counter-timer circuitry is useful for many applications, including counting the occurrences of a digital event, measuring digital pulse timing, and generating square waves and pulses. All these applications can be implemented by using three counter-timer signals: gate, source, and output. The *gate* is a digital input that is used to enable or disable the function of the counter. The *source* is a digital input that causes the counter to increment each time it toggles, and therefore the source provides the timebase for the operation of the counter. Finally, a *counter* can generate digital square waves and pulses at the output line.

The most significant specifications for the counter-timer are the resolution and the clock frequency. The resolution is the number of bits the counter uses to count with; higher resolution means the counter can count higher, a typical value being 65,536. The clock frequency determines how fast the digital source can be toggled. A higher frequency lets the counter increment more quickly and therefore detect higher-frequency signals on the input and generate higher-frequency pulses and square waves at the output.

6.8.7 Analysis Hardware

In some high-performance applications, the PC microprocessor cannot process data rapidly enough to respond to a real-world signal. Other applications have numerically intensive calculations that need to be performed on large data sets, requiring that the user wait a long period before seeing any results. Plug-in boards based on digital signal processors can solve these problems by making computations at even higher rates than the PC processor. In addition, these complex calculations can be downloaded to the analysis hardware so that computations are made simultaneously as the PC microprocessor executes the application program.

6.8.8 Software

Software transforms the PC and data acquisition hardware into a complete data acquisition, analysis, and display system. The lowest level of data acquisition software development is the programming of the board. At this level, you directly program registers in the hardware. You need to determine proper binary values to write to these registers. The programming language must be able to read and write data from the boards plugged into the computer. Many system developers, however, do not have the desire or time to program at this level.

Data acquisition software, which removes the complexities associated with register-level programming, is broken into two areas: driver-level software and application-level software. Driver software simplifies programming by taking care of the low-level hardware programming details and giving you high-level function calls that can be used with conventional programming languages. Application software packages add analysis and data presentation capabilities to the driver software and integrate instrument control (GPIB, RS-232, and VXI) into data acquisition. Application packages designed for the various PC types are available in several different levels of complexity and capability. One vendor's catalog[1] lists nine different software packages, ranging from the simplest to the most sophisticated. Some use a command-language approach while others use a graphical interface with menus and icons.

The GUI concept pioneered by Macintosh and later adopted in Microsoft Windows is available in a number of data acquisition software packages. An interesting feature of many of these is the concept of the *virtual instrument*. A virtual instrument is a software module packaged graphically to have the look and feel of a physical instrument, such as an oscilloscope. That is, the computer monitor displays an image which looks like the front panel of an oscilloscope, with the usual cathode-ray tube display, knobs for adjusting sensitivity, sweep speed, etc. Using the keyboard and/or a mouse, the user can set up this virtual instrument to behave essentially like the actual instrument. Several such virtual instruments (scopes, counters, digital thermometers, etc.) can be set up in a single system and displayed on the computer monitor at will. Another application of the GUI uses a block diagram of system interconnections to actually program and wire the system. That is, as one manipulates the system block diagram on the screen to achieve the functions desired, the necessary software programming and hardware connections are automatically established. Thus the displayed block diagram actually *becomes* the program, making interpretation of system operation simple, and modification quick and easy.

In concluding this section, we should point out that hardware and software for data acquisition and experiment control are available for almost any application, including control tasks such as motor control and general-purpose PID controllers. A complete and detailed discussion of the design and operation of an actual computer-aided experiment used

---

[1] *Catalog,* vol. 24, 1991, Keithley Metrabyte, 440 Myles Standish Blvd., Taunton, MA 02780, phone (508) 880-3000.

for testing of vane-type rotary pneumatic motors is available[1] in the literature. While this example uses a command language rather than the more user-friendly GUI methods, it does provide more detail on the overall design process than was given in this brief section.

BIBLIOGRAPHY

1. G. Pahl and W. Beitz, *Engineering Design: A Systematic Approach,* Springer-Verlag, New York, 1988.
2. N. Suh, *The Principles of Design,* Oxford, New York, 1990.
3. W. Trylinski, *Fine Mechanisms and Precision Instruments: Principles of Design,* Pergamon Press, New York, 1971.
4. A. H. Slocum, *Precision Machine Design,* Prentice-Hall, Englewood Cliffs, NJ, 1992.
5. P. H. Sydenham, *Mechanical Design of Instruments,* Instrument Society of America, Research Triangle Park, NC, 1986.
6. W. R. Moore, *Foundations of Mechanical Accuracy,* Moore Special Tool Co., Bridgeport, CT, 1970 (distributed by MIT Press).
7. L. E. Andreeva, *Elastic Elements of Instruments,* translation from Russian, Israel Program for Scientific Translation, 1966
8. P. Ricci, *Standards: A Resource and Guide for Identification, Selection, and Acquisition,* Bernards and Co., St. Paul, MN, 1989.
9. E. O. Doebelin, *Measurement Systems,* 4th ed., McGraw-Hill, New York, 1990.
10. E. O. Doebelin, *Control System Principles and Design,* Wiley, New York, 1985.
11. W. R. D. Manning and S. Labrow, *High Pressure Engineering,* CRC Press, Cleveland, OH, 1971.
12. I. L. Spain and J. Paauwe (eds.), *High Pressure Technology,* vol. 1: *Equipment Design, Materials and Properties,* Marcel Dekker, New York, 1977.
13. J. F. O'Hanlon, *A User's Guide to Vacuum Technology,* 2d ed., Wiley, New York, 1989.
14. A. Roth, *Vacuum Technology,* 3d ed., Elsevier, New York, 1990.
15. P. R. Yoder, Jr., *Opto-Mechanical System Design,* Marcel Dekker, New York, 1986.
16. D. C. O'Shea, *Elements of Modern Optical Design,* Wiley, New York, 1985.
17. W. K. Roots, *Fundamentals of Temperature Control,* Academic Press, New York, 1969.
18. J. R. Leigh, *Temperature Measurement and Control,* Peter Peregrinus, London, 1988.
19. S. Zinn and S. L. Semiatin, *Elements of Induction Heating: Design, Control, and Application,* American Society for Metals, Novelty, OH, 1988.
20. G. K. White, *Experimental Techniques in Low-Temperature Physics,* 2d ed., Oxford, London, 1968.
21. A. C. Rose-Innes, *Low Temperature Laboratory Techniques,* 2d ed., Crane, Russak and Co., New York, 1973.
22. W. P. Rodgers, *Introduction to System Safety Engineering,* Wiley, New York, 1970.
23. R. A. Wadden, P. A. Scheff, et al., *Engineering Design in the Control of Workplace Hazards,* McGraw-Hill, New York, 1987.
24. G. Marshall, *Safety Engineering,* Brooks/Cole, Monterey, CA, 1982.
25. J. H. Moore, M. A. Caplan, and C. C. Davis, *Building Scientific Apparatus,* 2d ed., Addison-Wesley, Reading, MA, 1989.
26. E. B. Magrab, *Computer Integrated Experimentation,* Springer-Verlag, New York, 1991.

---

[1] Doebelin, *Measurement Systems,* pp. 923–946.

CHAPTER
7

EXPERIMENT EXECUTION

7.1 APPARATUS CONSTRUCTION AND DEBUGGING

Once the apparatus design is "completed", the construction phase can commence. We put the word *completed* in quotation marks because we often find that design changes become necessary as construction and execution proceed. This is particularly true in studies which have a large component of novelty and thus are somewhat exploratory and tentative. More routine experiments will have a more predictable construction phase, and design changes will be few. We should thus expect the construction phase to be interrupted occasionally (or often in complex and original experiments) by the encountering of design gaps or flaws that will need attention before construction can proceed.

Most projects will have rather stringent constraints on time, so we must carefully plan the timing of delivery for purchased parts and the production of those that we make. Often a simple chart showing the planned occurrence of all the events in the construction sequence over time will be helpful. More sophisticated methods such as critical-path planning and Gantt charts[1] are available for more complex situations. For parts or construction operations whose performance is somewhat uncertain, we should provide alternative plans if possible, so that there will not be undue delays if assumed behavior fails to be realized.

In planning the construction sequence, we should be careful to identify those phases which can be performed in parallel and those which must be sequential. If two or more phases *can* be worked on simultaneously, we can speed up construction if we have sufficient personnel and facilities. However, if operation B depends on the prior successful completion of operation A, lack of recognition of this dependence can lead to waste of time and resources.

When a system is comprised of identifiable subsystems, it may be possible to construct and debug the individual subsystems more efficiently than if we try to do everything at once. We must, of course, be careful that any subsystem tests of operation are performed under conditions which duplicate or sufficiently approximate those operative in the complete system. For example, sensors whose electric output will ultimately be fed to a computerized data acquisition system, if tested as separate subsystems, should have their output terminated in an impedance which sufficiently approximates that of the data system. Tests of an oven's heat-up speed should be done with the oven loaded with material which approximates the thermal behavior of the actual specimens to be used.

If all or part of the apparatus is to be built "from scratch" rather than assembled entirely from purchased components, the apparatus design engineer will need to work with machine shop personnel and laboratory technicians. Clear communication among these groups is vital. Sketches and drawings must be clear and unambiguous. Every machine shop has its own peculiarities which affect how things are done, and the design engineer needs to become familiar with these. Furthermore, preliminary designs should be shown to machinists before they are finalized so as to reap the benefit of their unique knowledge of the manufacturing capabilities of that particular shop. Machinists and shop supervisors often have excellent suggestions for improving design details and avoiding problems of manufacture and assembly. Also, if they have been actively involved in the design stages, machinists and shop supervisors will consider the apparatus as partly "theirs" and give its construction more careful attention.

---

[1] J. C. Gibbings (ed.), *The Systematic Experiment,* Cambridge University Press, New York, 1986, pp. 19–27.

Each specific apparatus will require a debugging effort suited to its own peculiarities; however, we can give some overall guidelines. If our experimental plan requires that certain variables be exercised over specific ranges, we must make sure that these ranges can actually be reached by our equipment. We thus set our controls for both the lowest and highest desired values to see whether these extremes can be attained. If certain rates of change are necessary, these must also be checked. The accuracy of holding the desired values may need to be validated as well. We must check that no unforeseen dangerous conditions occur when the apparatus is operating at its designed limits.

Can all the adjustments built into the equipment actually be properly made? If there are several controlled variables, can all the desired *combinations* of values be achieved? When we are exploring individual, or combinations of, maximum values, safety considerations suggest that we "creep up" on these gradually, watching for the first indications of trouble, rather than going to them suddenly. Just as in construction, debugging should be done, insofar as possible, on sub-systems first, going to complete system operation only as subsystems are qualified. This applies also to procedures. Those subsystems thought to be most likely to encounter problems should be debugged first, if possible. If they require design changes, the changes may influence other subsystems, which may then require their own changes. If our experiment involves complicated procedures, we should try to break these down into simpler subtasks and debug these first, before attempting to debug the overall operation. If any standard codes or operating procedures (dis-cussed in Chap. 6 as they related to design) apply to the test, be sure to study these for any useful guidelines and warnings.

Since all but the simplest experiments nowadays include some aspect of computer management, we need to debug not only our experimental hardware but also the software. As discussed briefly in Sec. 6.8, software for laboratory automation can range from ready-made (where the experimenter's programming may be limited to simple fill-in-the-blanks operations), to detailed line-by-line programming in a fundamental language like BASIC, FORTRAN, or C. Since software engineering is a specialized and complex technology in its own right, with well-developed techniques for debugging, we cannot hope to deal with this here and must limit ourselves to some overall suggestions. Most important is the fact that since our software and hardware are interfaced, software errors can result in dangerous and/or expensive *physical* results. In pure software debugging, the *program* may crash, but this doesn't really cause any physical damage. Thus, insofar as possible, we should debug our software *before* we interconnect the computer with our apparatus. Finally, of course, the complete system must be tried out, errors detected, and corrections made. To minimize the chance of disastrous occurrences, we should again try to debug only small portions of our interfacing at one time.

7.2 MEASUREMENT SYSTEM CALIBRATION[1]

While the measurement system is part of the overall apparatus and could be included in the debugging discussed in the previous section, its importance and special needs justify a separate treatment. Basically, before we start to take serious data, we need to verify that all our instruments are "telling the truth." That is, they must be calibrated in one way or another against suitable standards. The most convincing calibrations are those performed *in situ,* i.e., with the instrument connected to the apparatus as it will be in actual operation. This ensures that any effects of the apparatus on the instrument are taken into account. Provisions for in situ calibration should be considered at the apparatus design stage, and decisions made as to whether this approach is technically and economically feasible. While desirable, in situ calibrations are often difficult or impossible to implement, and we must be satisfied with other methods. If in situ calibration is not feasible, we may be able to analytically correct (at least partially) for the effects of connection to the apparatus ("loading effects") by use of impedance methods.[2]

Whether performed in situ or with the instrument disconnected from the apparatus, calibration involves comparison of the instrument's readings with some kind of standard which is more accurate than the instrument itself. Standards can be known values, such as a standard weight (mass) for calibrating a force transducer, or they may simply be other instruments for measuring the given quantity that are known to be more accurate. For example, we can compare the readings of an electric pressure transducer with those of a mechanical piston gage (whose accuracy is much better) in order to calibrate the electric transducer. Static calibrations are the simplest and most common, but then we need to be aware that if we use the instrument to measure an *unsteady* quantity, dynamic errors may result.

In assessing loading effects (both static and dynamic) of transducers, one can sometimes use the following technique. Recall that we usually want the value of the measured quantity that would exist if the transducer were *not* present. If we attach a *second* transducer (identical to the first) in the "same" location on the apparatus, so that now *two* transducers are

---

[1] See also Chap. 3.

[2] E. O. Doebelin, *Measurement Systems,* 4th ed., McGraw-Hill, New York, 1990, pp. 74–88.

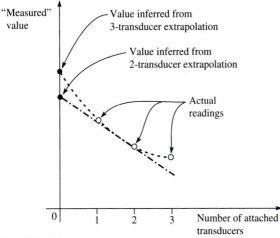

FIGURE 7.1
Determination of transducer loading effect.

present, then the effect of transducer attachment should be increased. Attachment of a third transducer would increase the loading effect even more. By graphing (see Fig. 7.1) the instrument reading versus the number of transducers, we can extrapolate this curve back to *zero* transducers and thus infer what the reading would have been if no transducers were present. If it is feasible to attach only two transducers, then the extrapolation to zero transducers would probably be along a straight line. In the rare case where three transducers can be used, any nonlinearity in the loading effect can be detected and a curved extrapolation employed. Clearly, this approach (even with only one added transducer) will not always be practical, but it is a possibility to consider.

When we calibrate individual sensors disconnected from our apparatus, we must always be careful to consider any extraneous effects which might change this calibration when the sensor is attached. A room-temperature calibration of a pressure transducer may be inaccurate if our apparatus test procedure results in working temperatures significantly higher or lower than the calibration value. A force transducer, calibrated with a carefully aligned uniaxial load, can give misleading data if the actual conditions of use involve some off-axis forces in addition to the intended on-axis component. Apparatus design, sensor installation, and apparatus operating procedures should attempt to minimize such errors. For the most exacting applications, it may be necessary to perform calibrations which *include* the undesired inputs as well as the main desired input. That is, the pressure transducer will have to be calibrated against *both* pressure and temperature, the operating temperature measured during apparatus operation, and the pressure reading corrected for the extraneous temperature input. When an instrument has many extraneous inputs which cannot be ignored, such calibrations can become quite complex. One can then formulate a mathematical model relating the instrument output (reading) to all the inputs, desired and extraneous. A calibration experiment then consists of exercising all the inputs over their range of interest and using multiple regression to find the best values of the coefficients in the model. The model is then used to correct instrument readings so that they are closer to the true value of the measured quantity. A good example of this complex procedure, as it relates to precision multicomponent load cells (force transducers), is available in the literature.[1]

7.3 DATA SHEETS AND LOGBOOKS

Even though data taking is today often computerized, it is still important to give careful consideration to the mechanics of actually laying out the forms used for this important operation. Let's look at the layout of a typical data sheet as used for manual data recording in a student lab experiment. Many of the features of such a form will also be usable if we are designing a data sheet for computer-aided recording, because most of the basic functions are the same.

An overriding principle in the design of an effective data sheet is that it should be understandable by engineers *other* than those who originally recorded the data. That is, it should be largely self-explanatory. If we are not careful to meet this basic criterion, then quite likely the information on the sheet may be misinterpreted even by those who *did* record the data, when they try to use it a few days or weeks after running the test. That is, things which may have seemed perfectly obvious and clear at the time of taking the data tend to become ambiguous and fuzzy as time passes and memory fades. We must protect ourselves from this pernicious phenomenon by taking sufficient care, when we originally lay out and use our data sheet, to make it as self-explanatory as we reasonably can.

For manual data taking, each sheet should include the names of the lab group members present and, in particular, the name of the individual who actually wrote the data on the sheet. It is vital that we later be sure who has responsibility for the data, should we need to consult with this person about any questions that may arise. The time, date, and location

[1] A. Bray, G. Barbato, and R. Levi, *Theory and Practice of Force Measurement,* Academic Press, New York, 1990, pp. 289–338.

of the experiment should also be noted. A main heading with a *descriptive* title for the experiment is desirable. If we later get results (based on the data sheet) that do not seem to be correct, we must be able to investigate possible sources of discrepancy. One such source is faulty instrumentation, malfunctioning recorders, sensors out of calibration, etc. To even have a chance of checking for such problems, we must know which specific instruments were used. The lab may have 20 oscilloscopes; we need to know which *one* was actually used if we want to check out its operation. Thus the data sheet must specifically identify every instrument used in our experiment. There are a number of ways to do this. Most organizations inventory equipment when it is purchased and affix a permanent tag with an inventory number. This number may be a quick way to identify things and takes little space on the data sheet. Any other scheme that serves the basic function of unambiguous identification would be acceptable. With regard to saving space on the data sheet, remember that paper is probably the cheapest item we use in the lab. It is much preferable to use more space or even additional sheets in our quest for clarity than to produce a crowded and unclear data sheet.

If we have a sequence of related data values, it is usually best to record these in a column rather than a row. This makes it easier to spot errors as we go, since the digits and decimal point can be kept aligned. Data forms with a *lightly* ruled grid of horizontal and vertical lines (about $\frac{1}{4}$-in spacing works well) expedite the layout of columns and also are an aid in making any apparatus sketches that might be needed on the data sheet. Each column of data should have a descriptive heading giving a word name for the quantity recorded, a letter symbol, and units. An example might be condenser inlet pressure P_{ci}, psia. When we choose letter symbols, we try to make them "self-defining," as in the condenser pressure example. For example, if we had two pressures to record, say, the condenser inlet and condenser outlet, we *could* just call them P_1 and P_2, but wouldn't P_{ci} and P_{co} be easier to remember and use?

It is good practice to make a *tentative* data sheet layout on scratch paper before making the final data sheet and actually taking data. This rough data sheet is useful for judging how much space to leave for various tables and sketches and for discussing the general layout and details with other members of the lab group. Group effort, here as in many other engineering tasks, can increase the number of good ideas produced, reduce the chance of oversights, and increase the commitment of each group member to a successful project. In academic contexts, some lab instructors insist on student groups making tentative data sheets as a mechanism for enforcing thoughtfulness at the beginning of an experiment. Also, an instructor's critique of the proposed data sheet gives opportunity to praise good features and suggest needed revisions. Actual data taking should, of course, be on the final data sheet, not the rough version. One should, with rare exceptions, never take data on scratch paper with the thought of later transcribing them onto the "good" data sheet. Such transferal of symbols from one sheet of paper to another is a potential source of error and should be avoided. One possible exception would be a consequence of data points occurring rapidly in time. Here, we might take the data with wide spacing on our scratch paper to allow rapid writing with legibility. This would be followed by careful transcription into neat, closely spaced columns on our final data sheet.

Many experiments produce data in some orderly sequence, and we have already mentioned our preference for data in vertical columns rather than horizontal rows, as a simple means of detecting errors. Even more useful for this purpose is the practice of plotting "running graphs." Here, as we take the data points, we immediately plot any appropriate graphs. Compared with a table, a graph generally is a much more sensitive and reliable detector of "odd" behavior. By keeping these running graphs updated as we add each reading, we can often detect problems and malfunctions quickly. Such graphs are usually *not* part of our final data sheet and will usually be done on scratch graph paper. Since such graphs are needed immediately, we plot, whenever possible, actual data values from our tables, rather than related quantities that might be the computed results of our experiment. That is, if we are recording water level in a standpipe and the discharge rate of an orifice at the bottom of the standpipe, for the purpose of producing a graph of flow rate versus pressure drop for the orifice, our running graph could use the raw water-level data in inches rather than require us to take the time to compute the pressure (pounds per square inch) from the water level.

Unless there is some good reason, the data sheet should *not* show any calculations but must rather be reserved for raw data only. It is especially bad practice to convert some raw data to a simply calculated result and then record on the data sheet *only* the calculated result. The data sheet *must* show *all* the raw data. Showing calculations on a data sheet also uses up valuable space, causes clutter, and may confuse later readers of the sheet as to what is data and what is calculation.

Apparatus *sketches* are almost always a vital part of a good data sheet. Well-planned and well-executed sketches can contribute greatly to the clarity of a data sheet and its ultimate utility. Unfortunately, skill in technical sketching has a low priority in many engineering curriculums, and students have little opportunity to develop it, leading to a serious gap in technical communication. This impacts engineering performance in a number of areas, such as design, experimentation, and technical writing. Reasons for this curricular underemphasis include a general reduction in the graphic communication component of engineering education. In the few remaining courses, an *over*emphasis on computer-aided methods leaves little time for unglamorous topics such as sketching.

A few broad guidelines and some specific examples will be offered on this topic. One of the most common data sheet errors of novice experimenters is the allocation of insufficient *space* on the data sheet for the sketch. There seems to be an overriding goal to get everything on *one sheet*. Here again, try out your sketch ideas on scratch paper, to get an idea of how much space and what shape of space is needed to accommodate the sketch. At this point students may be saying, "All this thinking and planning before we ever get started on the experiment means we will never get finished in the time allotted." Both instructors and students need to address this problem. Instructors need to assign less complex experiments which concentrate on fundamentals, and students need to spend more time thinking about experiments *before* they appear at the lab.

In the visual design of an apparatus sketch, we need to avoid the photographic approach. With rare exceptions, actual photographs of apparatus are usually *less* informative than well-thought-out line drawings. A photograph "shows everything," presenting an unorganized and confusing picture to the viewer. It is our task in sketching (1) to extract from the photographic view those essential features which we wish to explain and then (2) to organize them into the simplest sketch which will communicate the desired information most efficiently.

Most sketches can be broadly classified as *pictorial drawings, schematic diagrams,* or *block diagrams.* Pictorial drawings can be two-dimensional or three-dimensional and are a simplified "picture" of some real object or objects, often a piece of our apparatus or the entire apparatus. Critical portions of our equipment should be shown in detailed sketches, with important features and dimensions labeled. A word statement of a dimension, such as "fin length = 2.34 cm" for a heat exchanger, may be perfectly clear to the data taker at the time of taking this data, but can be quite ambiguous to a colleague trying to use the data a week later. Schematic diagrams are more abstract than pictorial drawings, showing symbolic elements and their interconnection to make clear the configuration and/or operation of a system. Electrical schematics, showing the familiar symbols for $R, C, L,$ op amps, etc., are the most familiar example, but mechanical and fluid system schematic diagrams are also widely used. Block diagrams are the least pictorial of all, showing only blocks representing components or subsystems and lines representing the flow of signal or energy between blocks. Both blocks and lines are suitably labeled to communicate the system functions. Figure 7.2 shows both a pictorial drawing and a block diagram for the same system. This system had been theoretically analyzed so that mathematical models (transfer functions) could be shown in the block diagram.

When making pictorial drawings, schematic diagrams, or block diagrams, we need to keep in mind a few basic principles of visual

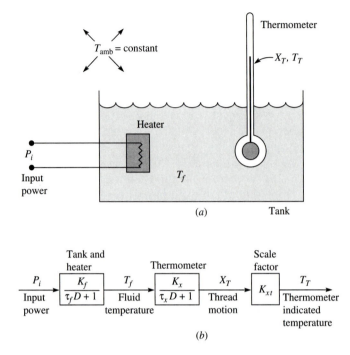

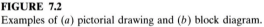

FIGURE 7.2
Examples of (*a*) pictorial drawing and (*b*) block diagram.

communication. Since none of these illustrations is "photographic," we always must decide how much, and in what way, we will simplify the actual forms of the apparatus for sketching purposes. Forms used in sketches should be kept as simple as possible *without* giving up any vital information. Detailed shapes need be shown only if these details are necessary for communicating the desired information. If the sketch includes letter symbols to represent dimensions or other quantities, again we should choose these with some care, so that the symbols themselves will jog our memory as to what they mean. This often leads to the use of subscripts for identifying different quantities of the same general class, such as three different temperatures. Some engineers and other writers avoid or minimize the use of subscripts since subscripts are more difficult to type in a report or paper. This leads to notation such as Tin and Tout rather than T_{in} and T_{out}, for example. We believe the use of carefully chosen subscripts is well worth the extra time and effort. We can recall many instances where the avoidance of subscripts has resulted not only in poor visual appearance and difficulty in reading but also in technical errors of interpretation.

   While data sheet sketches made under laboratory conditions cannot and need not achieve the precision of mechanical or computer-aided drawings, they do require sufficient care and neatness if they are to serve their function. The use of a simple straightedge and perhaps some kind of plastic template for drawing circles, etc., can both speed up the sketching and improve the overall quality. Again, do not skimp on space; paper is cheap, and allowing more space generally makes it easier to produce good results. When you think your sketch is complete, always stop and ask yourself whether anything needs to be added and whether anything can be deleted or simplified without a loss of information. Finally, be sure to leave sufficient clear *margins* on all sides of the data sheet, especially the left side, where the report will be bound. Most reports and data sheets will be reproduced, and the reproduction process may lose material that is too close to the paper's edges.

For lengthy projects (more than a few days) it is conventional to use a *logbook,* rather than one or several data sheets, to record the progress of the experiment. Most workers prefer that such books be permanently bound rather than the looseleaf type and that the size be generous, say, about 8 by 10 in. Data can be recorded directly in such a book, or if separate printed data forms are more convenient, these can be used and then later bound into the logbook for permanence. The logbook is, of course, used for more than simple data recording. It gives a chronological record of all the significant events observed by the experimenter during the experiment. Be sure to leave some blank pages at the front for a table of contents, which is updated as time goes by. It is vital to get your interpretations of observed phenomena into the logbook soon after they occur, before the passage of time begins to blur important details. Negative results may be as useful as positive ones, so be sure to record and discuss them. Include enough detail in all entries that the work can be replicated at a later date if necessary. If patent considerations are pertinent, be sure to date and have witnessed any relevant sections of the logbook. A well-kept logbook is a great aid in later writing reports or papers and is thus deserving of careful attention.

CHAPTER
8

DATA
ANALYSIS
AND
INTERPRETATION

8.1 INTRODUCTION

Once our data has been gathered, we want to interpret and analyze it to discover its meaning. Generally, tools available for such studies use *statistical, numerical,* and *graphical* methods. Specific tools often use a combination of these approaches. In terms of this text, a large portion of the topic of data analysis has already been discussed in Chap. 4, where we covered experiment planning. That is, in a planned experiment, the data analysis procedures are part of the plan. For those readers who might not be reading the book in sequence and have just turned to Chap. 8 for advice on data analysis, you need to go to Chap. 4 to find most of this material. However, now we review and elaborate on this material for the purposes of this chapter.

Prior to, and perhaps as part of, data analysis we should continue to be on the watch for invalid data, using whatever *data-checking* procedures are appropriate to our experiment. Weeding out of "bad" data needs to be a continuing effort at all stages of an experimental

454

project. Undetected spurious data, once they get into a sophisticated data analysis procedure, can bias results and interpretations in many subtle ways. When results are not following our expectations, this can be due to new (and possibly important) real effects in our apparatus or bogus effects caused by faulty data. Tools for rejection of spurious data that have already been covered in this text include the treatment of *outliers* in Chap. 4 and the use of *running graphs* in Chap. 7. As the experiment proceeds and we accumulate more data, we maybe want to employ more comprehensive data-checking tools. (As always, however, we need to exercise extreme care to not throw out *good* data just because it is unusual.)

The data-checking methods actually used will, of course, depend somewhat on the nature of the specific experiment; however, a few general suggestions can be made. An obvious, but often too expensive approach is simply to *repeat* part of or all those runs which appear questionable. If this *can* be justified, it will be most convincing if any random effects (such as the experiment operators, time of day, source of specimens) are allowed to be different from the original run when we replicate it. This philosophy can become quite expensive to implement since it may require that instruments be recalibrated and apparatus disassembled and reassembled. The decision as to whether to undertake a replication and how comprehensive it should be clearly will depend on the degree of doubt we have about the original run and how important the questioned results are to the success of our project.

When we graphically display data relating two variables, the extrapolation of the curve to the origin may be a useful checking method. Efficiencies of machines plotted against load on the machine should extrapolate to zero at zero load. When we study properties of mixtures of components, these properties should extrapolate to the properties of a component when the percentage of the other components goes to zero. We should always look for and use any such known relations to check the behavior of our data. Another useful tool employs conservation or balance principles to check whether data is consistent with itself. Conservation of energy tells us that if we have measured all the inputs and outputs of energy to a process operating at steady state, then the sum of all the inputs must equal the sum of all the outputs. If calculations of these inputs and outputs show a discrepancy, we may have some faulty data. Conservation of mass can be used in a similar way if we are measuring material inflow and outflow rates. In practice, such balance checks may not be possible or totally convincing if there are significant *unmeasured* "losses" such as heat leaks or fluid leaks in the system.

In general, we should carefully consider all the available theoretical information about our apparatus and then make use of those aspects which allow us to partially or totally check the validity of our data. If

Some important classes of experiments	
1	Estimation of parameter mean value
2	Estimation of parameter variability
3	Comparison of mean values
4	Comparison of variabilities
5A	Modeling the dependence of a dependent variable on several qualitative independent variables
5B	Modeling the dependence of a dependent variable on several quantitative independent variables
6	Accelerated testing

FIGURE 8.1
Classes of experiments.

possible, this should be done before we undertake any large-scale overall data analysis associated with the principal goals of the experiment.

8.2 STATISTICAL ANALYSIS AND INTERPRETATION

Figure 4.1 listed the major categories of experiments treated in that chapter, and Fig. 8.1 repeats this information. Since the seven categories all involve samples of size larger than 1, and since we assume our readings always have some element of randomness in them, all the data analysis methods discussed in Chap. 4 involve statistical considerations. All require numerical calculation, and most use graphs at various stages of evaluation and interpretation, so they are really a mixture of numerical, statistical, and graphical techniques. However, one usually finds these methods in books on statistics, thus the title of Sec. 8.2.

We said that data analysis aims at discovering *meaning* in our data, so let's interpret the seven categories of Fig. 8.1 in that context. The simplest experiment (category 1) is intended merely to establish the numerical value of some measurable quantity, so there is not a high level of analysis involved. The situation is well summarized in Fig. 4.3*b*. We must choose the percentage of the mean value that we wish our confidence interval to be and the confidence level we want. We also need an estimate of the coefficient of variation. Then Fig. 4.3*b* tells us how large a sample is needed. If no data have yet been taken, the estimate of the coefficient of variation must be based on past experience or, lacking this, outright guesswork. When we have tested two specimens, we can improve this estimate with real data. As we test more specimens (if we do), this estimate improves further with each additional test, and our belief in the predicted confidence interval increases. If the experiment was not *planned* with the aid of Fig. 4.3*b* (or equivalent methods) and we are simply faced with a set of raw data, "data analysis" consists merely of using Fig. 4.3*b* to determine the sizes of confidence intervals for chosen confidence levels. For example, if we have a sample size of 10, mean value of 6.8, and standard deviation of 0.47, Fig. 4.3*b* gives a 95 percent confidence interval of about 6.8 ± 0.33. Our best estimate of the measured value is 6.8, and we are 95 percent sure it is somewhere between 6.47 and 7.13.

When we are estimating parameter *variability* (category 2 of Fig. 4.1) rather than average value, we can use Fig. 4.7 in much the same way as we just used Fig. 4.3*b*. Categories 3 and 4 are used when we are trying to decide whether one material or process is better than an alternative in terms of having a different average value or standard deviation. Section 4.3 showed how to analyze the comparison of average values by using confidence intervals on the difference between the two values, perhaps augmented by overlap graphs. Our STATGRAPHICS software also gave us some useful graphical tools of interpretation in the double-histogram and box-and-whisker plots of Figs. 4.11 and 4.12. For comparing two standard deviations, Sec. 4.4 used their *ratio* and the *F* distribution to reach decisions as to whether the two values were about the same or significantly different. Recall from Chap. 4 that our approach for *all* the categories of experiments augmented the standard tests found in most statistics texts with computer simulations using widely available software for personal computers. These simulations are *very* helpful in making good decisions in the (many) cases where the standard tests give "fuzzy" predictions.

Once we leave categories 1 through 4, which only estimate a single number or compare two numbers to see if they are "different," data analysis becomes considerably more extensive. We are now looking for *relations* among several variables rather than just isolated numerical values. Since the possible number of relations (mathematical functions) between even two variables is infinite, our task is inherently more difficult. Once we speak of relations and functions, then it is natural to also think of graphs displaying these relations for our visual interpretation. In fact, in earlier times, data analysis of this class of experiment was often carried out by using manual graphical tools such as special kinds of graph paper (log-log, semilog, square root, etc.) or linear graph paper with distorted scales. Even today, preliminary studies using strictly graphical approaches may be valuable, and we will shortly devote a section to these. Our Chap. 4 discussion, however, emphasized the most common approach, which is today a numerical and statistical method using multiple regression or related tools. Let's now review this material.

In Sec. 4.5.1 we treated the problem of finding a relation between a single independent variable and a single dependent variable (simple regression). Preliminary data analysis should certainly include a simple plotting of the measured data, the pairs of x, y values. If we have a physical theory which predicts a definite mathematical relation between the two quantities, clearly we should check whether the graph is consistent with this theory. If visual evaluation does indicate that the data could follow the predicted straight line, parabola, sine wave, etc., then we can use the available regression software to find the specific curve, of the given family, that *best* fits the data. Functions that are linear *in the fitting coefficients* are the easiest to fit (linear regression) and should be used whenever possible. Nonlinear regression methods (Sec. 4.5.6) are available for other cases but are less straightforward to use.

When the data scatter is slight and the data curve follows closely a theoretical prediction, data analysis and interpretation using the best-fit curve from the regression are simple. If we have a theoretical prediction for the type of curve to be expected but the data show large scatter, then the interpretation becomes less clear-cut. If, in spite of the large scatter, we can still visually discern a trend in the data which does not contradict the theory, then of course we should get a best-fit regression, using the theoretically predicted type of curve. We can then graph the residuals for this fit to see if they are largely random, or if we can detect a trend. If no trend is apparent, we may have found the best model. If a trend is detectable, its nature may give us some clues about how to adjust the form of our model to get a better fit. A final choice of model is based on criteria such as visual goodness of fit, numerical value of R^2, numerical values of individual residuals relative to the practical importance of different regions of the curve, and the complexity of the model.

When there is *no* theoretical prediction and we have considerable scatter, a reasonable approach is to choose the simplest form of function which can fit the data (use functions linear in the coefficients if you can) and do a regression (least-squares curve fit). Often one uses a polynomial model and starts with the simplest, the straight line. Results from this simplest model can then be compared (by using the criteria just listed above) with those of successively higher-order polynomials, and a final choice made. Be sure to remember that an n-point data sequence can *always* be perfectly fitted with a polynomial of degree $n - 1$, but such fits are almost always *not* good models since they fit the "noise" in the data rather than the underlying trend. Also, when scatter is large and *no* trend is visually obvious, models extracted by regression methods, no matter how sophisticated, may really be quite meaningless for any practical use. We always need to carefully study the *consequences* of our using this model and decide whether it is to our benefit or not.

When we study the relation between one dependent and two

independent variables, graphical displays are still possible for preliminary data analysis, but we now deal with three-dimensional surfaces rather than plane curves. This makes visual analysis and interpretation much more difficult, and three-dimensional graphing is rarely used as a starting point of analysis. Usually we begin such studies with regression analysis. If a theoretical prediction of the type of functional relation to be expected is available, this is the logical starting point for the modeling process. Regression methods can be used to find the best-fit surface of the class specified, and the usual criteria for evaluating the quality of fit may be applied (see Sec. 4.5.2). As usual, data with large noise make the choice of the best model more difficult.

When we have no theoretical prediction to guide us, three-dimensional plots (see Figs. 4.28b and 4.59) of the raw data are possible but are almost useless for the purpose of choosing a model form if there are many points and/or there is much scatter. Rather, we undertake a systematic, stepwise study of increasingly complex surfaces, using multiple regression results to evaluate the quality of fit. Lacking any information to the contrary, we usually use polynomial-type surfaces, starting with the simplest, the plane. In addition to the usual evaluation criteria, we may now use the *component-effect* graphs (see Fig. 4.30) as aids in reaching a decision on the best model. All these comments on models with two independent variables also apply directly to those with n independent variables, except now, for $n > 2$, graphical display of a "surface" is no longer possible.

When we study models with more than a few independent variables, the concepts of factorial or fractional-factorial experiment plans, and orthogonal arrays, become useful (see Sec. 4.5.3). These topics are very important in experiment planning, but are not really data *analysis* methods. The data produced by these types of experiment plans can still be analyzed by using the same tool as for the simpler cases, i.e., multiple regression. In fact, as stated in Chap. 4, a *goal* of this text was to *simplify* the engineering application of statistics by applying the *same* tool to several classes of problems that are usually treated with *different* tools in statistics texts.

When our experiment deals with *qualitative* (nonnumerical) factors (independent variables), the classical data analysis method is called ANOVA (*an*alysis *of* *v*ariance). While use of ANOVA is very widespread (and therefore we did devote some text space to it), our main thrust was to show that ANOVA wasn't really necessary. We could in most cases accomplish the same (and sometimes more and better) results by just using multiple regression and/or ANOME (*an*alysis *of* *me*ans). The idea, of course, is that engineers (who are not professional statisticians and use statistics occasionally) will use statistical tools more proficiently if they learn a few "universal" ideas well, rather than several specialized ideas superficially. These concepts and methods are developed in detail in Sec. 4.5.4.

The methods of *stepwise regression,* useful in model building and evaluation, are presented in Sec. 4.5.5. These techniques use the same basic multiple regression methods that we have used all along, but apply them in a clever way to speed up the process of finding the best model. This section is also used to introduce the generally useful concept of *model validation,* wherein part of the data is used for model building and part is set aside for checking the model validity (see Fig. 4.49*b*).

Certain model forms require use of *nonlinear regression* tools. A brief introduction to these methods and the difficulties associated with them is given in Sec. 4.5.6. These methods do not use any new model analysis or interpretation tools. The difference lies just in the algorithm used to find the best model. *Response-surface* methods are discussed in Sec. 4.5.7. Here our usual multiple regression tool is just augmented with some additional techniques for locating peaks and valleys in the response of our model. This is useful in finding the combination of process inputs which produce a maximum quality or production rate, or a minimum cost.

Accelerated testing (Sec. 4.6) is a specialized but important topic in experimentation. It has its own unique data analysis techniques, which are explained in some detail. The presentation in Sec. 4.6 emphasizes graphical methods using special graph papers that are commercially available. While they are quite widely used in actual engineering practice, these graphical methods were chosen in Chap. 4 mainly because they allow the clearest presentation of the basic concepts. Equivalent computerized numerical techniques are also available. These do not require graphical methods for the data *analysis,* but would still use graphs as a valuable tool of data *presentation and display.*

This completes our review of the data analysis methods presented in detail in Chap. 4. Recall that we stated earlier that analysis should bring forth the *meaning* of our data. This meaning will, of course, have its own detailed interpretation for each specific example. We can, however, note the general meaning associated with each of the classes of experiment listed in Fig. 8.1. In classes 1 and 2, we are simply interested in getting a reliable estimate of the average value, or variability of some quantity. Typical quantities might be material properties such as modulus of elasticity or electrical resistance, component properties such as stiffness of a spring or gain of an amplifier, or system performance criteria such as efficiency of a hydraulic motor. The "meaning" of such data is defined by how we will use it. Material properties are used in design calculations for components. Component properties are used in design calculations for subsystems or systems. System performance criteria tell us whether overall system design goals have been met. Average values are used to calculate nominal design values; standard deviations are used to estimate variability around this design goal.

The "meaning" of the data developed by experiments in classes 3 and 4 is often related to decision making at various stages of the design process. We are evaluating alternative materials or design concepts so that we can choose the best. Such choices often require data on both the average values and the variability. If steels A and B have similar average strengths but B has significantly less variability, we may choose B, assuming other factors are comparable. If two alternative production processes show similar variability but different average values, we can often adjust the machinery to get the average value desired. Choice between the two systems would thus rest on other considerations, such as first cost or operating cost.

Experiment classes 5A and 5B are used to find relations (models) between process "outputs" such as quality, cost, and yield and the process inputs which affect these outputs. Once such relations have been discovered, we can use them to guide adjustments to the process aimed at improving operations. We may wish to shift the average value of an output to a desired value. Variability in an output may be reduced or minimized by suitable adjustments. Operating points which not only maximize some desirable output but also make it less sensitive to changes in input values can allow us to relax specifications on component parts or materials, reducing costs but still maintaining high quality. The "meaning" of data which defines a process model thus lies in our better understanding of how the process works.

Accelerated testing, the class 6 experiment, has a number of uses, as explained in Sec. 4.6. A major application related to the design process is the early detection of design flaws, so that product design can be improved before large-scale manufacture commences. This involves finding a sufficiently accurate model relating "stress" to life, so that short-time tests at high stress levels can be used to predict life at design stress levels. The "meaning" of accelerated testing data thus lies in this ability to reduce testing time and cost.

8.3 GRAPHICAL DATA ANALYSIS

We saw in Sec. 8.2 that graphs are regularly used as adjuncts to data analysis methods that are largely numerical and statistical. In this section we concentrate on methods in which the graph plays the central role. Some of these techniques are less widely used now that interactive software for personal computers makes regression methods so easy. Since graphs are still useful for the simpler problems and can be used when no computer is available, we want to discuss them briefly. Also, most of the computerized methods are related to the earlier graphical ones, so the ideas will be useful in both contexts.

Most of the methods are intended for exploring the relation between only two variables, by using a simple x, y graph. To speed up the work, special graph paper with various kinds of distorted scales (log-log, linear-log, gaussian probability, etc.) is often used. When such paper is not available, one can get the equivalent effect by using linear-scale paper and plotting not the raw data, but rather the desired *function* (such as a logarithm) on the graph. This is in fact what the more "modern" computerized versions of these methods do, using regression analysis to define the best line, rather than doing it "by eye." A variety of such specialized graph papers are available.[1]

As a simple and common example of these methods, let's consider the relation of the form

$$y = Cx^a$$

$$\log y = \log C + a \log x \tag{8.1}$$

$$Y = b + mX$$

By using the logarithmic transformation, the curve $y = Cx^a$ becomes the straight line $Y = b + mX$. We can graphically implement this transformation by plotting Y ($\log y$) and X ($\log x$) on linear-scale graph paper or y and x on log-log paper. ("Before computers," the special paper saved a lot of time.) The data analysis aspect of this procedure is as follows. We have gathered some x, y data and want to see whether they are well fitted by a relation of the form (8.1). For example, experiments which employ dimensional analysis to increase efficiency often use equations which have this form. Since the human eye and brain are not expert at judging whether a curve has a particular formula, but *are* well suited to judging whether points fall on a straight line, we choose to work with the transformed, rather than raw, data. We fit by eye the best straight line to our data and then decide by "visual judgment" whether the fit is acceptable. If the fit is acceptable, we get numerical values for the fitting constants C and a by graphical measurements. (If we were using the computerized version of this method, the best line, and its equation, would be automatically decided by a regression procedure and we would use the usual tools to decide if the fit were satisfactory.) If the fit is not acceptable, we need to study its faults to get some clues as to the next type of relation to try.

---

[1] TEAM Graph Papers, Box 25, Tamworth, NH 03886, phone (603) 323-8843. Codex Book Co., 74 Broadway, Norwood, MA 02062, phone (617) 769-1050.

Figure 8.2 shows a numerical example of such a procedure where we have used computerized regression methods. We use our usual approach of simulating a real experiment so that we know the "correct" answer and can judge how well the method is working. The x, y data in the table were generated from the formula

$$y = 8.37x^{1.63} \tag{8.2}$$

We then submitted these "perfect" data to the regression software and asked for the best straight line relating $\ln y$ to $\ln x$. (One can use either natural or base-10 logs in such transformations.) The software returned a straight line with slope 1.630 and intercept of 2.125 (2.125 is the natural logarithm of 8.37), essentially perfect results, and the graph of Fig. 8.2b. When we made the problem more realistic by adding some gaussian random noise to the y values, the fitted line (see Fig. 8.2c) had slope 1.87 and intercept 1.58 (1.58 is the natural log of 4.85). These numbers deviate from the known correct relation, but Fig 8.2c shows a good visual fit to the noisy data. Figure 8.2d compares the perfect relation YG (this would, of course, be unknown in a practical application), the noisy data (YG + NOISE), and our fit to the noisy data (YGFIT), using the *actual x, y* values rather than the logarithms used in the fitting process. This kind of check should always be done when the regression uses *transformed,* rather than actual, data values in the least-squares fitting process. That is, a fitted curve that is best for the transformed values may not be best for the actual values, so we always need to look at the fit quality for the actual values before deciding whether the fit is acceptable. The fit in Fig. 8.2d looks about as good as that in Fig. 8.2c, so the use of logarithms has apparently not created any serious problems in this case.

Further examples, using different transformations to rectify different functions, could be given at this point. Since the computerized methods using regression analysis are generally preferable and were discussed in Chap. 4, we choose to only give some references[1] on the older methods. These can sometimes be of use in finding a transformation that converts the regression problem from a nonlinear one (with all the difficulties discussed in Sec. 4.5.6) to a more tractable linear type.

For data relating a dependent variable to two or more independent variables, graphical data analysis is rarely effective.[2] Exceptions to this

---

[1] H. Schenk, Jr., *Theories of Engineering Experimentation* 2d ed., McGraw-Hill, New York, 1968, chap. 9. W. J. Worley, "Curves to Straight Lines," *Machine Design,* November 10, 1960, pp. 173–180. C. Daniel and F. S. Wood, *Fitting Equations to Data,* Wiley-Interscience, New York, 1971, pp. 19–24.

[2] Daniel and Wood, ibid., pp. 50–53.

x	y	Noise
3	50.17	−17.03
4	80.18	−16.34
5	115.36	−8.35
6	155.28	7.83
7	199.63	−2.48
8	248.18	−18.94
9	300.71	−4.71
10	357.05	−15.39
11	417.06	−13.74
12	480.61	10.77

(a)

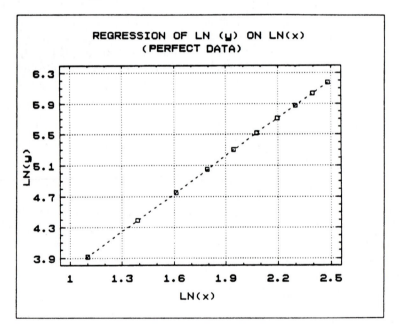

(b)

FIGURE 8.2
Linearizing transformations in data analysis.

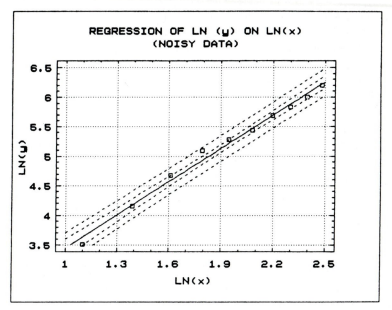

(c)

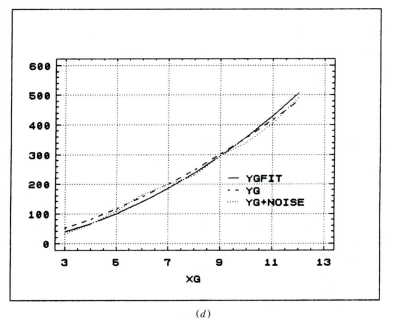

(d)

FIGURE 8.2
(Continued).

general rule are found when the experiment design includes runs where all variables except one are held constant while that one is varied over some range. Recall that such experiment plans are considered (and rightly) by statisticians to be *inefficient* compared with the factorial or fractional-factorial plans discussed in Chap. 4. They do, however, allow useful plotting of the dependent variable versus the one independent variable allowed to change. Such graphs can give clues as to the function shapes that might be tried in regression studies. The problem, of course, is that the shape that we observe in a single such plot may change radically when the other independent variables take on different values. Another exception occurs when the dependent variable is influenced *mainly* by one of the independent variables, the others having only slight effects. A sequence of graphs of the dependent variable versus each of the independent variables will, of course, reveal which independent variable is dominating and what the shape of this dependency looks like.

While most analysts, such as Daniel and Woods, agree that multivariable models are better explored by regression than by pure graphical approaches, others continue to search for ways to make graphing more useful in such cases. One such proposed approach is called the *scatterplot matrix*[1] and is in principle applicable to models with any number of variables. If the total number of variables (independent and dependent) is k, then one can make an ordinary x,y graph between any two of them, there being a total of $k(k-1)/2$ such graphs. The idea of making such graphs is, of course, not novel; the contribution of the scatterplot matrix concept lies in how the graphs are organized and presented. Cleveland[2] uses the actual air pollution data of Fig. 8.3 to explain this graphical data analysis method. Measurements of ozone concentration, wind speed, solar radiation, and air temperature were made on 111 days from May to September of 1973 at sites in the New York metropolitan area. In a scatterplot matrix display, the graphs are arranged in rows and columns such that each row or column has all the graphs relating a certain variable to all the others. In our example, the upper row has graphs of solar radiation plotted against each of the other three variables. The third column from the left plots solar radiation, ozone, and wind speed against temperature. All the other rows and columns are similarly interpreted.

Such a display will always show *twice* as many graphs as computed

---

[1] W. S. Cleveland, *The Elements of Graphing Data,* Wadsworth, Belmont, CA, 1985, pp. 210–215.

[2] Cleveland, op. cit.

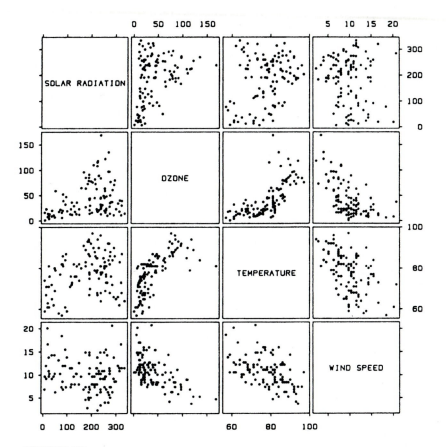

FIGURE 8.3
Scatterplot matrix for multivariable graphical data analysis. (*From* The Elements of Graphing Data, *by W. S. Cleveland. Copyright © 1985 by Bell Telephone Laboratories, Inc. Reprinted by permission of Brooks/Cole Publishing Company, Pacific Grove, CA 93950.*)

from $k(k-1)/2$, since each graph appears twice, with horizontal and vertical axes interchanged. This redundancy is not suppressed since it sometimes helps in the analysis. Combining some qualitative knowledge of pollution chemistry with information from the various graphs allows one to reach some useful conclusions about this pollution process (see the reference for details). The reference also discusses an extension of the scatterplot matrix concept called *brushing*. This technique takes advantage of interactive computer graphics to get even more information out of the graphs. Actual production of a scatterplot matrix, with a few simple commands, is provided in some commercial statistics software packages, including the STATGRAPHICS used in this text. In STATGRAPHICS, however, it is called "the draftsman's plot."

8.4 NUMERICAL DATA ANALYSIS

As mentioned earlier in this chapter, most data analysis methods employ a combination of statistical, graphical, and mathematical (numerical) tools. In this section we focus mainly on those numerical techniques which are not of a statistical nature. Since these calculations are all executed on a digital computer, and since software can be written to perform any calculation we might need, the scope of such methods is essentially unlimited. Also, each experiment will have its own unique set of calculations to suit the detailed nature of the data and the type of results desired. For these reasons, our treatment here is limited to giving an overview of the *general-purpose* numerical analysis routines that are commercially available with lab software packages.

As our illustrative example, we use the NB-DSP230X Series Analysis Library.[1] This is a very comprehensive and complete library of routines designed for Macintosh computers which have been equipped with a high-performance plug-in board for digital signal processing (DSP). These DSP boards are used to unload the host computer from the tasks which require the highest computing speeds, allowing "real-time" processing even for very fast data such as acoustics and vibration. The routines are 32-bit single-precision, floating-point analysis functions that can be called from an application program written for the particular experiment. One can use conventional text-based languages, such as C, to write the application program, or higher-level, more user-friendly languages such as LabVIEW 2. LabVIEW 2 uses graphic software modules called *virtual instruments* (VIs) that have front-panel user interfaces and block-diagram programs. By connecting icons for these modular VIs in a block diagram, you can easily create, modify, combine, and exchange VIs to form more sophisticated VIs. A compiler is used to generate machine code so that VIs execute on the Macintosh at speeds comparable to compiled C programs.

The routines available are classified into seven categories:

Fourier and spectral analysis
Digital filters and windows
Waveform analysis
Waveform generation
Statistical analysis

---

[1] *1992 Catalog,* pp. 3–157, National Instruments, 6504 Bridge Point Parkway, Austin, TX 78730.

Vector and matrix algebra
Numerical analysis

Fourier and spectral analysis includes

Real FFT and inverse
Complex FFT and inverse
Fast Hartley transform and inverse
Fast Hilbert transform and inverse
Power spectrum
Cross power spectrum

Digital filters and windows include

Adaptive filter: LMS
Nonlinear filter: Median
IIR filters (low-pass, high-pass, bandpass, bandstop)
    Butterworth
    Chebyshev I
    Chebyshev II
    Elliptic
FIR filters (low-pass, high-pass, bandpass, bandstop)
    Eqi-ripple
Parks-McClellan *FIR* filter design
Smoothing windows
    Hanning
    Hamming
    Bartlett
    Blackman, exact Blackman, Blackman-Harris
    Kaiser-Bessel
    Flat-top
    General cosine
    Force
    Exponential

Time-domain analysis includes

Convolution

Deconvolution

Autocorrelation

Cross-correlation

Integration

Differentiation

Peak detection

Pulse parameters

Decimation

Interpolation

Shift

Clip

Waveform generation includes

Impulse

Impulse train

Ramp

Triangle train

Sawtooth

Sinusoid

Sinc

Square

Random pattern

Uniform white noise

Gaussian white noise

Statistical analysis includes

Mean

Standard deviation

Variance

Root mean square

Moments about the mean

Median
Histogram
Mode
Linear, exponential, and polynomial curve fitting
Mean square error

Vector algebra and matrix algebra include

Dot product
Vector and matrix normalizations
Matrix multiplication
Matrix inversion
Determinant
Transpose
Trace
Linear system of equations

Numerical analysis array functions include

Complex add, subtract, multiply, and divide
Complex ln, log, and exponential
Scaling functions
Rectangular-to-polar transformations
Phase unwrapping
Magnitude-to-decibel conversion

Many of the above functions are simple and self-explanatory, but a few comments on some of the less obvious ones may be helpful. The various fast Fourier transform (FFT) routines are widely used to transform data from the time domain to the frequency domain (or vice versa). The direct Fourier transform accepts a time-varying function as input and computes the frequency spectrum of this signal. From a single such analysis one can get the power spectrum of a signal, which reveals the frequency content. If we transform two signals, one at the input of a (linear) system and one at the output, we can also get the cross-spectrum between the two signals and then the transfer function which defines the dynamic behavior of the system. If the system analyzed is a *measurement system,* this transfer function can be used to *correct* measured data for the dynamic imperfections of the measurement system. Once the measurement has been corrected in the frequency domain, we can use the inverse FFT to recover the (corrected) time-domain signal. When the tested system is not a measurement system but some "product," such as an automobile suspension system, the transfer function is used to judge whether dynamic behavior is acceptable or not. In research and development work, such information is useful in evaluating and improving designs. In production work, transfer functions of individual products are compared with benchmark values to make accept-or-reject decisions.

The various digital filter functions are useful for processing time-varying signals so as to keep or remove selected portions of the frequency content of the signals. Low-pass filters retain frequency content below a selected value (say, 20. Hz) and reject all frequencies above. High-pass filters remove the average value (dc) and low frequencies while passing high-frequency components unchanged. Bandpass filters reject all frequency components except those in a chosen narrow band of frequencies. Bandstop filters pass all frequency components except those in a narrow band. All these functions are also available in analog (hardware) versions, and sometimes this approach must be used (e.g., antialiasing low-pass filters are used *before* sampling). When we can use the digital (software) versions, we gain the advantages of accurate parameter setting, no drift, easy adjustability, and programmability. Smoothing windows of various types are usually used in conjunction with FFT operations to control some of the undesirable effects associated with sampling and discrete mathematics.

The time-domain operation of autocorrelation involves averaging the product of the time signal and a time-shifted version of this signal. When the time shift is zero, this gives the mean-squared value of the signal. As the time shift is incremented to larger values, if there is any periodicity in the signal, the autocorrelation function will show relative peaks at time shifts which are integer multiples of the period. This can be used to extract small periodic signals buried in random noise. In cross-correlation, we average the product of two signals, one of which is time-shifted relative to the other through a range of values. It has been used, for example, in flowmetering by injecting a signal (say, a thermal signal) into the flowing fluid and then detecting this signal downstream a known distance. When we cross-correlate the downstream signal with the injected signal for different time shifts, a peak will occur at a time shift equal to the known distance divided by the fluid velocity. Knowing this time shift and the distance, we can compute the unknown fluid velocity.

Numerical differentiation and integration are well-known operations so we will only remind the reader that integration (even successive integrals) is generally a benign operation while differentiation is generally difficult and sometimes impossible, due to its noise-accentuating behavior. Waveform generation is often used to stimulate a system into responding. By measuring stimulus and response, we can use FFT techniques to discover the system's dynamic behavior (transfer function).

Various waveforms should be available so that we can select the most appropriate one for each application.[1]

BIBLIOGRAPHY

1. J. M. Chambers, W. S. Cleveland, B. Kleiner, and P. A. Tukey, *Graphical Methods for Data Analysis,* Wadsworth, Belmont, CA, 1983.
2. C. Daniel and F. S. Wood, *Fitting Equations to Data,* Wiley-Interscience, New York, 1971.
3. J. W. Tukey, *Exploratory Data Analysis,* Addison-Wesley, Reading, MA, 1977.
4. G. E. P. Box, W. G. Hunter, and J. S. Hunter, *Statistics for Experimenters,* Wiley, New York, 1978.
5. C. Lipson and N. J. Sheth, *Statistical Design and Analysis of Engineering Experiments,* McGraw-Hill, New York, 1973.
6. M. S. Phadke, *Quality Engineering Using Robust Design,* Prentice-Hall, Englewood Cliffs, NJ, 1989.

---

[1] E. O. Doebelin, *System Modeling and Response,* Wiley, New York, 1980, chap. 6 (most recent printing available from Long's Bookstore, 1836 N. High St., Columbus, OH 43201).

CHAPTER
9

TECHNICAL COMMUNICATION

9.1 INTRODUCTION

The importance of technical communication skills for engineers cannot be overemphasized. If we cannot explain our work to others, it will detract greatly from its recognition and utility, no matter how technically brilliant it may be. Technical communication in written, oral, or graphic forms is, of course, necessary in both theoretical and experimental work. Inclusion of some material on this subject in a text on experimentation is largely a matter of historical convention. Most engineering curriculums have for many years concentrated the emphasis on report writing in various laboratory courses. Even though this is the accepted practice, some engineering educators (myself included) feel that the cultivation of clarity in writing and speaking should be exercised in *all* the courses, so that this vital skill is continually honed and perfected. This of course requires that some time be stolen from the purely technical material and that all instructors accept some responsibility for critical evaluation and development of communication skills as they grade and comment on student writing and speaking, no matter what the context, including mundane documents such as homework papers.

Because I wish to promote clear writing and speaking "across the curriculum" rather than just in laboratory courses, the presentation in this chapter for the most part will address the *general* problem of producing clear technical communications. While there are specialty B.S., M.Sc., and Ph.D. degrees in the subject of technical communication,[1] most technical writing is still done by the engineers and scientists who performed the technical work. In her article,[1] Prof. Steininger makes a case for turning over technical writing to those with degrees in that field, stating, "English/journalism graduates can write, but have no technical expertise. Engineers and technicians have technical expertise but can't write.... Don't ask an engineer to be an author or ask a journalist to critique a tractor design." I would agree that certain kinds of technical writing are best done by writing specialists, but would insist that many others are better carried out by professional engineers with proper training in writing.

Although engineering graduates *are* regularly criticized for poor communications skills when they first enter industry, engineering educators should not give up on this important task. More curricular time and faculty interest simply needs to be devoted to developing writing and speaking. Such an effort not only improves communications but also often enhances technical performance. That is, when we strive for clarity in explaining ideas and results to someone else, we usually clarify and deepen our own understanding as well. In my opinion, most students are quite capable of becoming good writers if faculty would provide only a modest increase in emphasis on this topic. While a certain division of labor is clearly needed in today's complex world, we also need to avoid overspecialization; doing engineering and explaining it should *not* really require two separate degrees!

Engineers are called upon to produce many different types of technical communications:

Letters	Memos	Proposals
Reports	Papers	Articles
Talks	Manuals	Films
Video tapes	Audio tapes	Computer presentations
Brochures	Books	Patent disclosures

This brief chapter concentrates on reports, papers, articles, and talks. References are given for topics not covered in detail.

---

[1] J. Steininger, "Hire the Right Person for the Write Job," *Hydraulics and Pneumatics,* July 1988, p. 84. A list of colleges which give degrees in technical communications can be obtained from the Society for Technical Communication, 815 15th St., NW, Washington, DC 20005, phone (202) 737-0035. The society publishes a quarterly journal called *Technical Communication.*

9.2 REPORTS, PAPERS, AND ARTICLES

Technical writing for engineers probably arises most often in the context of presenting to colleagues or superiors a report on some assigned project. In this section we also treat papers and articles since these kinds of documents are closely related to reports. The distinction between a paper and an article may not be obvious or universally defined, so we need to say a few words here. By a *technical paper* here we mean a document submitted to some technical society or industry group for printing in a journal and/or oral presentation at a conference. Such papers are usually reviewed by a group of referees who recommend acceptance or rejection. An *article,* usually considered to be at a lower technical level than a paper, generally appears in a trade journal rather than a society's transactions and is not subjected to peer review.

9.2.1 Audience Analysis

In planning *any* technical communication, we should always begin by considering the nature of the audience receiving our information. We have a natural tendency to write for ourselves rather than someone else, and we must overcome this. Students in school (or recently graduated) may have another bad habit in this regard—writing for the instructor. In school we know that our reports will be read by instructors skilled in the subject, and we tend to leave out material which we assume they will supply from their background. Because of problems such as these, we must *actively* force ourselves to analyze the audience and then tailor our presentation to *their* needs, not ours. This may not always be easy since the audience may be diverse or not familiar to us, but we must nevertheless do the best we can in appraising their needs and capabilities.[1]

9.2.2 Overall Organization: Use of Outlines

While we do not want to restrict our writing by a single rigid structure, most writers find it helpful to have in mind some kind of outline for the entire report as they begin writing. For certain classes of documents,

---

[1] G. H. Mills and J. A. Walter, *Technical Writing,* 5th ed., Holt, Rinehart and Winston, New York, 1986, pp. 19–30. T. E. Pearsall, *Audience Analysis for Technical Writing,* Glencoe Press, Westerville, OH, 1979.

such outlines can be relatively fixed and specific. Many technical journals require authors to follow a *style manual* provided by the journal. These style manuals may be quite specific about how papers for that journal are to be written. Larger companies may similarly require that reports follow a standard company format. If you are writing in such a situation, be sure to follow the guidelines provided by the journal or the company. If not, you may select a format that you feel is appropriate.

Outlines for reports might differ from those for papers. The major difference lies in the sequencing of topics according to their importance. In a paper, usually intended for reading by the author's peers, we often use an outline which starts with explanatory details and gradually builds up to important conclusions. This form of organization is well suited to an audience expected to read the entire paper. In a report intended for superiors, it may be better to present the main results early and leave the details for later sections. This format better suits the reader who may be managing several projects at once and needs to scan quickly for problem areas that need attention.

Next we present a sample outline which is well adapted to reports of experimental projects. With a few modifications, this will also serve nicely for most papers and reports, whether they are experimental, theoretical, or both. After presenting and discussing this outline, we also consider how to organize each paragraph and each sentence in the report to achieve maximum clarity. By concentrating on these organizing principles at all three levels (total report, paragraph, sentence), we establish a framework and method for achieving effective communication.

9.2.3   Report Outline for Experimental Projects

One of several possible good ways to organize a report on an experimental project is as follows:

1. Title
2. Table of contents
3. Lists of figures and tables
4. Definitions of symbols
5. Objectives
6. Summary of results
7. Equipment tested
8. Test method
9. Presentation and discussion of results
10. Conclusions and recommendations

11. References and Bibliography
12. Appendix
    A. Testing apparatus
    B. Sample calculations
    C. Raw data sheets

    The choice of a suitable *title* for the report should not be treated lightly. When other engineers are searching for information on a certain topic, because of the vastness of the technical literature in most fields, they often must decide, based only on the title, whether to pursue a particular reference. We thus try to come up with the most descriptive concise title that we can. Unless there are explicitly stated limits on the length of the title, we are more concerned with clarity than with conciseness. If the database in which our document will be catalogued allows both a title and an abstract, then we consider shortening the title, relying on the abstract to clearly guide the literature searcher. If possible, the title should clarify whether the work is theoretical, experimental, or computer simulation and whether it is original or a review of previous work. Put yourself in the place of the literature searcher, and try to give a title which would be most useful for this purpose.

    The *table of contents* should clearly locate all the important sections of the report by page number. These sections include the items listed above in our outline but often show additional detail, if necessary. That is, the test method, e.g., might have several sections of its own that we would want to appear in the table of contents. If sections and subsections in the report are numbered according to some scheme, it is best to keep such numbering simple. That is, avoid having subsection numbers such as 2.2.3.3.8. Such awkward and confusing proliferation of subsection numbers can usually be avoided by simply not numbering the minor subsections. The name of the subsection (perhaps indented, underlined, or capitalized to make it stand out) appears in the text and in the table of contents, but we do not assign it any number.

    Lists of figures and tables are provided to help the reader locate these features when they are referred to in a text location remote from the feature location. That is, if figure 12 appears on page 23 but is later referred to on page 134, the reader wants to be able to quickly locate it. *Definitions of symbols* serve a similar purpose. They include both a word description and the proper units. Depending on the audience, you may give more than one set of units, such as British and SI. As a detail of notation, I prefer to use the symbol $\triangleq$ rather than $=$ in making lists of symbols, since I like to keep clear the difference between equality ($=$), identity ($\equiv$), and definition ($\triangleq$).

    In the *objectives* section we state clearly and concisely the main purposes of the study. What is it that we are trying to accomplish? Since clarity requires some attention to detail and since conciseness implies brevity, these two attributes are opposed and we must always strike a proper compromise. One cannot give a blanket rule to use a certain number of sentences or paragraphs to state the objectives; each project must be considered individually. Because this outline is organized in order of *decreasing* importance and *increasing* detail, we keep the objective statement short, to facilitate rapid scanning by a busy manager, our boss. For students writing in an academic environment, most instructors want you to treat the experiment as a "real-world" professional task, *not* an educational exercise. Thus the objective should be stated not as an educational one ("to teach the student . . ."), but rather as a technical one.

The *summary of results* is a section where we give the *major* results corresponding to the stated objectives. Again we are trying here to give our boss a quick but accurate picture of how the project is doing, so we don't give *all* our results (there is a later section where we can do this); we give only those that we judge to be of prime interest for the manager's needs. This section is the last where we have sacrificed detail in order to facilitate quick scanning. Many times, the boss will read no further, so we must provide accurate summaries of all the information needed for the manager's decision-making process. From here on in this report outline, we get into greater and greater detail. Such detail is not usually of interest to our boss but must nevertheless be included for other important reasons.

The hardware associated with any experiment can be divided into two general groups: *equipment tested* and testing apparatus. If we are obtaining speed/torque curves for an internal-combustion engine, the engine is the *equipment tested* and the instruments used to make the needed measurements are the *testing apparatus.* Equipment tested is of major importance, so it appears fairly early in the main body of the report. Testing apparatus is usually of secondary interest and is relegated to the appendix. Under "equipment tested" we give a detailed description of this hardware, using appropriate sketches and diagrams to augment the text. In deciding which features of the equipment to discuss and which to leave out, we emphasize those features which pertain most closely to the tests actually run.

Having just described what was tested, we can now explain how the testing was done, in the *test method* section. If we used a standard test method or code that is described in complete detail in a standards document (such as one of the ASTM standards), we refer the reader to that document or provide a copy in our report. Otherwise, we give a complete and detailed explanation of our testing procedures. If alternative test methods were considered and rejected, we explain why the chosen method is superior. While the boss generally will not read details such as test methods, they *are* necessary in our report. Any unusual or unexpected results may be challenged, and we might need to search for explanations. Bad results can arise from many sources. These include bad data, calculation errors, faulty test methods, and inappropriate theory. Our report must document all these operations so that we can later check them for possible errors. A detailed description of the test method allows us or others to evaluate it to make sure it is suitable for our purposes. Also, if we have to *repeat* some of or all the experiment, such a description is again vital. Diagrams and sketches are used where appropriate to augment and clarify the text.

While the major results have already been briefly presented earlier, the section on *presentation and discussion of results* provides a more complete treatment of all the results and their significance. We first emphasize that this discussion centers on the *results,* not on general discussion of the subject area. (A common fault of student report writers is to use this section for discussion of extraneous background material obtained from textbooks or other references, rather than as a critical examination of *their own results.*) While the nature of the discussion in this section will clearly depend somewhat on the specific experiment, we point out some general features.

Results are usually a combination of verbal statements, numerical values, tables, and graphs. In some cases we may give the same result in both tabular form and graphical form. The graph is useful for quickly showing overall behavior and trends, whereas the table is better for giving accurate numerical values, if these are judged important. (Because graphs and tables are so important in their own right, we devote an entire section to discussing them shortly.) If there is a theoretical prediction for a result which was measured, we usually compare theory and experiment and discuss their agreement or disagreement. The accuracy analysis methods of Sec. 3.5 should be applied to *both* the theoretical calculation and the measured result, giving nominal values and standard deviations. These accuracy calculations will often use estimates of the uncertainties of the measured quantities rather than being based on actual data from repetitions of the experiment. When the experiment is not repeated one or more times, a decision as to whether theory and experiment agree or disagree is a judgment based on how close the two nominal values are and on the overlap of their respective uncertainty bands, say, $\pm 2\sigma$ or $\pm 3\sigma$. If the experiment is repeated, then the more quantitative methods of Sec. 4.3 might be used.

In addition to comparing theory and experiment, we compare and relate our results to those of other workers in the field, if such results are known to us. If there are significant discrepancies, we explain these in terms of differences in apparatus, test methods, operating conditions, or calculation schemes. We also evaluate and explain the practical importance of our results to company products or processes in terms of performance improvement, cost reductions, and productivity increases.

In the section on *conclusions and recommendations* we draw together in our mind all the results and explain what conclusions we have reached based on these results. These conclusions must be justified in terms of logical arguments, past experience, and personal judgments. It is important to make clear which statements are based on facts and which on opinion. Personal opinions and judgment calls are very often important parts of practical engineering work, and we should not denigrate them; however, in our writing we must be clear as to when we are arguing factually and when we have moved on to opinion.

When we conclude that some action should be taken (or not taken) as a result of the findings of our study, we make recommendations. These may be related to the objectives stated in the front of the report, or the recommendations may be actions that became clear only after the study was completed. In sponsored research projects, there is often a recommendation for further work, prolonging the life of the project and the jobs of the participants. Based on my own reading of many such reports, it would be refreshing to occasionally find a recommendation that work be terminated. Of course, it is true that *any* topic can be explored ad infinitum, and human curiosity is an admirable trait; but in a world of limited resources, the fact that something *can* be done does not always mean that it *should* be done.

With regard to *references and bibliography,* several formats are possible. I personally prefer the method used in this book, where a complete citation appears as a footnote on the same page as the text reference. Thus the reader does not have to flip continually to the back of the report to a reference section. I also like to make references *complete* (including the title), so that the reader at least has some idea as to what the reference is about. Some footnote schemes don't give the title of the reference at all, leaving the reader in the dark as to the reference's general content. Gathering all the references together in a single section at the rear of the report does have the advantage of giving a quick visual overview of the pertinent literature. This advantage is, in my opinion, outweighed by the convenience, in my recommended method, of immediately seeing the reference title and date at the bottom of the page during the reading of the report. One could of course use *both* methods if report space is not at a premium. Recall that a *bibliography* is used to list literature that is pertinent to the general subject area of the report. Sometimes it includes items that were also referenced. Both the references and the bibliography should give *complete* data on the entry, so that the reader will have no difficulty in identifying the source.

Appendices (some prefer *appendixes*) are useful for documenting information of lesser importance or for removing from the main text those items which, perhaps due to complexity, would needlessly disrupt the smooth progress of the reader. Detailed mathematical derivations whose end result is the main interest, may, for example, be deferred to an appendix. Items which are routinely found in the appendix are shown in our earlier report outline.

The *testing apparatus* appendix is simply a careful listing and complete identification of all the instruments used in studying the equipment tested (item 7 in the outline). Such identification is vital if we encounter "bad results" and need to do some detective work to uncover the sources of error. If our lab has 13 oscilloscopes, we must know exactly which one of these was used in our experiment if we need to check it for proper operation and calibration.

The *sample calculations* appendix serves a similar purpose in allowing the checking of our calculation methods for possible errors. The word *sample* implies that we must show *one typical* calculation of every different kind that was used in the study. If we calculated horsepower 10 times, we need only show *one* typical horsepower calculation, not all 10. Such calculations should include a descriptive subtitle, a formula with letter symbols, a complete set of numerical values, and a definition of each symbol and its units. For example,

Air-spring stiffness

$$K_s = \frac{A^2 p}{V} = \frac{(1.50)^2 (50.0)}{3.20} = 35.2 \frac{\text{lb}_f}{\text{in}} \tag{9.1}$$

where $K_s \triangleq$ air-spring stiffness, lb_f/in
  $A \triangleq$ piston area, in^2
  $p \triangleq$ air pressure, lb_f/in^2
  $V \triangleq$ chamber volume, in^3

If any of the items A, p, or V are not raw data but were themselves calculated from raw data, these calculations must also be shown. That is, *every step* in the calculations from raw data to final results must be completely documented so that it can be checked by someone who has *not* run the experiment or done the calculations.

In this report outline, the *raw data sheets* are the very last item in the report. If all has gone well, they probably will never be looked at again. If, however, we get unexpected results and need to search for error sources, it is vital that the raw data have been accurately preserved for careful scrutiny or perhaps recalculation. In experiments which are highly automated and computerized, it is important that the functional equivalents of the raw data sheets be similarly preserved, if possible. While we may strive for a paperless environment in this electronic age, actual paper hardcopy (in addition to magnetic disks or tapes) of critical data may still be a reassuring safety factor.

9.3 PRINCIPLES OF CLEAR EXPRESSION

Having just given a general framework for organizing a report on an experimental project, we now present some techniques for "filling in the details" that will maximize communication with our readers. These principles of clear expression actually apply to all kinds of written or oral technical communication, so our discussion will have wide applicability rather than being limited to reports on experiments. Many books and articles have been written on the subject of technical writing and speaking, so the principles and guidelines presented here are hardly original.

9.3.1 Write for the Intelligent but Uninformed Reader

We have already mentioned the need for audience analysis in all kinds of technical communication. The writer of a report or paper is usually so familiar with the material that it takes a conscious effort to question each sentence and paragraph to make sure that sufficient explanatory detail is included. Unless we are sure that the audience has a higher level of familiarity, it may be wise to assume a level of preparation equal to that of a senior engineering student in the subject area. Clearly, adding explanation to a document will increase its length, so we must make our best judgment on this tradeoff. However, "long writing makes short reading" is a catch phrase with some relevance here. If, in an attempt to save *space,* we cause the puzzled reader to waste *time,* are we really being efficient?

9.3.2 Use the First Person and the Active Voice

In scientific writing, it used to be *required* that the author use the third person and passive voice: *The equipment was calibrated,* rather than *We calibrated the equipment.* The idea was to make the writing sound "scientific," i.e., objective and detached from any personal influences. Today, many style manuals for high-level technical journals encourage a less formal and more conversational tone by using the active voice. Sentences written in this way often read more easily and use fewer words to express ideas with equal or greater clarity. Unless there are specific requirements to the contrary, we recommend the use of the active voice and first person as the usual form. This does not mean that passive voice is *never* to be used. When it seems to fit better, use it.

Some writers recommend, "Write the way you talk, then polish."[1] Another text[2] implements this concept by recommending that our normal mode of writing be based on first *dictating* documents into a tape recorder and using professional typists to produce paper copies for editing. Many advantages are claimed for this method, but it might not work well for material which includes many equations. Engineers who regularly use a personal computer for technical and word processing tasks often find advantages in doing their own typing, even though their typing speed is modest. There are real advantages to having complete personal control over the entire process and immediate access to your document.

9.3.3 Use a Three-Step Sequence for Important Explanations

When you are going to explain an important and perhaps complex idea, consider using the following three-step plan:

1. Tell the audience what you are going to tell them.

2. Tell them.

3. Tell them what you told them.

Step 1 prepares the reader by warning that some important information is about to be presented. Step 2 actually presents the new information, and step 3 recapitulates the concept, perhaps in different words or from a different viewpoint. While this somewhat complex structure is not

---

[1] Matt Young, *The Technical Writer's Handbook,* University Science Books, Mill Valley, CA, 1989, p. 4.

[2] M. J. Murray and H. Hay-Roe, *Engineering Writing,* 2d ed., PennWell Books, Tulsa, OK, 1986, pp. 113–121.

warranted for simple concepts, it has been found to greatly enhance the understanding of complicated ones.

9.3.4 Organize Sentences according to Reader Expectations

Just as experienced report readers expect an entire report to show some kind of structure or organization such as our sample outline, there is also a more subtle and subconscious effect of the structure of individual sentences. This effect has recently been put on a scientific basis[1] and explains some well-known writing rules. Information is interpreted more easily when it appears where most readers *expect* it to occur. We can use this fact to improve our sentence and paragraph structure.

Readers expect the subject of a sentence to be followed fairly closely by the verb. When this does not happen, clarity suffers because, without the verb, we don't know what the subject is doing and thus what the sentence is all about. We should thus avoid separating the subject and verb too much and watch for this problem when we proofread rough drafts. Another rule deals with the material that we place in the *topic position* (beginning of the sentence) and the *stress position* (end of the sentence). Readers naturally expect that material at the beginning of sentence will provide linkage (looking backward) into the previous sentence and context (looking forward) into the new sentence. Thus the topic position should be used for such transitional material which provides a smooth flow between sentences. The topic position is also where we place the person or thing whose "story" is being told by the sentence. Readers expect a sentence to build to an emphasis at the end. Thus we should strive to place the most important material in that location, the stress position.

Another way of looking at this is in terms of "old" information and "new" information. Familiar, previously introduced material is old information while material making its first appearance is new information. Writers have a natural, but unfortunate tendency to place new information in the topic position (early in the sentence), because they want to "capture the thought" before it escapes. Readers, on the other hand, expect the new information to be in the stress (late) position. Once we are aware of these contradictions, we can look for and correct this kind of problem during proofreading.

Suitable punctuation can create secondary stress positions, allowing

---

[1] G. D. Gopen and J. A. Swan, "The Science of Scientific Writing," *American Scientist,* November–December 1990, pp. 550–558.

several pieces of stress-worthy information to be clearly presented in a single sentence. This is most commonly done with a semicolon, which creates essentially two "subsentences," each with its own stress position. Such constructions can lead to rather long sentences which, nevertheless, are clear and easy to read. Some writing manuals insist that sentences never exceed a certain number of words (usually about 30). While lengthy sentences should not be the norm, length alone is not the problem. A 15-word sentence can be quite unintelligible while a properly designed 40-word sentence can be perfectly clear.

9.3.5 Structure Paragraphs Properly

Most writers prefer to begin a paragraph with a topic sentence. This sentence presents the idea to be developed in the rest of the paragraph. It is usually best to have only one main idea in each paragraph; however, simple ideas may require only one or a few sentences, leading to overly short paragraphs. We should then try to treat several ideas in one paragraph by showing some relation among them. Complicated ideas can lead to overly long paragraphs. Long, unbroken sections of text repel the reader's eye and overload the memory, so we need to *force* paragraphs in such cases, even though we are still developing the same main idea. This can usually be done by using suitable transitional words at the paragraph break. For single-spaced text with the usual margins, try to have one or more paragraph breaks per page.

9.3.6 Choose Words Carefully

In addition to providing the optimum *structure* for the entire report, the paragraphs, and the sentences, it is vital that the words we choose be the best possible. A good vocabulary comes mainly from wide general reading, but other tools are available for those who need to improve this background in limited time. A good word processor will have a rather extensive thesaurus, and regular use of this aid will gradually improve your word choice. Some writing manuals[1] have alphabetical lists of misused words. *The Technical Writer's Handbook* by Matt Young is especially useful for technical writers, and I would recommend it even to those with some experience and expertise.

We need to avoid use of vague words, unfamiliar words, needlessly

---

[1] W. Strunk and E. B. White, *The Elements of Style*, 3d ed., Macmillan, New York, 1979. Young, op. cit.

complicated words, and overworked words and phrases (cliches).[1] Technical jargon is acceptable only if we are sure that the audience is familiar with it. When we proofread, we should always be looking for unnecessary words and delete them. We now give a few examples related to the several categories of word choice just listed, to make these recommendations more concrete. For a much more complete treatment of these areas, you should consult a writing text such as Matt Young's.

Vague word or phrase	*Specific word or phrase*
periodically	hourly, daily, weekly
fastened	bolted, welded, spliced
electricity	voltage, current, power
machined	turned, milled, ground

The words listed as vague are of course perfectly acceptable when properly used. They become vague only when we use them in place of a more correct specific word that is available.

Unfamiliar or complex word	*Simpler word*
components	parts
conflagration	fire
initiate	begin
compensation	pay

9.3.7 Removal of Unnecessary Words

Wordy: During the time that the pieces are connected together, their temperature is in the neighborhood of 450 degrees.

Better: While the pieces are joined, their temperature is near 450 degrees.

Wordy: Through the use of numerical integration, we are able to effect a solution of the equation.

Better: We solved the equation by using numerical integration.

When writing the first draft, we often are somewhat wordy since we

---

[1] Mills and Walter, *Technical Writing*, pp. 37–44.

try to quickly record the ideas "as they occur to us" and before we forget them. By careful proofreading, with a checklist of bad habits in mind, we can polish the text to make it clearer and more concise.

9.3.8 Selected Examples of Word Misuse

While we have made reference to some excellent writing texts for more details, a few errors are so common that we want to deal with them even in this short discussion.

absolute words	*Perfect, essential, unique,* and *certain* are examples of absolute words, which do not normally take modifiers. That is, we don't say "very perfect," "most essential," "less unique," or "absolutely certain."
adverse, averse	To be averse to something means to be opposed to it. Adverse consequences are unfavorable consequences.
affect, effect	To affect means to have an effect on. *Affect* is usually a verb; *effect,* a noun.
complementary, complimentary	Complementary means completing or making up what is lacking. Complimentary remarks are praise.
criteria, criterion	*Criteria* is plural; *criterion,* singular.
critique, criticism	A critique is a careful analysis. Criticism is the act of finding fault.
data	Some insist *data* is the plural form while *datum* is the singular form. Many writers (myself included) use data as either singular or plural.
discreet, discrete	Discreet means prudent. Discrete means made up of distinct parts.
ensure, insure	Ensure means to make certain. Insure means to "take out a policy."
fewer, less	Use *fewer* for countable objects, *less* for an amorphous *quantity.* (Fewer beads, less sand.)
he/she	To avoid sexist language, the use of *he/she* has become common. I find this awkward and prefer to dodge the problem by using plurals, which allows use of the nonsexist form *they.*

irregardless	This is *not* an acceptable written form. Use *regardless.*
lend, loan	*Lend* is a verb; *loan* is a noun.
man-made	Use *synthetic* or *artificial* to avoid sexism.
orientate	Orientate means to turn to the east. Orient means to arrange in position.
phenomena, phenomenon	*Phenomena* is the plural form.
principal, principle	A principal is a person of presiding rank or a sum of money on which interest is paid. A principle is a basic law or truth.

9.4 USE OF GRAPHS, FIGURES, AND TABLES

Most engineering reports, papers, or articles include graphs because graphs can be a great aid to understanding. Reference material relating to graphs can be classified into two main types. One type[1] concentrates on the principles guiding the direct visual communication between the graph and the reader. The other[2] deals with methods used to *analyze* the meaning that may be hidden to the casual eye. In Sec. 8.3 we have given a brief treatment of this data analysis aspect, so this section focuses on the mechanics of designing the layout of graphs for clarity of data presentation.

While graphs can be used for many detailed purposes, there are two major classes that I want to mention here. These are called *presentation graphs* and *working graphs*. Working graphs are not just looked at; they are used to pick off numerical values for some purpose. Because these days it is more convenient and accurate to use a computerized table, formula, or curve fit for such needs, working graphs are much less important than they used to be. The major differences between a presentation graph (used to present or display data in a written or oral report or in a paper) and a working graph lie in the graph size and the

---

[1] W. S. Cleveland, *The Elements of Graphing Data,* Wadsworth, Belmont, CA, 1985. E. R. Tufte, *The Visual Display of Quantitative Information,* 1984, and *Envisioning Information,* 1991, Graphics Press, Cheshire, CT.

[2] J. M. Chambers, W. S. Cleveland, B. Kleiner, and P. A. Tukey, *Graphical Methods for Data Analysis,* Wadsworth, Belmont, CA, 1983.

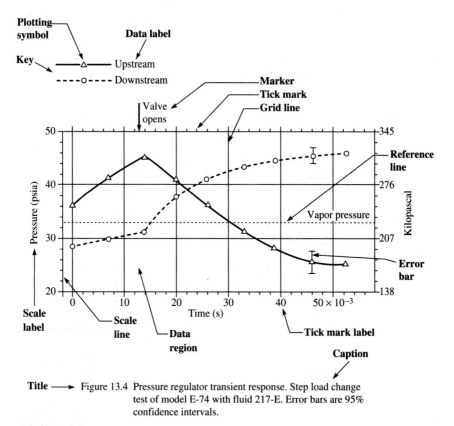

Figure 13.4 Pressure regulator transient response. Step load change test of model E-74 with fluid 217-E. Error bars are 95% confidence intervals.

FIGURE 9.1
Basic elements of a graph; definitions.

spacing of the grid lines. For accuracy in picking off values, a working graph needs to be quite large and have closely spaced grid lines. Since our main interest is in presentation graphs, the remainder of our discussion will emphasize this type.

We start with some definitions of the various basic elements which make up a graph, using Fig. 9.1 to augment the text discussion. Graphing terminology has not been standardized, so be prepared for some variations when you read about graphs or use commercial graphing software. *Scale lines* (also called *axes*) form the horizontal and vertical borders of a rectangle called the *data region.* While one vertical scale line at the left and one horizontal scale line at the bottom of the data region is a layout that is sometimes used, we prefer (and most commercial graphing software has as default) a scheme which uses two horizontal and two vertical scale lines to completely enclose the data region in a box.

Along the scale lines we place *tick marks* (also called *ticks*), which locate numerical values along the scales. Try to keep the data region *uncluttered.* This important general principle of graphing leads us to recommend that tick marks be *outside* the data region rather than inside. Commercial graphics software may not offer a choice on this, and in many cases inside tick marks will not be a serious defect, but if there is a choice, outside tick marks are preferred because they don't interfere with data points that lie close to the scale lines. Tick marks should not be too closely spaced, and not all need to be labeled with numbers, again to avoid graphical clutter (called *chartjunk* by Tufte[1]).

Grid lines extend across the data region from the tick marks and may or may not be used. Most presentation graphs have sparse tick marks and *no* grid lines. We want our *data* to stand out, so we avoid cluttering the data region with any unnecessary features. If a few grid lines are felt to be helpful for some specific reason, they should be light (relative to the bolder data curve) and perhaps dashed rather than solid. Sometimes one or more light *reference lines,* placed in strategic locations (not necessarily at a tick mark) and extending across the data space, can serve our needs better than a complete set of grid lines. When you attach numerical values (called *tick mark labels*) to selected tick marks and/or grid lines and a power-of-10 notation is desired, be careful how you set up the *scale label* (sometimes called the *axis legend*) which names the variable plotted along a scale line. If the scale label, for example, is given as "Time, seconds, $\times 10^{-3}$, more than one interpretation is possible. That is, a tick mark numbered 5 *could* mean either 5×10^{-3} sec or 5000. sec. Avoid such ambiguity by either labeling the scale as "Time, milliseconds" or numbering the rightmost tick mark with the 10^{-3} symbol.

A *key* (sometimes called the *legend*) is usually placed outside the data region and identifies some feature of the data within the data region. Most often, it shows the symbols used for plotting each of several curves in the data region and includes descriptive names, called *data labels,* for each curve. *Plotting symbols* are used to locate the individual data points on a curve. We need a selection of plotting symbols when we show several curves in one panel. A *panel* is a single data region defined by one set of scale lines. Sometimes our graphical display will have several related panels, all shown together on one page. When we display several curves in one panel, the curves are *superimposed.* When it is clearer to separate the curves into several panels located next to each other, the curves are *juxtaposed.* When curves are juxtaposed, the data labels are

---

[1] E. R. Tufte, op. cit.

usually placed inside the data region in a location where they do not interfere with the curve or other features, perhaps near a corner.

The choice of appropriate plotting symbols involves several considerations. Symbol form and size must be such as to make the points clearly visible. For hand-plotted student graphs that I regularly see, the most common error is the use of *tiny* solid dots. These become nearly invisible when a bold curve is drawn through them. When several curves appear in one panel, we distinguish between them by using different plotting symbols and/or different types of lines (solid, dashed, dot/dash, etc.). The symbols need to be sufficiently different to be easily distinguished. Another possible problem is symbol overlap when several points lie close to each other. Some shapes of symbols tolerate overlap better than others, allowing us to discern individual points when they are closely grouped. When superimposed curves in a single panel are too similar and/or too numerous to clearly distinguish, we may need to use juxtaposed curves in adjacent panels.

Markers are brief notes, usually outside the data region, with arrows pointing to specific locations on a scale line that we wish to highlight. The graph *title* is usually placed below the graph and includes both a figure number and a word title. A *caption* (sometimes called a *legend*) which adds some explanatory material to the title may also be found in this location. Often, the terms *caption* and *title* are used interchangeably and no separate statements are distinguished. If both a brief title and a more detailed caption are used, the title might be in all capital letters with the caption in lowercase except for the usual first-letter-of-sentence capitalization.

In choosing numerical scales for our scale lines, it is *not* necessary that the scale include zero, as recommended in some books.[1] The referenced book was aimed at the general public, rather than professional engineers, and it tried to alert this audience to graphical "tricks" that might be used to mislead them. For an audience of engineers, we usually choose our numerical scales so that the data curve nearly fills the data region, giving the best visual resolution in the total space available, regardless of where zero is located. Since we have recommended use of *two* scale lines for each axis, this allows us to provide, when appropriate, useful features such as one scale with British units (say, pounds per square inch) and another with SI units (pascals). Use of nonlinear scales, such as logarithmic, gaussian probability, and reciprocal, has been discussed elsewhere in this text.

---

[1] D. Huff, *How to Lie with Statistics*, Norton, New York, 1954, pp. 64 65.

When data have been subjected to statistical analysis which provides both expected values and measures of uncertainty, we may want to include in the graphs some indication of the uncertainty in the plotted points. *Error bars* are the most common means of doing this, and methods for superimposing them on basic data plots are provided in most commercial software. When the error-bar option is invoked in your commercial software, be sure to include in your report text or graph caption the precise *meaning* of the displayed error bars, so that your readers will not need to guess. Common definitions are ±50 percent or ±95 percent confidence intervals (sometimes we may show both). We also need to make clear whether we are showing *confidence intervals* or *prediction intervals* (see text discussion of Fig. 4.19*a*).

While *line plots,* such as our Fig. 9.1, are most common in engineering reports, there are of course many other forms of graphical displays that we may find useful. Figure 9.2*a* shows a *scatterplot* based on a data set provided in the statistics software STATGRAPHICS.[1] Scatterplots show only points, not any connecting lines. This data set has information on 155 automobiles regarding their fuel consumption, country of manufacture, number of engine cylinders, weight, etc. In Fig. 9.2*a* we show how fuel consumption depends on vehicle weight for all 155 cars. While there is the expected statistical scatter, the general decrease in miles per gallon with increasing weight is clear. This graph also illustrates the problem of plotting symbol overlap, mentioned earlier. There are two or three areas in the graph where we cannot be sure how many points are actually plotted; however, the problem is not severe, and the graph is certainly useful as it stands. When we have so many points (155), choice of a different symbol probably will not give much improvement; using a larger graph is about the only way to avoid the overlap.

The same array of points can be *coded* in various ways to communicate more information than can be perceived in Fig. 9.2*a*. In Fig. 9.2*b* we have used code numbers 1, 2, 3 to show the country of manufacture. Now the overlap problem is severe, perhaps unacceptable. If we insist on using this type of display, we may have to simply use a larger graph to improve clarity. Figure 9.3*a* uses coding to display another attribute—number of engine cylinders. Again, overlap is a problem. In addition to using coded letters, numbers, or symbol shapes to communicate information, we can exploit *color.* While color monitors, printers, and copiers are all available, the routine use of color in engineering

---

[1] STSC, Inc., 2115 E. Jefferson St., Rockville, MD 20852.

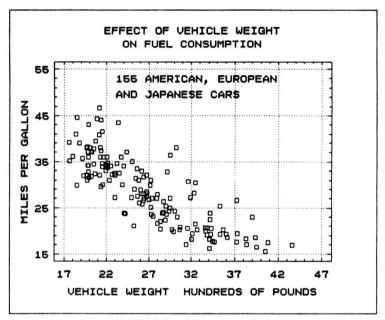

(*a*)

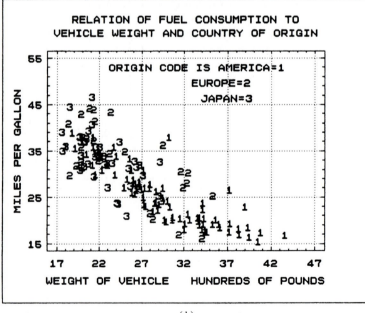

(*b*)

FIGURE 9.2
(*a*) Basic scatterplot and (*b*) scatterplot with coded second variable.

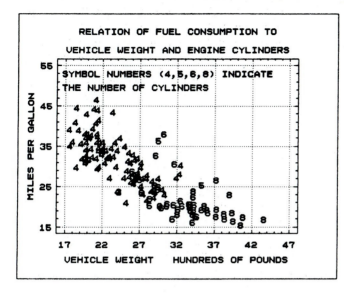

(*a*)

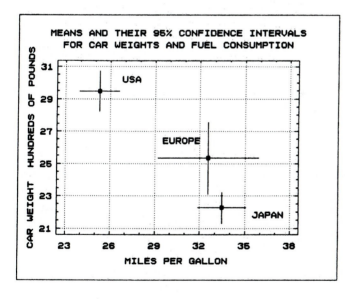

(*b*)

FIGURE 9.3
(*a*) Coded scatterplot and (*b*) data analysis to reduce clutter.

reports or journal articles is still somewhat expensive and thus limited. Certain applications, of course, benefit greatly from color, and then we make every effort to use it well. Examples include visualization of fields such as temperature distributions in heated objects, stress and deflection in loaded machine parts, and complex fluid flows. As a simple example of color use, in Fig. 9.2*b* we could communicate both the country of origin and the number of cylinders by printing the 1s, 2s, and 3s in four different colors, coded to represent the four different cylinder numbers. Now, however, overlap would likely be quite serious, requiring use of a larger graph area. Also, we need always to avoid graphical clutter and information overload. If we try to pack *too much* information into a single graph, the viewer may just get confused.

By doing some simple data analysis on the raw data, other types of graphs become available. In Fig. 9.3*b* we have computed separately the mean values and their confidence intervals for the car weights and miles per gallon. These values can then be plotted as shown to summarize the data for the three countries of origin. Three-dimensional graphical displays have become more popular since many software packages make the production of impressive-looking graphics quite quick and easy. Here we need to be a little critical because the beauty of the display may overshadow its real utility in communicating information. When tempted to use a dazzling three-dimensional display, always stop and make sure that there is no simpler form that does a better job. Figure 9.4*a* shows in one graph the effect of car horsepower and weight on fuel consumption for only U.S.-made cars. While there is some overlap problem, this graph is actually quite useful. The three-dimensional histograms shown in Fig. 9.4*b* (and similar bar graphs), especially when done in color, have become quite common in many magazines because they are visually impressive. Unfortunately, they usually do not present the information in the most useful way, and they should be replaced by displays which communicate the data more clearly. The multiple, two-dimensional graphs of Fig. 9.5 use more space, but provide clearer and more detailed information. Of course, certain types of information are best displayed in three dimensions, and we should then use this mode of display as effectively as possible. The response-surface graphs of Figs. 4.54 and 4.55 are good examples.

Bar graphs are a well-known and useful form that has several versions. In addition to the simplest version we have the *clustered-bar* and *stacked-bar* types. Figure 9.6*a* shows a clustered bar graph used to display expenditures on three research projects over the four quarters of a year. If more accurate dollar values are desired, we can add labels with the exact dollar amounts to each bar. The same basic data can also be displayed in the stacked-bar version of Fig. 9.6*b*. Here the *total* expenditure is easier to see, but the individual project contributions are

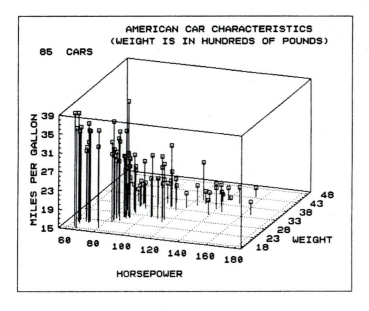

(*a*)

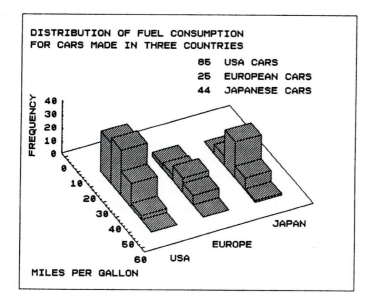

(*b*)

FIGURE 9.4
(*a*) Use and (*b*) misuse of three-dimensional displays.

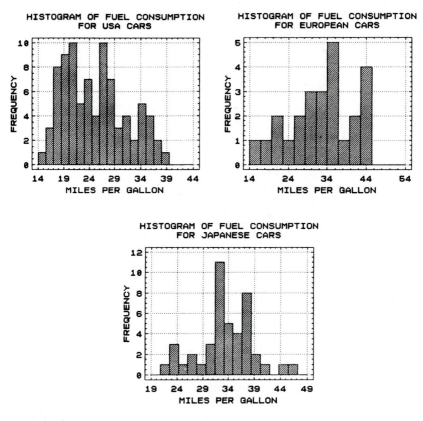

FIGURE 9.5
Figure 9.4*b* data better displayed in two dimensions.

less clearly compared. Note the addition of some grid lines to aid in picking off numerical values. *Pie charts* are in common use for displaying the contributions of several parts to the whole, even though some graphics experts say that there are always better ways to communicate such data. One defect of pie charts is the poor angle-measuring capabilities of the human eye and mind; we have trouble distinguishing between angles unless they are very different. This can usually be easily overcome by simply labeling the pieces of pie with their percentage contributions, as in Fig. 9.7. Note that the two "pies" are different sizes. The software made the areas proportional to the total expenditures for each quarter. Just as for angles, the eye is not good at judging the numerical ratio of two areas, even for simple figures like circles. If it is important for the viewer to access such numerical data, use some labels on the pie charts or else go to an alternative type of graph.

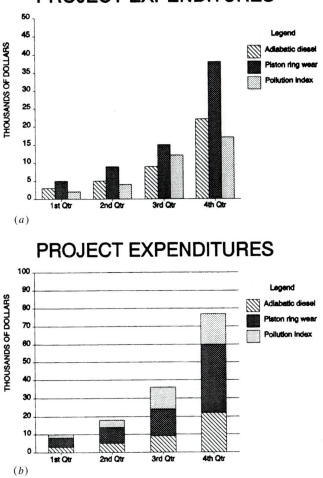

(a)

(b)

FIGURE 9.6
(a) Clustered and (b) stacked bar graphs.

Turning now from graphs to figures, we consider the term *figure* to include line drawings, computer solid modeling, photographs, and text charts. We emphasize line drawings because they are by far the most common in reports and papers. Line drawings can be classified into pictorial drawings, schematic diagrams, and block diagrams. Pictorial drawings can be two-dimensional or three-dimensional and are a simplified "picture" of some real object or objects, often a piece of our

PROJECT EXPENDITURES

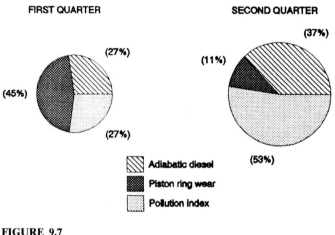

FIGURE 9.7
Pie charts.

apparatus or the entire apparatus. Schematic diagrams are more abstract than pictorial drawings, showing symbolic elements and their interconnection to make clear the configuration and/or operation of a system. Electrical schematics, showing the familiar symbols for R, C, L, op amps, etc., are the most common example, but mechanical and fluid system schematic diagrams are also widely used. Block diagrams are the least pictorial of all, showing only blocks, representing components or subsystems, and lines, representing the flow of signals or energy between the blocks. Both blocks and lines are suitably labeled to communicate the system functions. Figure 9.8 shows all three kinds of line drawings, using a simple example from the area of vibrating systems. The pictorial drawing presents a simplified view of a machine mounted on a beam. By making suitable assumptions, this system can be represented schematically as a mechanical model, using the standard symbols for masses, springs, and dampers. Analysis of this model by using Newton's law and differential equations allows us to get a relation (transfer function) between the input force F_i causing the vibration and X_o, the vibratory displacement of the mass. We can then draw a block diagram which shows this relation explicitly.

When you make pictorial drawings (either sketches for a data sheet or "mechanical" drawings for the report), give enough thought to this process that your illustration shows all that is needed but not more. The

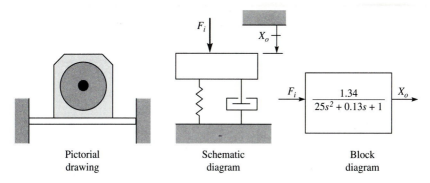

Pictorial drawing   Schematic diagram   Block diagram

FIGURE 9.8
Examples of pictorial drawing, schematic diagram, and block diagram.

advantage of a drawing over a photograph is that we have complete control over which parts of the photographic view we choose to retain in the picture and which we leave out. This usually requires that we take time to ask ourselves, What is the real purpose of including this drawing? Figure 9.9 shows a pictorial drawing of a computer-controlled dynamometer for speed/torque testing of rotary air motors in an instrumentation lab at The Ohio State University. The drawing's purpose is to show all the important components of this computer-aided machine and how these components relate to one another in the complete system. To do this without making the drawing too cluttered requires that we make some decisions about what details to include and how to represent them.

If you were standing in front of the actual apparatus, with your eye-brain system recording a "photographic view," you could then decide how to make your drawing. Each engineer faced with this task would probably come up with a unique interpretation, different from that of other engineers. Figure 9.9 shows the drawing that I personally made. At the left I showed the rotary air motor in its mounting bracket, but both the motor and the bracket shape are shown much simplified, because their actual detail forms are not useful in any analysis that will be done. The air supply for the motor is only indicated by the label "0 to 60 psig air." In the actual apparatus there is also a filter/lubricator/regulator, but I chose not to even show this component because it is of secondary interest and would begin to clutter a drawing that is already rather "busy." These air supply components *do* need to be mentioned and explained somewhere in the report, but are not vital to the purpose of the present drawing. Also not shown are shaft-support bearings and a misalignment coupling between the motor and the pneumatically actuated friction brake used to load the motor. (We will see shortly that some of the bearing details *are* vital for later analysis, but are better shown in a separate drawing which emphasizes these features.)

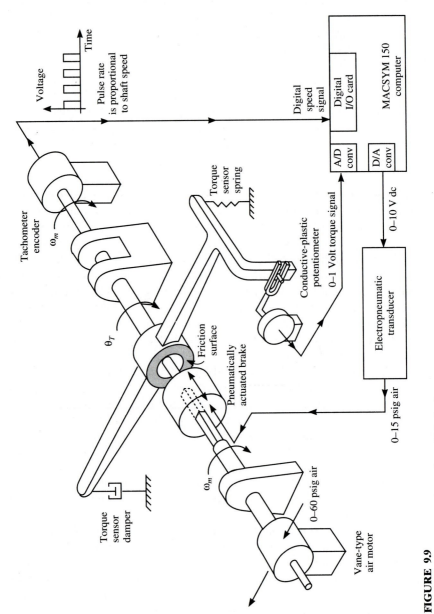

FIGURE 9.9
Pictorial drawing with schematic and block elements.

The friction brake itself is quite complicated, as we see in the cross-sectional drawing (Fig. 9.10) obtained from the manufacturer, which would be included somewhere in our report. In Fig. 9.9 we show the brake much simplified, using a square, sliding shaft to represent how varying pneumatic pressure can be applied to control brake friction torque while the entire assembly rotates. The "stationary" part of the brake housing is restrained from rotation by a homemade, spring-type torque sensor. The actual design uses a cantilever beam for the spring, but our drawing represents this with the simpler "standard" spring symbol. This spring transduces the brake friction torque (the motor's load torque which we wish to measure) into a related rotary displacement θ_T (which we label) and which is in turn measured with a potentiometer. An adjustable air-piston type of damper is used to improve torque sensor dynamics and is again shown in simplified schematic-symbol form.

Note that we have combined pictorial and block-diagram representations when we include the electropneumatic transducer and computer in Fig. 9.9. The internal workings of the electropneumatic transducer are somewhat complex but are of no direct interest in this project, so we don't show them pictorially; however, we do label the input and output signals. The D/A, A/D, and digital I/O interface cards need to be "called out" in our picture, but block-diagram representations are again sufficient. Speed is measured with a tachometer encoder, which we represent in very simple graphical form. We do, however, remind the reader of its operating principle by showing a graph of its output waveform with a short explanatory note. Since it is not graphically obvious that all the rotating shafts shown run at the same speed, we label both the motor shaft and the tachometer shaft with the same symbol ω_m. Both the tachometer encoder and the potentiometer require electric power supplies, but showing these or even noting their presence would clutter the picture without adding any pertinent information. Figure 9.9 and the associated discussion are intended as an example of good practice in constructing a simplified but informative picture of an apparatus. While two-dimensional views are often sufficient and easier to draw, here a three-dimensional view makes possible a clearer communication of the desired information.

In addition to overall views such as Fig. 9.9, we may need drawings of smaller details when this information is required for some explanatory or analysis purpose. To do an accuracy analysis of the torque-sensing method, some internal details of the friction brake and its mounting bearings are needed. In particular, the location of several bearings is critical to an evaluation of bearing friction effects on torque measurement. Figure 9.10 has some of this information in it, but, as with most conventional assembly drawings, it has too much detail to be very useful as an analysis diagram. We thus provide Fig. 9.11 to show only the

PARTS LIST

ITEM NO.	DESCRIPTION	NO. REQD
1	Set Screw	3
2	Hub Assembly with Disc	1
●3	Facing Screws	6
●4	Friction Facing	1
5	Piston	1
●*6	Shoulder Bolts	6
●*7	Springs	6
8	Bearing Housing	1
●9	"O" Ring	1
●*10	"O" Ring	6
●11	"O" Ring	1
12	Internal Retaining Ring	1
13	External Retaining Ring	1
●14	Bearing	1
**15	Bushings	1
**16	Key	1

*Only Three (3) are required in FWB
**BE SURE TO STATE MODEL NUMBER AND BORE SIZE DESIRED.
●Repair Kit Items.

REPAIR KITS

FWB	Prod. No. 8473
LWB	Prod. No. 8474
MWB	Prod. No. 8475
HWB	Prod. No. 8476

FOR REPLACEMENT PARTS CONTACT YOUR LOCAL HORTON DISTRIBUTOR.

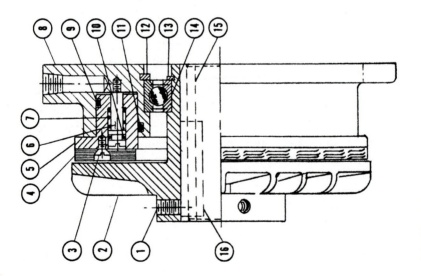

FIGURE 9.10
Assembly diagram of pneumatic friction brake.

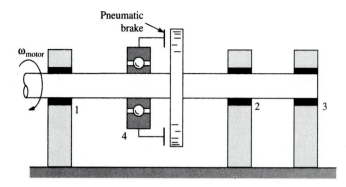

FIGURE 9.11
Analysis diagram of brake for bearing friction error study.

features essential to a study of bearing friction. The brake in this figure is shown completely disengaged (friction surfaces not touching), as it would be when the torque sensor is calibrated by applying known deadweights at a known lever arm on the spring-restrained brake arm. Students are asked to compare two alternative methods of doing this calibration. In the first method the motor is not rotating, and in the second it *is* rotating at a speed equal to the average of the highest and lowest speeds used during a speed/torque test.

If the motor is not rotating during calibration, as we apply weights to the brake arm (which is rigidly attached to the ball bearing outer race), shaft-support bearings 1, 2, and 3 hold the shaft fixed, since their combined static friction happens to be greater than that of the ball bearing. Before the brake arm begins to deflect the sensor spring, the applied calibration weights must overcome the ball bearing static friction torque. This causes a calibration error because an unknown portion of the applied deadweight torque is taken up by the friction and thus is not applied to the spring. (How would the situation change if the ball bearing had greater friction than the shaft-support bearings?)

In the alternative calibration scheme (motor rotating), the friction in the three shaft-support bearings has already been broken loose before any calibration weights have been applied. Now, however, the ball bearing exerts a running-friction torque on the brake arm, and thus on the spring, causing a small deflection with no weights yet applied. This deflection is *not* a calibration error since we take this condition as the zero-weight, zero-deflection starting point of our calibration curve. Also, the calibration condition (motor running) is closer to the actual measuring condition during speed/torque testing, and the slight vibration associated with running tends to reduce all friction coefficients. This second calibration method thus appears to be the better one, even though there are still some remaining error sources in both the calibration and the measuring situations. Thoughtful consideration of all these questions is greatly aided by the clarity of Fig. 9.11. A less well-designed drawing might make it quite difficult to sort out the various frictional effects.

We have mentioned that line drawings of the entire apparatus are usually more informative than photographs; however, there are cases where photographs are desirable or essential. Surface details of metallurgical samples or failed specimens may be vital to proper interpretation. Many flow visualization methods use photographic records for data presentation and analysis. Scientific and engineering photography is a specialty field that we are not able to treat in any detail in this text, so we list only a few references.[1] For "casual" photographs that might be appropriate in a report we can give a few broad guidelines. To center the viewer's interest on the subject of the photograph, keep the background uncluttered. If the background clutter cannot be physically removed, try to arrange the lighting so that the offending background is dark. Sometimes the background can be intentionally thrown out of focus by using larger apertures to decrease the depth of field. For closeups which require sharpness over the whole field of view, use small apertures and long exposures to get maximum depth of field. Use a tripod whenever possible, especially for long exposures.

The design of *tables* usually presents less difficulty than we have with drawings and diagrams. *Ruled tables* are most common, but we also can use *text tables* and *bulleted lists*. A text table is usually short and is not numbered, titled, or ruled into rows and columns; it is considered part of the text.

TRANSDUCER TYPE	DC RESPONSE?	STATIC CALIBRATION?
Piezoelectric	No	Usually not

Bulleted lists generally have only one column and are not usually numbered:

TYPES OF DISPLACEMENT TRANSDUCERS

- Potentiometer
- LVDT
- Capacitance
- Piezoelectric
- Eddy current
- Optical
- Pneumatic

---

[1] A. A. Blaker, *Handbook for Scientific Photography,* 2d ed., Focal Press, Stoneham, MA, 1988. C. L. Tucker, *Industrial and Technical Photography,* Prentice-Hall, Englewood Cliffs, NJ, 1989.

Ruled tables are major tables which are titled, numbered, and ruled into rows and columns:

TABLE 5
Results of three alloying treatments

Property	Alloy 1	Alloy 2	Alloy 3
Hardness, Rockwell C	54	43	49
Specific weight, lb_f/in^3	0.253	0.289	0.222
Tensile strength, psi	70,600	56,900	63,500

Note that this table could also have been arranged differently:

	Hardness, Rockwell C	Specific weight, lb_f/in^3	Tensile strength, psi
Alloy 1	54	0.253	70,600
Alloy 2	43	0.289	56,900
Alloy 3	49	0.222	63,500

This second arrangement would probably be preferred since it makes the numbers for the various properties a little easier to compare. As with any graphics in our reports, tables should be proofread, not only to check the numbers but also to weed out clutter that detracts from the main message.

With respect to the *location* of graphics in a report, most writers and readers prefer to find a graphic located close to its first mention in the text. Graphics are usually discussed in the text, and it is convenient to see the graphic and read the associated text without turning pages. Of course, there will be situations where this convenience is not possible, but we should *not* routinely force all graphics to a common location at the back of the report, as sometimes has been the practice. The *size* of graphics should also be carefully chosen to maximize clarity without wasting space. It is becoming more common for engineers to produce their own reports (except for very lengthy ones or those that require the talents of professional illustrators), using word processors on their personal computers. Thus, when choosing software for this task, we must be sure that the word processor has equation-writing, drawing, and graphing capabilities, either built-in or conveniently accessible. This gives the engineer complete control over report production, without the delays and errors associated with the use of professional typists.

9.5 ALTERNATIVE FORMS OF OUTLINES FOR TECHNICAL COMMUNICATIONS

Because this is a text on laboratory experimentation, we have emphasized in this chapter the use of an outline suitable for laboratory reports to be read by a supervisor. Other forms of outline are desirable for other types of communications, and we now briefly consider these so as to make this chapter of more general usefulness.

The form of outline discussed in Sec. 9.2 is that of *decreasing importance,* also called the *inverted-pyramid* style. We start with the major objectives and results and gradually work our way down to details of lesser and lesser importance. An outline form that is essentially the reverse of this, sometimes called the *pyramid* style, can also be useful. (Remember, too, that we need not constrain our writing rigidly to *any* specific outline form; we can combine various forms if that seems best.) Using *increasing* importance as our outline guide, we would begin with details and gradually build to a climax of important results and conclusions. Many technical papers (as contrasted with reports to superiors) are written in this style, especially if they are rather short. For longer papers, it might be better to *combine* two standard outline forms by giving a short statement of important results in an introduction and then revert to the increasing importance style for the rest of the paper. This gives the readers an early insight into where they are being led, which piques their interest and may also help them to understand the intervening details.

Many other logical ways of organizing or outlining a technical communication are available.[1] Weisman explains 10 such forms which we can use, singly or in combination, to help us organize the document. A *chronological* form of outline organizes the material in a time sequence, which may be appropriate when the historical order of events is particularly significant or the steps in a procedure must follow a specific time pattern. *Geographic or spatial* organization patterns might be

---

[1] H. W. Weisman, *Basic Technical Writing,* 4th ed., Merrill, Columbus, OH, 1980, pp. 138–140. D. E. Zimmerman and D. G. Clark, *The Random House Guide to Technical and Scientific Communication,* Random House, New York, 1987, chap. 7.

appropriate in a report discussing a proposed reallocation of space in a manufacturing plant. A *functional* form of outline could be used in breaking down the operation of a complex machine or process into simpler basic functions. We have already discussed the two *order-of-importance* (increasing and decreasing) outlining schemes. When arguing for a new approach to some problem, we could use the method called *elimination of possible solutions.* Here we propose, and then "shoot down," all the alternative solutions that we *don't* like, before presenting our preferred solution.

In the *deductive* (general to particular) organization, we start with general principles and then branch off into specific applications. A general discussion of laser material-processing principles would be followed by applications to cutting, drilling, welding, and heat-treating. *Inductive* (particular to general) organization just reverses this concept. We might start with some examples of practical problems involving forces and motion and then show how a general principle, Newton's law, allows us to solve all of them. In the *simple-to-complex* plan, we begin with simple and/or known situations and then apply this understanding to a new, more complex problem. In electric circuits, we could develop understanding of the voltage/current behavior of R, C, and L as isolated devices and then discuss the analysis of complete circuits. The *pro-and-con* type of outline is useful in studies which recommend that a particular course of action, rather than some alternative, be pursued. For each alternative, we list and discuss its good and bad features. Finally, in the *cause-and-effect* scheme, we begin with a set of facts or causes and then proceed to the consequences or results of these. The converse process may also be useful (reasoning from results to probable causes).

9.6 ORAL PRESENTATIONS

Preparing for an oral presentation involves many of the same techniques used in producing a written document, so we need not repeat a detailed discussion of these. The audience must be analyzed, outlines of some kind are generally useful, paragraphs and sentences need to be properly structured, the best words need to be chosen, and visual aids should be effectively utilized. In a talk, clarity and simplicity are even more important than in writing, because the audience has no chance to "reread" an obscure statement.

In presenting a talk, we could use one of three approaches. We could write down every word of the talk, polish this written document in the usual ways, and then read the final document to the audience. This is the approach used when the President of the United States presents the annual State of the Union speech, but it is rarely appropriate for a technical talk. Only when the result of misspeaking ourselves is truly catastrophic should we rely on such a presentation. A second approach is to again prepare a verbatim written document, but now we *memorize* it so that we can present it without any written aids. This approach is both undesirable and difficult or impossible for most people. The third approach, which is the usual and recommended one, employs a relatively short set of notes outlining the main ideas of the talk in the desired sequence. These notes are used by the speaker to maintain the desired train of thought and to ensure that major topics are not left out. We rely on the speaker's inherent familiarity with the topic to "fill in the gaps" of the sparse notes. (Most audiences really expect the speaker to be knowledgeable on the announced subject.) If the talk includes overhead transparencies, slides, flip charts, or a computer-monitor presentation, these can often also serve as the notes, cuing the speaker as to the proper sequence of topics and major content.

Since it is *extremely* easy to lose an audience during a technical talk, we should strive to present the *minimum* necessary information in the clearest and most connected way. Carefully planned *repetition* of important or difficult ideas may be necessary. Try to avoid display of complicated equations by giving them in explanatory *word* form rather than in symbols. As we suggested for written communications, use a three-stage process for complicated ideas:

1. Tell the audience what is about to be explained.

2. Give the explanation.

3. Reiterate what you just told them.

Normal speaking rates are about 120 to 180 words per minute. Use this as a guide in estimating the total time if the number of words is fixed, or the number of allowable words if time is limited. To gain confidence, verify timing, and obtain critical comments, rehearse your speech in front of a colleague if possible. Figure 9.12 shows a rating sheet used for student oral presentations in a lab course at The Ohio State Department of Mechanical Engineering. The sheet lists many aspects of a talk that contribute to its effectiveness. Both the instructor and each student in the audience rate the performance by placing a mark next to each item. A cross means "needs improvement," a checkmark means "better than average," and no mark means "average performance." You should now examine this form to get familiar with these attributes of a speech.

Visual aids that accompany a talk should conform to the same principles of clarity required in written communications; however, some

ORGANIZATION

Good opening statement
Good overview
Good development
Good closing statement

CHOICE OF TOPIC

Too simple
Too complex
Interesting
Appropriate

USE OF VOICE

Volume
Speed
Pitch
Pauses
Pronunciation
Mumbling

USE OF LANGUAGE

Choice of words
Sentence clarity

VISUAL AIDS

Print/sketches too small
Too complex
Too simple
Too many
Too few
Meshing with talk
Blackboard usage
Props

TIMING

Too long
Too short
Rushed pace
Dragging pace
Comfortable pace

USE OF HUMOR

Proper
Overdone

USE OF NOTES

Too much "reading"

"BODY LANGUAGE"

Eye contact
"Talked to the chalkboard"
Gestures
Facial expression
Posture
Appearance ("neatness")
Fidgeted

HANDLING OF QUESTIONS

Questions encouraged?
Looked at person asking question?
Repeated the question if not clearly heard?
Gave answer to the whole audience?
Made sure audience understood question?
Answered competently?

IMPRESSIONS/ATTITUDE

Enthusiasm
Sincerity
Confidence
Nervousness

FIGURE 9.12
Checklist and evaluation sheet for oral presentations.

additional considerations need to be mentioned. Because slides, overhead transparencies, and flip charts are viewable by the audience only for a short time during a talk, they must be exceptionally easy to understand. Also, the text and graphics must be *large enough* to be easily visible from the last row of the audience. The violation of this one simple rule is perhaps the most common graphics problem in technical talks. Letters on a screen need to be about 25 mm high for every 10 m of viewing distance. Another useful rule is that the maximum viewing distance should not exceed 6 times the width of the projected image (24 ft would be maximum for a 4-ft-wide image). If you are using an 8.5-in by 11-in piece of paper to prepare copy for a 35-mm slide, a 16-point typeface (72-point type is 1 in high) is a reasonable choice. A sans serif typeface such as Helvetica bold (or equivalent) gives good readability. Use all capitals for emphasis, such as in headings; but use text-style lettering (mix of capitals and lowercase) for most of the text, because it is easier to read. In designing text for a slide, state the key idea of the slide in a bold heading, followed by a bulleted list of descriptive phrases giving more information. Phrases are preferable to complete sentences because the audience understands them more quickly.

When you actually present the talk, match the spoken words to those on your slides. Using different words forces the audience to deal with *two* presentations, one visual and the other oral, causing confusion. While we recommend simplicity, some slides are necessarily complex. You should then take time to lead the audience through a step-by-step analysis, using a pointer. Optical pointers rather than "sticks" are most effective. We also try to anticipate major questions that might come from the audience and prepare answers for them. Since we often intentionally leave out some details that our audience may ask about, it may be wise to prepare some extra slides that address these issues, should they come up.

Several "special effects" are possible with overhead transparencies. Using suitable writing instruments, we can "ad lib" material onto the transparency while it is being viewed, thus adapting it to audience questions or comments as they occur. Overlays can be used to build up a complex illustration in easy steps. *Masking* allows us to reveal each part of a multistep process as we discuss it. *Billboarding* uses cutouts in colored transparent sheets to emphasize a portion of the display that we are discussing.

9.7   SOFTWARE AND HARDWARE AIDS FOR TECHNICAL COMMUNICATION

We have already briefly mentioned the utility of a suitable word processor in producing technical communications. Most engineers find a general-purpose word processor with a good equation capability adequate for their work. Specialized technical word processors (T^3 is one example[1]) exist and may be helpful for the more complicated documents.

Graphing software and drawing software are also very useful and come in several forms. In terms of the underlying software details, two broad classes exist: *bit-mapped* and *vector*. Bit-mapped graphics, usually used in the simpler programs, represent images as a series of dots called *pixels*. Vector graphics, sometimes called *object-based*, work with complete objects, such as circles and triangles. Each pixel in a bit-mapped image can be edited, giving editing versatility; however, when a bit-mapped image is moved or enlarged, the resolution deteriorates and the image often appears jagged. Vector graphics generally gives considerably better and smoother images and allows convenient editing of entire objects, so it is used in the more advanced graphics software. A comprehensive comparison of the two techniques is difficult in a single

---

[1] TCI Software Research Inc., 1190 Foster Rd., Las Cruces, NM, 88001, phone (800) 874-2383.

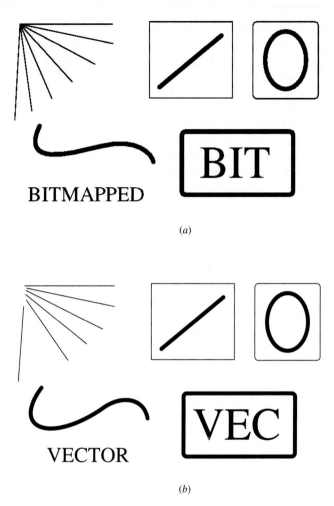

FIGURE 9.13
Comparison of (*a*) bit-mapped and (*b*) vector graphics.

example, but Fig. 9.13 shows some of the differences. Figure 9.13*a* was made with the (bit-mapped) Paintbrush program which is part of Microsoft Windows 3.0, while Fig. 9.13*b* used the (vector) software called Drawperfect. The size of Fig. 9.13 may make the differences nearly undetectable, so you might want to make such a comparison on your own computer and printer.

Software from a single company which provides word processing, drawing, and graphing capabilities in a single integrated package might seem to be ideal from the standpoint of ease of use, but other considerations may lead you to use separate packages. (If we decide to use separate packages, they must allow *easy* "importing and exporting" of data between the various programs so that the overall document can be assembled efficiently.) Software designed mainly for drafting tasks (CAD-type programs) can be used to produce figures for reports, but may be too complex and/or expensive for routine document preparation. Most drawing programs intended for illustration work rather than drafting also provide at least some graphing capability. Such programs may also be classified as *business graphics* or *technical graphics.* This is particularly true for those which emphasize *graphs* rather than drawings.

For engineering and scientific communications, we may want to invest in "high-powered" graphing software specially designed for this application area, rather than trying to "get by" with a business-type graphing program. Such technical graphing software[1] usually provides powerful data *analysis* tools (curve fitting, regression, spectrum analysis, etc.) in addition to graph plotting. Since engineers will have individual needs and preferences, many satisfactory combinations of software packages are possible. I happen to use an IBM-type personal computer rather than a Macintosh, and I do most of my work with the WordPerfect word processor (which handles equations and tables nicely), the companion drawing package Drawperfect (which allows easy export of its figures into WordPerfect), and the graphing capabilities of my statistics package STATGRAPHICS. I originally got STATGRAPHICS for my statistics work, but found that its excellent graphics capability could be used for my general-purpose graphing, since it imports into Wordperfect fairly easily. Drawperfect has some basic graphing capacity, and I use it for the simpler graphs and for bulleted lists. It also has a large library of "clip art,"[2] a common feature of drawing programs. *Clip art* is ready-made images, some quite complex and detailed, which you can import from your disk at will and insert into a drawing. Drawperfect's clip art is classified into the categories of animals (24 birds, fish, insects, etc.), arrows (24 versions), 48 business articles, 24 pieces of computer equipment, 48 flags, 24 flowchart symbols, 48 graphic "devices" (banners, backgrounds, borders, stars, etc.), 48 maps, 24 pieces of military equipment, 48 "objects" (cameras, bottles, ladders, etc.), 24 people, 24

---

[1] Two examples are SIGMAPLOT, Jandel Scientific, 65 Koch Rd., Corte Madera, CA 94925, phone (800) 874-1888, and AXUM, Trimetrix Inc., 444 N. E. Ravenna Blvd., Suite 210, Seattle, WA 98115, phone (206) 527–1801. The magazine *Scientific Computing and Automation* is a good source for surveying the available technical graphics software at any time. Surveys of this area also appear from time to time in the various personal-computer magazines.

[2] L. Simone, "Clip Art," *PC Magazine*, May 14, 1991, pp. 203–273.

special occasions, 24 sports items, 48 symbols, and 24 pieces of transportation equipment. These are intended more for business than technical applications, but I occasionally find them useful.

While many engineering oral presentations use 35-mm slides or overhead transparencies for their graphics, another medium useful for this task is the *computer* slide show. Here the graphics are not only prepared on the computer, but a computer monitor is used to display them during the actual presentation. Many graphics packages provide a slide-show feature which allows presentation of graphic images in any sequence desired. To allow this technique to be used with any but the smallest groups, special monitors called *presentation monitors*[1] are used. They have screens with diagonal measurements of 25 to 35 in. When even larger displays are needed (for larger groups), video projectors or projection panels which can be placed on an overhead projector are available.[2] Some of these can project both computer data and graphics and video.

Production of 35-mm slides has also been facilitated by various technical tools. By using Polaroid[3] film in any 35-mm camera and a special Polaroid processor and slide mounter, slides can be produced in just a few minutes. With one of several available films, this technique can be used for producing images of ordinary objects, documents, or computer screens. For higher-quality slides of computer screens, but with an equipment cost about 10 times that of the direct photographic approach just discussed, film recorders[4] are available. For any computer graphic, one can also use commercial slide services.[5] These services accept the digital data file, either on a floppy disk that you ship them or directly over a phone line by modem, and they process this to produce the slide.

Electronic copyboards[6] perform the functions of a classical chalkboard during meetings and conferences, with the added advantage of producing paper copies of all or part of what is written on the board when a copy request is entered. They are available wall-mounted or on casters in sizes up to about 45 in by 67 in. Although most engineers will not be engaged in producing filmstrips, films, or videos, we wanted to

---

[1] A. Poor, "Presentation Monitors," *PC Magazine,* May 14, 1991, pp. 347–375.

[2] Visualon, 9000 Sweet Valley Drive, Cleveland, OH 44125, phone (216) 328–9000.

[3] Polaroid Corporation, 575 Technology Square, Cambridge, MA 02139, phone (800) 345-5000.

[4] A. Poor, "Film Recorders," *PC Magazine,* May 14, 1991, pp. 305–336.

[5] K. S. Betts, "Slide Services," *PC Magazine,* May 14, 1991, pp. 277–301.

[6] Quartet Ovonics, 5700 Old Orchard Rd., Skokie, IL 60097, phone (708) 965-0600.

mention these means of technical communication and list some references.[1] Videotaping of a trial run of an oral presentation is an excellent means of discovering any errors or peculiarities that should be improved. Filmstrips, or their computerized versions, are quite effective in technical education because the learner can advance to the next topic, or return to earlier ones that remain unclear, at will. Films and videos, on the other hand, proceed at *their* speed, not the learner's, although they (especially videos) can, of course, be rewound to a desired location and rerun. While extremely sophisticated computer animation techniques have been developed, easy-to-use software for simpler animation tasks is available.[2] The referenced software does not require special video motion boards or data compression hardware and is quite inexpensive. Using a simple 286-based PC running at only 12 MHz, one can replay VGA (640×480) images with 16 colors at about 15 frames per second. Faster systems will replay images proportionally faster.

BIBLIOGRAPHY

Journals

1. *Journal of Technical Writing and Communication*
2. *Technical Communication*
3. *IEEE Transactions in Professional Communication*
4. *Computer Graphics*
5. *Pixel*
6. *AV Communication Review*
7. *Technical Photography*

Books

1. D. E. Zimmerman and D. G. Clark, *The Random House Guide to Technical and Scientific Communication,* Random House, New York, 1987.
2. H. M. Weisman, *Basic Technical Writing,* 4th ed., Merrill, Columbus, OH, 1980.
3. M. Young, *The Technical Writer's Handbook,* University Science Books, Mill Valley, CA, 1989.
4. M. J. Murray and H. Hay-Roe, *Engineered Writing,* 2d ed., PennWell Books, Tulsa, OK, 1986.

---

[1] J. Mercer, *The Informational Film,* Stipes Publishing, Champaign, IL, 1981. E. Dale, *Audiovisual Methods in Teaching,* 3d ed., Holt, Rinehart and Winston, New York, 1969. L. Herman, *Educational Films,* Crown Publishing, New York, 1965. A. Strasser, *The Work of the Science Film Maker,* Communication Arts Books, Hastings House, New York, 1972.

[2] Micro-Movies, Eclectic Systems, 8106 St. David Ct., Springfield, VA 22153, phone (703) 440-0064.

5. W. Strunk and E. White, *The Elements of Style,* 3d ed., Macmillan, New York, 1979.
6. N. Enrick, *Handbook of Effective Graphic and Tabular Communication,* R. E. Krieger Publ., New York, 1980.
7. A. Bishop, *Slides: Planning and Producing Slide Programs,* Kodak Publication S-30, Rochester, NY, 1984.
8. W. S. Cleveland, *The Elements of Graphing Data,* Wadsworth, Belmont, CA, 1985.
9. E. R. Tufte, *The Visual Display of Quantitative Information,* Graphics Press, Cheshire, CT., 1983.

INDEX